Universitext

W0106890

Universitext

Editors (North America): J.H. Ewing, F.W. Gehring, and P.R. Halmos

(continued after index)

Alexander J. Hahn

Quadratic Algebras, Clifford Algebras, and Arithmetic Witt Groups

Springer-Verlag

New York Berlin Heidelberg London Paris
Tokyo Hong Kong Barcelona Budapest

Alexander J. Hahn
Department of Mathematics
University of Notre Dame
Notre Dame, IN 46556
USA

Mathematics Subject Classifications (1991): 15A66, 16H05, 13A20

Library of Congress Cataloging-in-Publication Data
Hahn, Alexander, 1943–
    Quadratic algebras, Clifford algebras, and arithmetic Witt groups/
Alexander J. Hahn.
        p.   cm. — (Universitext)
    Includes bibliographical references.

    ISBN-13: 978-0-387-94110-3   e-ISBN-13: 978-1-4684-6311-8
    DOI: 10.1007/978-1-4684-6311-8

    1. Clifford algebras.   2. Commutative rings.   3. Forms, Quadratic.
I. Title.   II. Series.
QA199.H34   1993
512′.57 — dc20   93-5139

Printed on acid-free paper.

Production managed by Francine McNeill; manufacturing supervised by Vincent Scelta.
Camera-ready copy prepared by the author using Microsoft Word.

9 8 7 6 5 4 3 2 1

To my parents
with great affection and esteem

Preface

This volume had its starting point with certain questions about the structure of the Clifford algebra over a commutative ring, in particular with the existence of certain "special elements" in the centralizer of the even subalgebra. These — see Chapter 7 of Hahn-O'Meara *The Classical Groups and K-Theory*, Springer-Verlag 1989 — play a role in the isomorphism theory of linear groups of small rank. When I realized that the structure theory of Clifford algebras over commutative rings could be developed around special elements and the closely related concept of quadratic algebras, these early investigations had developed into the theme for this volume. The flow of ideas was refined, expanded, and finalized in the course of discussions, lectures, and seminars at the University of Notre Dame. In this context, I would like to express a warm thank you to my colleagues at the University of Notre Dame, especially to Timothy O'Meara, who introduced me to the theory of quadratic forms years ago and has been a source of inspiration since; to Stephan Stolz and Gudlaugur Thorbergsson for contributing their considerable expertise to Chapter 15, especially for providing Chapters 15F and 15G, respectively; to Ken Grant, who provided valuable assistance with the number theory required for the table in Chapter 14E; and to Warren Wong for the many discussions on the topics of this text. A word of gratitude goes also to my colleagues at the Mathematisches Institut der Universität Innsbruck, particularly to Ottmar Loos and Ulrich Oberst, for their hospitality both during the academic years 1987–88 (toward the beginning of the present project) and again in 1992–93 (at its conclusion). Thanks, too, to my student Zhang Qi for her diligent proofreading, and to the staff at Springer-Verlag, New York, for its courtesy and professionalism. To my wife Marianne: many thanks for your understanding throughout this enterprise! Last, but certainly not least, I wish to thank the National Security Agency for supporting my efforts with a two-year research grant awarded under its Mathematical Sciences Program.

Innsbruck, Austria

Contents

Introduction

The goal of this volume is to introduce the reader in an elemental and accessible way to the large and dynamic area of algebras and forms over commutative rings. Quadratic algebras and their analysis give the volume its direction. Indeed, its defining moment is the fact that quadratic algebras lie at the heart of the theory of quadratic forms and Clifford algebras over commutative rings.

The first two chapters set out the main themes: algebras over a commutative ring R, involutions on algebras, gradings and tensor products of algebras, and separable algebras. These concepts are illustrated in the concrete case of free quadratic algebras, i.e., algebras of the form $R[X]/(X^2 - aX - b)$. The third chapter gathers the isomorphism classes of separable free quadratic algebras into a group, the "free quadratic group," and studies its main properties. This chapter ends the introductory phase of the book. It is very elementary, but gives a flavor of the material that is investigated later.

Chapter 4 quickly recalls the localization of rings and modules, introduces bilinear and quadratic modules, and develops the group of "discriminant modules," i.e., the group of isomorphism classes of projective bilinear modules of rank 1. The elementary properties of the Clifford algebra of a quadratic module are collected in Chapter 5 mostly without proof. The chapter that follows investigates algebras with "standard" involution. The most important examples are the quaternion algebras, i.e., Clifford algebras of quadratic modules of rank 2. They provide concrete examples of the basic concepts and constructions in the theory of Clifford algebras.

The third part of the book interrelates the first two. It begins in Chapter 7, with the study of the Arf algebra, i.e., the centralizer of the "even" part of the Clifford algebra in the full Clifford algebra. This algebra is quadratic and establishes the connection between quadratic algebras, Clifford algebras, and quadratic forms. Properties of the Arf algebra, and in particular certain "special elements" in it, have impact on the representation theory of Clifford algebras and clarify the connection between the tensor and graded tensor products. This is taken up in Chapter 8. In Chapter 9, the structure of the Clifford algebra $C(M)$ and its even part $C_0(M)$ for a general finitely generated projective nonsingular quadratic module M is analyzed. Both are shown to be separable. If the rank of M is even, then $C(M)$ is central, and if the rank of M is odd, then the even part is central. Chapter 10 proves that special elements exist whenever M is free and nonsingular. Chapter 11 applies the results of Chapter 8 and develops the matrix theory of Clifford algebras over fields, including the real and complex numbers, and local and global fields. It includes the periodicity phenomena. Chapter 12 returns to a commutative ring, constructs and investigates the "projective" version of the free quadratic group, and studies its connection with the group of discriminant modules.

The last component of the book consists of applications of the earlier material. Chapter 13 recalls (with proofs of only those facts that are not easily tracked in the literature) the Brauer group, its graded version, the Witt groups of quadratic and symmetric bilinear forms, and the important invariants of forms. Chapter 14 uses the facts from Chapter 12, analyzes the Witt group of quadratic forms over arithmetic Dedekind domains, and studies the quotient of the symmetric bilinear Witt group over the quadratic Witt group. The concluding chapter concerns itself with the fact that Clifford algebras and their representations play a significant role in a number of areas of pure and applied mathematics, and is a collection of short "essays" designed to give a glimpse of that role, especially in differential geometry and topology.

The preceding discussion also describes the pedagogical "flow" of the book. It is very elementary initially, requiring only elements from a basic algebra course. The degree of difficulty increases gradually. The first twelve chapters are − with the exception of the basics about Clifford algebras and standard facts from the theory of separable algebras − self-contained. The last three chapters are more advanced, and blend in the literature in a substantial way. There are many exercises throughout the book. There is great variation in the degree of difficulty. "Hints" are supplied .

New material in this book includes: the structure theorem for algebras with standard involution in Chapter 6A; the results concerning the representations of Clifford algebras in Chapters 8B, C, and D; the fact, proved in Chapter 10 that any free nonsingular quadratic module has a special element, and all the consequences of this fact when it is combined with the results of Chapter 8; the connection between $Qu(R)$, $Dis(R)$, and $Dis(R/4R)$ that is developed in Chapter 12, and the arithmetic results about the quadratic Witt group developed in Chapters 14B, C, and D. Another new aspect is the approach to the structure theory of the Clifford algebra over a commutative ring via the systematic use of quadratic (and quaternion) algebras.

The Bibliography is not intended to be complete, but does list much of the important literature related to the themes of this book. The text makes reference to the literature in the bibliography simply by stating the author's name (and the year of publication in ambiguous cases). For the most part the most accessible references were used. Little effort was made (in retrospect perhaps unfortunately) in singling out the originators of a particular theory, i.e., to questions of priority.

Notation and Terminology

Any ring A in this book is associative with identity 1 (for emphasis this is sometimes written as 1_A). A subring has the same identity, and ring homomorphisms preserve 1s. The possibility that $A = \{0\}$ is allowed. Obviously in this case, $1 = 0$. If $A \neq \{0\}$, then $1 \neq 0$; for otherwise, $a = a \cdot 1 = a \cdot 0 = 0$ for any a in A. If $A \neq \{0\}$ and $ab = 0$ implies that either $a = 0$ or $b = 0$, then A is a *domain*. If $A \neq \{0\}$, and $\{0\}$ and A are the only two-sided ideals of A, then A is *simple*. For a subset S of A, the *centralizer of* S *in* A is the subring $\mathrm{Cen}_A S$ of A defined by

$$\mathrm{Cen}_A S = \{a \in A \mid as = sa \text{ for all } s \in S\}.$$

The *center* of A is $\mathrm{Cen}\, A = \mathrm{Cen}_A A$. If $\mathrm{Cen}\, A = A$, then A is commutative. An invertible element of A is a *unit* of A. The group of units is denoted A^*. The ring of integers is denoted by $\mathbb{Z}$ and the quotient ring $\mathbb{Z}/k\mathbb{Z}$, $k \geq 0$, by $\mathbb{Z}_k$. Observe that there is a unique ring homomorphism $\mathbb{Z} \to A$. Since the kernel is an ideal of $\mathbb{Z}$, it equals $k\mathbb{Z}$ with $k \geq 0$. So there is an induced injective ring homomorphism $\mathbb{Z}_k \to A$. This integer k is the *characteristic* char A of A. If $k \neq 0$, we will also use $\mathbb{Z}_k$ to denote the finite cyclic group of order k (since the additive group of the ring $\mathbb{Z}_k$ is such a group). The infinite cyclic group is denoted $\mathbb{Z}$.

Let M be an A-module. The scalars will normally be written on the right and it is always assumed that $x1 = x$ for all x in M. If $A \neq \{0\}$ is commutative and M is free and finitely generated, then any two bases of M have the same finite cardinality. This number is the *rank* of M and is denoted rank M. It follows that any such module is isomorphic to the external direct sum $A \oplus ... \oplus A$ (rank M copies). If $A = \{0\}$, observe that $M = \{0\}$ is the only A-module. In this case, $M = A = A \oplus A = A \oplus ... \oplus A$, so that M is free of any rank. When A is a field or a division ring, then modules will be called vector spaces and rank becomes dimension.

In this book, R will generally denote an arbitrary commutative ring. When required, explicit additional assumptions will be made. Let A be an algebra over

3

R. Thus A is a ring with 1_A that is equipped with a ring homomorphism f from R into Cen A. Note that A has the structure $ra = ar = (fr)a$ for r in R and a in A, of an R-module. Observe that $f : R \to$ Cen A is injective if and only if A is *faithful* as R-module, i.e., if $r = 0$ is the only element such that $ra = 0$ for all a in A. If f is injective and surjective, then A is *central*. Our notation will, for the most part, suppress reference to f. In the process, $fr = (fr)1_A$ is denoted by r (with appropriate caution, particularly since f is not necessarily injective) and $R1_A$ denotes the image of R in Cen A. The maps $\mathbb{Z} \to A$ and $\mathbb{Z}_k \to A$ with k = char A, already mentioned, and the inclusion Cen A $\to$ A make any ring A into algebras over $\mathbb{Z}$, $k\mathbb{Z}$ and Cen A. Note that A is central in the last case. Of course, R is an R-algebra with f the identity map.

Properties that an R-algebra A has as a ring and as an R-module will also be ascribed to A as an algebra. For example, A is a commutative or simple if it is so as a ring; and A is finitely generated, projective, or free if it is so as an R-module; a basis of A is a basis of the R-module A. In this last situations, an element of caution is necessary, since the concepts free and finitely generated have (never in this book, but in the literature) also a ring theoretic and therefore a different meaning.

Let M be an R-module. The Abelian group $M \otimes_R A$ has the structure of an A-module that satisfies $(x \otimes a)b = x \otimes ab$ for all x in M and a and b in A. For an ideal $\mathfrak{a}$ of R, $M\mathfrak{a} = \{ \sum_{\text{finite}} xa \mid x \in M, a \in \mathfrak{a} \}$ is a submodule of M.

Let A and B be two R-algebras. Clearly, $A \times B$ is an algebra over $R \times R$ and, via the diagonal map $R \to R \times R$, also over R. In the latter case $A \times B$ will be denoted by $A \oplus B$. Note that the underlying R-module structure in this case is the external direct sum of the R-modules A and B. An (A-B) *bimodule* is an additive Abelian group M, which is both a left A-module and a right B-module and satisfies for all r in R, a and b in A, and x in M, $a(xb) = (ax)b$ and $rx = xr$.

For a set S, the identity map on S is denoted id_S (also 1_S, id, or 1, when the context is clear), and the restriction of a function f on S to a subset T is written $f|_T$. The cardinality of S is card S. For an Abelian group G, the *torsion* subgroup, i.e., the subgroup of elements of finite order is denoted by G_{tor}, and G_k denotes the subgroup of elements of order $\le k$. So G_2 is the subgroup of involutions of G. If $G = G_k$, we say that G has *exponent* k.

1
Fundamental Concepts in the Theory of Algebras

Overview

Let R be a commutative ring. This first chapter recalls some fundamental concepts and constructions from algebra, e.g., involutions and gradings on R-algebras, and tensor products and graded tensor products of R-algebras. These are also the general Leitmotifs of this book. In addition, this chapter introduces free quadratic R-algebras, i.e., algebras of the form $R[X]/(X^2 - aX - b)$, and some of their basic properties. These algebras provide concrete illustrations of the basic concepts and constructions just referred to; more significantly, they will be of importance in virtually every chapter of this book.

A. Free Quadratic Algebras

For an ideal $\mathfrak{a}$ of R, the quotient ring $R/\mathfrak{a}$ is given the structure of an algebra over R by taking the quotient map $R \to R/\mathfrak{a}$. The R-algebra $R/\mathfrak{a}$ is of course commutative; it is faithful only if $\mathfrak{a} = \{0\}$. More generally, if A is a commutative R-algebra and $\mathfrak{a}$ is an ideal of A, then the composite $R \to A \to A/\mathfrak{a}$ gives $A/\mathfrak{a}$ the structure of a commutative R-algebra.

An important class of algebras of the preceding form are the free quadratic algebras. Let $R[X]$ be the polynomial algebra in the indeterminate X over R. An R-algebra of the form $R[X]/(X^2 - aX - b)$, where $X^2 - aX - b$ is a monic quadratic polynomial in $R[X]$ and $(X^2 - aX - b)$ is the ideal it generates, is a *free quadratic algebra over* R. Let $S = R[X]/(X^2 - aX - b)$ be free quadratic. The identity of S is $1_S = 1 + (X^2 - aX - b)$ and $r \to r1_S$ is injective. So S is faithful and $R \subseteq S$. Also, $\{1, v = X + (X^2 - aX - b)\}$ is a basis of S as R-module and $v^2 = b + av$. Refer to Exercise 3 for the fact that any R-algebra S that is free of rank 2 is a free quadratic algebra over R.

The quadratic algebra $S = R[X]/(X^2 - X)$ is isomorphic to $R \oplus R$, i.e., the Cartesian product ring $R \times R$ made into an R-algebra via the diagonal map

$R \to R \times R$. This is easy to see. First note that $\{1 = (1, 1), (0, 1)\}$ is a basis of $R \oplus R$. Define $\varphi : R \oplus R \to S$ by $\varphi(1,1) = 1_S$, $\varphi(0, 1) = v$, and by extending linearly. Since φ takes a basis to a basis it is bijective, and since $(0, 1)^2 = (0, 1)$, φ is an isomorphism of algebras. The quadratic algebra $R[X]/(X^2 - X)$ is therefore called the *trivial* quadratic R-algebra. The quadratic algebra $R[X]/(X^2)$ is the *algebra of dual numbers* over R.

We now turn to the question of when two free quadratic R-algebras are isomorphic. The answer, perhaps not surprisingly, will turn out to be dependent on arithmetic properties of R. We begin with an easy, but important, isomorphism criterion.

(1.1). *Let* $R[X]/(X^2 - aX - b)$ *and* $R[X]/(X^2 - cX - d)$ *be quadratic algebras over* R. *Then*

$$R[X]/(X^2 - aX - b) \cong R[X]/(X^2 - cX - d)$$

if and only if there exist elements r *in* R *and* u *in* R^* *such that*

(i) $c = ua + 2r$ *and* (ii) $d = u^2 b - rua - r^2$.

Proof. Let $v = X + (X^2 - aX - b)$, and let $w = X + (X^2 - cX - d)$. Let an isomorphism

$$\varphi : R[X]/(X^2 - cX - d) \to R[X]/(X^2 - aX - b)$$

be given. Put $\varphi w = r + uv$, with r and u in R. Since $\{1, r + uv\}$ is a basis of $R[X]/(X^2 - aX - b)$, u is in R^*. Since

$$(\varphi w)^2 = r^2 + 2ruv + u^2 v^2 = (r^2 + u^2 b) + (2ru + u^2 a)v,$$

on the one hand, and $\varphi((w)^2) = \varphi(cw + d) = (rc + d) + ucv$, on the other, we get $uc = 2ru + u^2 a$ and $rc + d = r^2 + u^2 b$. The first equality provides (i), and plugging (i) into the second gives (ii). In the other direction, take the given r and u and define

$$\varphi : R[X]/(X^2 - cX - d) \to R[X]/(X^2 - aX - b)$$

by $\varphi(1) = 1$, $\varphi(w) = r + uv$, and by extending linearly. Since $\{1, r + uv\}$ is a basis of $R[X]/(X^2 - aX - b)$, φ is bijective, and by the preceding equations it is an algebra isomorphism.   QED.

If $X^2 - aX - b$ has a root, say γ, in R, then by an easy computation,

$X^2 - aX - b = (X - \gamma)(X - (a - \gamma))$, so that $a - \gamma$ is another root in R. If $a - \gamma = \gamma$, then γ is a *double root* of $X^2 - aX - b$. If γ is a root of $X^2 - aX - b$, then $(2\gamma - a)^2 = a^2 + 4b$; so if γ is a double root, then $a^2 + 4b = 0$.

Let γ in R be a root of $X^2 - aX - b$. Set $c = 2\gamma - a$. Taking $r = \gamma$ and $u = -1$ in (1.1) shows that $R[X]/(X^2 - aX - b) \cong R[X]/(X^2 - cX)$. If γ is a double root, then $c = 0$ and $R[X]/(X^2 - aX - b) \cong R[X]/(X^2)$. Suppose conversely that $R[X]/(X^2 - aX - b) \cong R[X]/(X^2 - cX)$ for some c. Then the element $\gamma = -ru^{-1}$ provided by criterion (1.1) is a root of $X^2 - aX - b$. If $c = 0$, then $ua + 2r = 0$, so $a - 2\gamma = 0$, and $\gamma = a - \gamma$ is a double root.

We have proved (2) and (3) of the following proposition. Part (1) is a trivial application of (1.1).

(1.2).　(1) *For* $t \in R^*$, $R[X]/(X^2 - aX - b) \cong R[X]/(X^2 - taX - t^2b)$.

　　　　(2) $R[X]/(X^2 - aX - b) \cong R[X]/(X^2 - cX)$ *for some* $c \in R$ $\Leftrightarrow$ $X^2 - aX - b$ *has a root in* R.

　　　　(3) $R[X]/(X^2 - aX - b) \cong R[X]/(X^2)$ $\Leftrightarrow$ $X^2 - aX - b$ *has a double root in* R.

Example 1. Let $R = \mathbb{C}$ be the complex numbers. An easy application of (1.2) shows that any quadratic algebra over $\mathbb{C}$ is isomorphic to either the trivial algebra $\mathbb{C}[X]/(X^2 - X)$ or the algebra of dual numbers $\mathbb{C}[X]/(X^2)$. By (1.1), these two algebras are not isomorphic.

(1.3).　*If* $R[X]/(X^2 - aX - b) \cong R[X]/(X^2 - cX - d)$, *then* $c^2 + 4d = u^2(a^2 + 4b)$ *for some* $u \in R^*$. *The converse holds under any one of the following assumptions:*

　(1)　$2 \in R^*$.

　(2)　R *is an integrally closed domain with* char R $\neq 2$.

　(3)　2 *is not a zero divisor of* R *and* 2R *is a prime ideal.*

Proof. The initial assertion is a routine consequence of (1.1). For the converse, suppose that there is some $u \in R^*$ such that $c^2 + 4d = u^2(a^2 + 4b)$. We prove (1). Put $r = 2^{-1}(c - ua)$. A computation that is completely routine shows that $d = u^2b - rua - r^2$. So $R[X]/(X^2 - aX - b) \cong R[X]/(X^2 - cX - d)$ by (1.1).

Now to case (2). Let F be the field fractions of R. Applying the proof just given to F shows that the element $r = 2^{-1}(c - ua)$ of F satisfies the equality

$d = u^2b - rua - r^2$. So r is a root of the monic polynomial $X^2 + uaX + (d - u^2b)$. It follows that $r \in R$. Since criterion (1.1) is satisfied with r and u, $R[X]/(X^2 - aX - b) \cong R[X]/(X^2 - cX - d)$.

Assume (3). Since $c^2 - u^2a^2 = 4u^2b - 4d$, $(c - ua)(c + ua) \in 2R$. Therefore, either $c - ua$ or $c + ua$ is in $2R$. Replacing u by $-u$ if necessary, we assume that $c - ua = 2r$ for some r in R. In view of (1.1), it remains to show that $d = u^2b - rua - r^2$. Since 2 is not a zero divisor, it suffices to show that $4d = 4(u^2b - rua - r^2)$. But this is easy: multiply out and substitute $2r = c - ua$.   QED.

Remark. Refer to Exercises 10, 15 and 18 for examples relevant to the preceding proposition. Suppose that R is a domain with char $R \neq 2$ and that F is its field of fractions. Refer to the proof of the proposition in case (2) and observe that the condition that any monic polynomial of degree 2 over R that has a root in F has a root in R is sufficient for the conclusion.

Examples 2 through 4 that follow are routine consequences of Proposition (1.3).

Example 2. Assume that one of the conditions (1), (2), or (3) of (1.3) holds for R. Then $R[X]/(X^2 - aX - b)$ is isomorphic to the trivial quadratic algebra if and only if $a^2 + 4b \in (R^*)^2$; and $R[X]/(X^2 - aX - b)$ is isomorphic to the algebra of dual numbers over R if and only if $a^2 + 4b = 0$.

Example 3. Consider the three quadratic algebras

$$\mathbb{R}[X]/(X^2 - X) \qquad \mathbb{R}[X]/(X^2) \qquad \mathbb{R}[X]/(X^2 + 1)$$

over the real numbers $\mathbb{R}$. They are, respectively, the trivial algebra, the dual numbers, and the complex numbers. No two of these algebras are isomorphic, and for an arbitrary quadratic algebra $S = \mathbb{R}[X]/(X^2 - aX - b)$:

$$S \cong \begin{cases} \mathbb{R}[X]/(X^2), & \text{if } a^2 + 4b = 0 \\ \mathbb{R}[X]/(X^2 - X), & \text{if } a^2 + 4b > 0 \\ \mathbb{R}[X]/(X^2 + 1), & \text{if } a^2 + 4b < 0. \end{cases}$$

Example 4. Let F be any field with char $F \neq 2$. Let D be a set of representatives of the group $F^*/(F^*)^2$. So $F^*/(F^*)^2 = \{d(F^*)^2 \mid d \in D\}$. Then the algebras

$$F[X]/(X^2) \qquad\qquad F[X]/(X^2 - d), \ d \in D$$

are a complete set of representatives (with no duplication) of the isomorphism classes of quadratic algebras over F. This implies that there are infinitely many isomorphism classes of quadratic algebras over the rational numbers. See Exercises 4 and 5.

For an analysis of the isomorphism classes of quadratic algebras over fields of characteristic 2, see Exercises 6 and 7 in this chapter and Exercise 19 of Chapter 3.

Example 5. Let $\mathbb{Z}[X]/(X^2 - aX - b)$ be an arbitrary quadratic algebra over $\mathbb{Z}$. By (1.1),

$$\mathbb{Z}[X]/(X^2 - aX - b) \cong \begin{cases} \mathbb{Z}[X]/(X^2 - n), & \text{if } a \text{ is even} \\ \mathbb{Z}[X]/(X^2 - X - n), & \text{if } a \text{ is odd.} \end{cases}$$

The listing involves no duplication.

B. Involutions on Algebras

Return to an arbitrary R-algebra A. Let α be a homomorphism or anti-homomorphism of A. So $\alpha : A \to A$ is R-linear, and (using exponential notation): $1^\alpha = 1$ and $(cd)^\alpha = (c)^\alpha(d)^\alpha$ for all c and d in A in the first case, and $(cd)^\alpha = (d)^\alpha(c)^\alpha$ for all c and d in A in the other. If α is an automorphism or antiautomorphism and $\alpha^2 = \text{id}_A$, then α is an *involution* on A. Note that if $\alpha : A \to A$ is a homomorphism or antihomomorphism of R-algebras such that $\alpha^2 = \text{id}_A$, then α is injective and surjective, and thus is an involution on A. If an involution is an antihomomorphism, we will, when emphasis requires, call it an *anti-involution*. If A and B are R-algebras with fixed involutions α and β, then an algebra homomorphism $\varphi : A \to B$ *preserves involutions* or is a *homomorphism of algebras with involution* if the diagram

$$
\begin{array}{ccc}
A & \xrightarrow{\varphi} & B \\
\downarrow{\alpha} & & \downarrow{\beta} \\
A & \xrightarrow{\varphi} & B
\end{array}
$$

commutes. In this case we also say that α corresponds to β under φ. If there is an algebra isomorphism from A to B that preserves involutions, then A and B are *isomorphic as R-algebras with involution.*

Let $S = R[X]/(X^2 - aX - b)$ be a free quadratic algebra. Define $\sigma : S \to S$ in the basis $\{1, v = X + (X^2 - aX - b)\}$ by $(r + tv)^\sigma = (r + ta) - tv$, for all r and t in R. It is not hard to check that σ is an R-algebra homomorphism. Since $(1)^\sigma = 1$ and $(v)^{\sigma^2} = (a - v)^\sigma = a - a + v = v$, $\sigma^2 = id_S$. So σ is an involution on S. Note that both v and $a - v$ are roots (in S) of the polynomial $X^2 - aX - b$ and that σ is the unique automorphism which interchanges these two roots. We call σ the *conjugation* of $R[X]/(X^2 - aX - b)$. In the case of the complex numbers $\mathbb{C} = \mathbb{R}[X]/(X^2 + 1)$, check that σ is complex conjugation. Let $s = r + tv \in S$. The trace and norm of an element s of S are respectively given by $s + s^\sigma$ and ss^σ. Let $s = r + tv$ with r and t in R. Show that $s + s^\sigma = 2r + ta$ and $ss^\sigma = r^2 + rta - rtv + trv + t^2av - t^2v^2 = r^2 + rta - t^2b$. Both are elements of R.

Let $S = R[X]/(X^2 - aX - b)$ and $T = R[X]/(X^2 - cX - d)$ be free quadratic algebras. Let σ be the conjugation of S and let τ be that of T. Let $\varphi : T \to S$ be any algebra isomorphism. Refer to the proof of (1.1) and its notation. In particular, let r in R and u in R^* be the scalars that φ determines and note that $c = ua + 2r$. Since $(\varphi w)^\sigma = (r + uv)^\sigma = r + ua - uv = c - r - uv$ and $\varphi(w^\tau) = \varphi(c - w) = c - (r + uv)$, it follows that $\sigma\varphi = \varphi\tau$. We have shown that any algebra isomorphism $\varphi : T \to S$ preserves the conjugations τ and σ.

C. Gradings on Algebras

We introduce gradings on algebras. Let A be an algebra over R. For each i in the group $\mathbb{Z}_2 = \{0, 1\}$, let A_i be an R-submodule of A and suppose that $A = A_0 \oplus A_1$. If $R1 \subseteq A_0$ and $A_iA_j \subseteq A_{i+j}$ for any i and j in $\mathbb{Z}_2$, then A is a *graded algebra* and the decomposition $A = A_0 \oplus A_1$ is a *grading* of A. The set $A_0 \cup A_1$ is denoted A_{hom} and its elements are the *homogeneous elements* of A. For $a \in A_{hom}$, the *degree* ∂a of a is equal to 0 or 1 according to whether a is in A_0 or A_1. The fact that $\partial 0$ can be either 0 or 1 is of no consequence. There is a more general notion of a grading of an algebra (this allows more than just two components), but the present "$\mathbb{Z}_2$-grading" is all that will be needed in this book.

Evidently, if $A = A_0 \oplus A_1$ is a grading of A, then A_0 is a subalgebra of A and A_1 is a A_0-module. Of course, any algebra A always has the grading obtained by taking $A = A_0$ and $A_1 = \{0\}$. This is the grading *concentrated in degree zero* or the *trivial* grading. The algebra A with this trivial grading will be denoted by <A>. Let B with grading $B = B_0 \oplus B_1$ be another graded algebra over R. An algebra homomorphism $\varphi : A \to B$ is said to *preserve the gradings* or to be *graded* if $\varphi A_0 \subseteq B_0$ and $\varphi A_1 \subseteq B_1$.

Consider a free quadratic algebra $S = R[X]/(X^2 - aX - b)$. When S is supplied with the trivial grading we will say that it is *even* and denote it by S^{even}. If $a = 0$, then the element $v = X + (X^2 - b)$ satisfies $v^2 = b$, so that S also has the grading $S = R1 \oplus Rv$. In this case we say that S is *odd* and write S^{odd}. Consider the ring $\mathbb{Z}_2 = \{0, 1\}$. For $\varepsilon \in \mathbb{Z}_2$, S^ε will mean S^{even} if $\varepsilon = 0$ and S^{odd} if $\varepsilon = 1$. So $S^0 = S^{even} = <S>$ and $S^1 = R1 \oplus Rv$, where it is understood that $a = 0$. Observe that the conjugation σ of S preserves the grading in either case. In the case of S^0 this is clear since the grading is trivial, and in the case of S^1 just note that $v^\sigma = -v$.

D. Tensor Products and Graded Tensor Products

Let A and B be two algebras over R and form the R-module $A \otimes_R B$. By use of the basic properties of the tensor product of modules, $A \otimes_R B$ can be made into an R-algebra with the property that $(a \otimes b)(a' \otimes b') = aa' \otimes bb'$.

Now let $A = A_0 \oplus A_1$ and $B = B_0 \oplus B_1$ be graded algebras. Consider the R-module $A \otimes_R B$ and the natural R-module isomorphism

$$A \otimes_R B \cong (A_0 \otimes_R B_0) \oplus (A_1 \otimes_R B_1) \oplus (A_0 \otimes_R B_1) \oplus (A_1 \otimes_R B_0).$$

Let C_0 and C_1 be the inverse images in $A \otimes_R B$ of

$$(A_0 \otimes_R B_0) \oplus (A_1 \otimes_R B_1) \text{ and } (A_0 \otimes_R B_1) \oplus (A_1 \otimes_R B_0)$$

respectively under the natural isomorphism above. Check that $C_0 \oplus C_1$ is a grading of $A \otimes_R B$.

We now put a second graded algebra structure on the R-module $A \otimes_R B$. This is done by twisting the algebra structure of $A \otimes_R B$ as follows: Let a and a' in A_{hom} and b and b' in B_{hom}, define

$$(a \otimes b) \cdot (a' \otimes b') = (-1)^{\partial b \partial a'}(a \otimes b)(a' \otimes b') = (-1)^{\partial b \partial a'}(aa' \otimes bb'),$$

and extend this product linearly to all of $A \otimes_R B$. This R-algebra is the *graded tensor product algebra* of A and B. It is denoted

$$A \overset{\wedge}{\otimes}_R B .$$

A moment's reflection shows that the decomposition $C_0 \oplus C_1$ also provides $A \overset{\wedge}{\otimes}_R B$ with the structure of a graded algebra. Notice that if $A = <A>$ or $B = <B>$, then $A \overset{\wedge}{\otimes}_R B = A \otimes_R B$ as graded algebras.

Suppose that $\varphi : A \to A$ and $\rho : B \to B$ are algebra homomorphisms. The properties of the tensor product provide a unique R-module homomorphism $\varphi \otimes \rho : A \otimes_R B \to A \otimes_R B$ that satisfies $(\varphi \otimes \rho)(a \otimes b) = \varphi a \otimes \rho b$ for all a in A and b in B. Note that $\varphi \otimes \rho$ is an algebra homomorphism. Check that if φ and ρ are both graded, then $\varphi \otimes \rho$ is a graded algebra homomorphism on both $A \otimes_R B$ and $A \overset{\wedge}{\otimes}_R B$. In particular, if φ and ρ are involutions on A and B, which preserve the gradings, then $\varphi \otimes \rho$ is a grading-preserving involution on both $A \otimes_R B$ and $A \overset{\wedge}{\otimes}_R B$.

We conclude this paragraph by analyzing the tensor and graded tensor product algebra in the special case where S and T are free quadratic algebras over R.

Put $S = R[X]/(X^2 - aX - b)$ and $T = R[X]/(X^2 - cX - d)$. Since S and T are commutative, the algebra $S \otimes_R T$ is also. Let $v = X + (X^2 - aX - b)$ and $w = X + (X^2 - cX - d)$. Since $\{1, v\}$ and $\{1, w\}$ are bases for S and T respectively, $\{1 \otimes 1, v \otimes 1, 1 \otimes w, v \otimes w\}$ is a basis for $S \otimes_R T$. It is left to the reader to provide a multiplication table for $S \otimes_R T$ in this basis.

Now supply S and T with the gradings S^ε and T^η. The following observations are easy: If $\varepsilon = \eta = 0$, then the gradings of S^ε and T^η are both trivial, and the grading of $S \otimes_R T$ is also trivial. If $\varepsilon = 0$ and $\eta = 1$, then S^ε has the trivial grading and $T^\eta = R1 \oplus Rw$; so the grading of $S^\varepsilon \otimes_R T^\eta$ is $(S \otimes_R R1) \oplus (S \otimes_R Rw)$. If $\varepsilon = 1$ and $\eta = 0$, the grading is $(R1 \otimes_R T) \oplus (Rv \otimes_R T)$, and if $\varepsilon = \eta = 1$, it is

$$((R1 \otimes_R R1) \oplus (Rv \otimes_R Rw)) \oplus ((Rv \otimes_R R1) \oplus (R1 \otimes_R Rw)).$$

Next turn to $S^\varepsilon \overset{\wedge}{\otimes}_R T^\eta$. If the grading of either S^ε or T^η is trivial, then $S^\varepsilon \overset{\wedge}{\otimes}_R T^\eta = S^\varepsilon \otimes_R T^\eta$ as graded algebras. So only the case $\varepsilon = \eta = 1$, i.e., $S^1 = R1 \oplus Rv$ and $T^1 = R1 \oplus Rw$ remains. This is the most interesting case. The grading of $S^1 \overset{\wedge}{\otimes}_R T^1$ is the same as that of $S^1 \otimes_R T^1$, so the focus is on the multiplicative structure of $S^1 \overset{\wedge}{\otimes}_R T^1$. Put $H = S^1 \overset{\wedge}{\otimes}_R T^1$ and label the basis of H by $1 = 1 \otimes 1$, $z_1 = v \otimes 1$, $z_2 = 1 \otimes w$ and $z_3 = v \otimes w$. Noting that a and c are both zero in the present situation, we find that

$$z_1^2 = (v \otimes 1)(v \otimes 1) = v^2 \otimes 1 = b(1 \otimes 1) = b1$$

$$z_2 z_1 = (1 \otimes w)(v \otimes 1) = (-1)^{1 \cdot 1}(v \otimes w) = -z_3$$

$$z_2 z_3 = (1 \otimes w)(v \otimes w) = (-1)^{1 \cdot 1}(v \otimes w^2) = -d(v \otimes 1) = -dz_1, \text{ and}$$

$$z_3^2 = (v \otimes w)(v \otimes w) = (-1)^{1 \cdot 1}(v^2 \otimes w^2) = -bd(1 \otimes 1) = -bd1.$$

It is left to the reader to complete these computations to the multiplication table

	1	z_1	z_2	z_3
1	1	z_1	z_2	z_3
z_1	z_1	$b1$	z_3	bz_2
z_2	z_2	$-z_3$	$d1$	$-dz_1$
z_3	z_3	$-bz_2$	dz_1	$-bd1$

of H. Observe, if we take $R = \mathbb{R}$ and $b = d = -1$, then H is Hamilton's classical quaternion division algebra.

A final comment concludes this chapter. Let σ and τ be the respective conjugations of S and T. Since $\sigma : S^\varepsilon \to S^\varepsilon$ and $\tau : T^\delta \to T^\delta$ are both graded algebra homomorphisms, it follows by observations already made that $\sigma \otimes \mathrm{id}_T$, $\mathrm{id}_S \otimes \tau$, and $\sigma \otimes \tau$ are involutions of both $S \otimes_R T$ and $S^\varepsilon \hat{\otimes}_R T^\delta$, which preserve the gradings.

E. Exercises

Unless otherwise specified, R is an arbitrary commutative ring.

1. Let S be a ring with char $S = k \neq 0$. In particular, S can be any finite ring. If k is odd, then $2 \in S^*$, and if k is even, then 2 is a zero divisor of S.

2. Let $\mathfrak{a}$ be an ideal of R. Let A be an R-algebra. The subset $\mathfrak{a}A =$ $\{\sum_{\text{finite}} ra \mid r \in \mathfrak{a}, a \in A\}$ of A is a two-sided ideal of A. Check that the ring homomorphism $R \to A$ induces a ring homomorphism $R/\mathfrak{a} \to A/\mathfrak{a}A$ which provides $A/\mathfrak{a}A$ with the structure of an $R/\mathfrak{a}$-algebra. The ring $A \otimes_R (R/\mathfrak{a})$ is an algebra over $R/\mathfrak{a}$ with the map $R/\mathfrak{a} \to A \otimes_R (R/\mathfrak{a})$ given by $s \to 1 \otimes s$. Show that the algebras $A/\mathfrak{a}A$ and $A \otimes_R (R/\mathfrak{a})$ are isomorphic.

3. Suppose that A is an R-algebra which is free of rank 2. Let $\{x, y\}$ be a basis of A. Set $1_A = \alpha x + \beta y$ and let $\mathfrak{a}$ be the ideal of R generated by $\{\alpha, \beta\}$. Show that $A/\mathfrak{a}A = \{0\}$ and therefore that $\mathfrak{a} = R$. Choose γ and δ in R such that $\alpha\gamma + \beta\delta = 1$ and show that $\{1_A, z = \delta x - \gamma y\}$ is a basis of A. Put $z^2 = b1_A + az$ and show that $A \cong R[X]/(X^2 - aX - b)$.

4. Let $R = \mathbb{Q}$ be the field of rational numbers. Show that if p and q are distinct primes, then $p(\mathbb{Q}^*)^2$ and $q(\mathbb{Q}^*)^2$ are distinct elements of $\mathbb{Q}^*/(\mathbb{Q}^*)^2$. So $\mathbb{Q}^*/(\mathbb{Q}^*)^2$ is a countably infinite elementary Abelian 2-group. It follows that there are infinitely many isomorphism classes of free quadratic algebras over $\mathbb{Q}$.

5. Let a be an odd positive integer. Show that there is an infinite sequence of positive integers $b_1, b_2, \ldots$ such that no two of the quadratic algebras $\mathbb{Q}[X]/(X^2 - aX - b_i)$ over $\mathbb{Q}$ are isomorphic.

6. Let K be a field of characteristic 2. Show that $K^2 = \{\alpha^2 \in K \mid \alpha \in K\}$ is a subfield of K and that K is a vector space over K^2. If K is finite show that $K = K^2$. For a finite K let F be the field of fractions of the polynomial ring $K[X_1,\ldots, X_n]$. Show that F^2 is the field of fractions of $K[X_1^2,\ldots, X_n^2]$ and $\{1, X_1,\ldots, X_n\}$ is a basis of F over F^2.

7. Let F be a field of characteristic 2. Consider any quadratic algebra of the form $F[X]/(X^2 - b)$. Show that $F[X]/(X^2 - b) \cong F[X]/(X^2) \Leftrightarrow b \in F^2$. If b and d are nonsquares, show that $F[X]/(X^2 - b) \cong F[X]/(X^2 - d) \Leftrightarrow$ the planes $F^21 \oplus F^2b$ and $F^21 \oplus F^2d$ in F are equal. With C the set of isomorphism classes of algebras of the form $F[X]/(X^2 - b)$ show: card $C = 1 \Leftrightarrow \dim_{F^2} F = 1$, card $C = 2 \Leftrightarrow \dim_{F^2} F = 2$, and card $C = \infty \Leftrightarrow \dim_{F^2} F > 2$. Exercise 6 provides examples.

8. Let A be a ring with char $A = k$. Let z be any element of A. The injection $\mathbb{Z}_k \to A$ has a unique extension to a homomorphism $\mathbb{Z}_k[X] \to A$ that maps X to z.

9. Show that the trivial quadratic algebra $\mathbb{Z}_2[X]/(X^2 + X)$, the dual numbers $\mathbb{Z}_2[X]/(X^2)$, the Galois field of four elements $\mathbb{Z}_2[X]/(X^2 + X + 1)$, and the ring $\mathbb{Z}_4$, is a complete listing of rings of order 4.

10. The quadratic algebras $\mathbb{Z}_2[X]/(X^2 + X)$ and $\mathbb{Z}_2[X]/(X^2 + X + 1)$ are not isomorphic and hence provide a counterexample in the context of Proposition (1.3). Construct more counterexamples over $\mathbb{Z}_4$.

11. Show that $\mathbb{Z}[X]/(X^2 - aX - b) \cong \mathbb{Z}[X]/(X^2 - nX)$ with $n \geq 0$ if and only if $X^2 - aX - b$ has a root in $\mathbb{Z}$.

12. Consider the exercise on page 618 in Bourbaki [Algebra 1972]: Show that any quadratic algebra $\mathbb{Z}[X]/(X^2 - aX - b)$ over $\mathbb{Z}$ is isomorphic to precisely one in the list: $\mathbb{Z}[X]/(X^2 - n)$, $n \geq 0$; $\mathbb{Z}[X]/(X^2 - nX)$, $n \neq k^2$; $\mathbb{Z}[X]/(X^2 - X - n)$, $n \neq k(k - 1)$. Is this assertion correct?

13. Let d be an integer which is square free. Show that the free quadratic algebra $\mathbb{Z}[X]/(X^2 - d)$ over $\mathbb{Z}$ is isomorphic to the $\mathbb{Z}$-subalgebra $S = \{n + m\sqrt{d} \in \mathbb{C} \mid n, m \text{ in } \mathbb{Z}\}$ of $\mathbb{C}$. Show that the conjugation of $\mathbb{Z}[X]/(X^2 - d)$ corresponds to the map σ of S given by $\sigma(n + m\sqrt{d}) = n - m\sqrt{d}$.

14. Let $R = \{n + mi \in \mathbb{C} \mid n, m \text{ in } \mathbb{Z}\}$, where $i = \sqrt{-1}$, be the ring of Gaussian integers. Show that $R[X]/(X^2 - X - r) \cong R[X]/(X^2 - iX + r)$, for any r in R. Next, check that $C = \{0, 1, i, i+1\}$ is a set of representatives of $R/2R$. Let $R[X]/(X^2 - aX - b)$ be a quadratic algebra over R. Choose c in C such that $a \in c + 2R$ and show that $R[X]/(X^2 - aX - b) \cong R[X]/(X^2 - cX - r)$ for some $r \in R$. Determine that $R[X]/(X^2 - r)$, $R[X]/(X^2 - X - r)$, and $R[X]/(X^2 - (1 + i)X - r)$, with r ranging over R, form a complete list of isomorphism classes of free quadratic algebras over R.

In Exercises 15 to 17, F is a quadratic number field. So $F = \mathbb{Q}(\sqrt{d})$, where $d \in \mathbb{Z}$ is an integer which is square free, i.e., has has no square factors. Refer to

Samuel, Lang, or Ribenboim for the basic facts about the ring of algebraic integers of F, i.e., the integral closure of $\mathbb{Z}$ in F, used later.

15. If d is not congruent to 1 mod 4, then

$$R = \{n + m\sqrt{d} \mid n \text{ and } m \text{ in } \mathbb{Z}\}.$$

If d is congruent to 1 mod 4, then

$$R = \{\tfrac{1}{2}(n + m\sqrt{d}) \mid n, m \text{ in } \mathbb{Z} \text{ both even or both odd}\}.$$

In the first case, show that $\{0, 1, \sqrt{d}, 1 + \sqrt{d}\}$ is a complete set of representatives of R/2R. In the second case, show that

$$\{0, 1, \sqrt{d}, 1 + \sqrt{d}, \tfrac{1}{2}(1 + \sqrt{d}), \tfrac{1}{2}(1 - \sqrt{d})\}$$

is a complete set of representatives of R/2R. Is 2R a prime ideal in either situation?

16. Suppose the quadratic number field F is *complex*, i.e., that $d < 0$. The group R^* has the description: $R^* = \{\pm 1\}$ if $d \neq -1$ and -3; $R^* = \{\pm 1, \pm i\}$ if $d = -1$; and $R^* = \{\pm 1, \tfrac{1}{2}(1 \pm \sqrt{-3}), \tfrac{1}{2}(-1 \pm \sqrt{-3})\}$ if $d = -3$. Show that $(R^*)^2$ is respectively equal to $\{1\}$, $\{\pm 1\}$, and $\{1, \tfrac{1}{2}(-1 \pm \sqrt{-3})\}$ and compute $R^*/(R^*)^2$ in each case.

17. Use the results of Exercises 15 and 16 to give a complete list of representatives of free quadratic algebras over R in the complex case.

18. Consider the ring $R = \{n + m\sqrt{5} \in \mathbb{R} \mid n, m \text{ in } \mathbb{Z}\}$. Show that the quadratic algebras $R[X]/(X^2 - X)$ and $R[X]/(X^2 - \sqrt{5}X + 1)$ are not isomorphic. Is this a counterexample to parts (2) and (3) of Proposition (1.3)?

19. Let $S = R[X]/\mathfrak{a}$ where $\mathfrak{a} = (X^2 - aX - b)$, and $T = R[X]/\mathfrak{b}$ where $\mathfrak{b} = (X^2 - cX - d)$, be two free quadratic algebras. Consider the R-algebra $R[X,Y]/\mathfrak{c}$, where $R[X,Y]$ is the polynomial algebra in the commuting variables X and Y and $\mathfrak{c}$ is the ideal generated by $X^2 - aX - b$ and $Y^2 - cY - d$. Show that the R-algebras $S \otimes_R T$ and $R[X,Y]/\mathfrak{c}$ are isomorphic.

Hints:

2. Show that $A \times (R/\mathfrak{a}) \to A/\mathfrak{a}A$ given by $(a, (r + \mathfrak{a})) \to ra + \mathfrak{a}A$ defines an algebra homomorphism $A \otimes_R (R/\mathfrak{a}) \to A/\mathfrak{a}A$ and that the inverse $A/\mathfrak{a}A \to A \otimes_R (R/\mathfrak{a})$ is given by $(a + \mathfrak{a}A) \to a \otimes (1 + \mathfrak{a})$.

4. Use the fact that $\mathbb{Z}$ is a unique factorization domain. For the last statement apply Example 4 or (1.3).

5. Choose c relatively prime to a. Since a is odd, the arithmetic progression $a^2 + 4c, a^2 + 2(4c), a^2 + 3(4c), \ldots$ contains infinitely many primes by Dirichlet's theorem.

6. If K is finite, then $K \to K^2$ defined by $a \to a^2$ is an isomorphism.

9. By Exercise 8, either char $R = 2$ or $R \cong \mathbb{Z}_4$. Now use (1.2).

12. It is wrong. The algebras $\mathbb{Z}[X]/(X^2 - n)$ with $n < 0$ are not on the list.

19. Let $x = X + \mathfrak{a}$, $y = X + \mathfrak{b}$, $x' = X + \mathfrak{c}$, and $y' = Y + \mathfrak{c}$. Show that there exists an algebra isomorphism $S \otimes_R T \to R[X,Y]/\mathfrak{c}$ that satisfies $1 \otimes 1 \to 1$, $x \otimes 1 \to x'$, and $1 \otimes y \to y'$.

2
Separable Algebras

Overview

This chapter introduces the important class of separable algebras. The concept of separability in this context is closely related to, and is in fact a generalization of, the classical concept of separable field extensions. The substance of the chapter is devoted to an analysis of separable free quadratic algebras and the conjugation automorphism. Throughout, R will be an arbitrary commutative ring.

A. Separability of Algebras

Let A be any R-algebra. Let A° be an R-module that is isomorphic to A and fix an isomorphism $\circ : A \to A^\circ$. Define a multiplication in A° by $a^\circ b^\circ = (ba)^\circ$. In this way, A° is an R-algebra, the *opposite algebra* of A. Suppose that A has an antiautomorphism α defined on it. Then $\alpha : A \to A$ is an isomorphism of R-modules and $(a^\alpha)(b^\alpha) = (ba)^\alpha$. We can therefore use α to construct A°. The *enveloping algebra* A^e *of* A is defined to be $A^e = A \otimes_R A^\circ$. Let M be an (A,A)-bimodule. So M is both a left A module and a right A module and for any x in M, a, b in A, and r in R, $a(xb) = (ax)b$ and $rx = xr$. It is easy to check that M is a left A^e-module with action $(a \otimes b^\circ)x = axb$; and that conversely, any left A^e-module is an (A,A)-bimodule in the natural way. Observe that A is a left A^e-module via $(a \otimes b^\circ)c = acb$ and that there is a unique homomorphism

$$\phi : A^e \to A$$

of left A^e-modules which satisfies $\phi(a \otimes b^\circ) = ab$. When appropriate, ϕ will be denoted by ϕ_A. Since $\phi(a \otimes 1^\circ) = a$, ϕ is surjective. So the sequence

18

$$0 \to \ker \phi \to A^e \xrightarrow{\phi} A \to 0$$

of left A^e-modules is exact. Note that if A is commutative, then $A = A^\circ$, $A^e = A \otimes_R A$, and ϕ is given by multiplication. In this case, ϕ is a homomorphism of R-algebras.

The algebra A is a *separable* R-algebra if the exact sequence

$$0 \to \ker \phi \to A^e \xrightarrow{\phi} A \to 0$$

splits, i.e., if there is an A^e-homomorphism $\theta : A \to A^e$ such that $\phi \theta = \mathrm{id}_A$. Note that if this is the case, then $A^e = A \otimes_R A^\circ = \ker \phi \oplus \theta A$. Consider the special case $A = R$. Here $R^\circ = R$ and $R^e = R \otimes_R R$. It is easy to check that $\phi : R^e \to R$ is an isomorphism whose inverse $\theta : R \to R^e$ is given by $\theta(r) = r \otimes 1$. In particular, R is a separable R-algebra.

The ring of $n \times n$ matrices $\mathrm{Mat}_n(R)$ over R is an R-algebra via $r \to rI$, where I is the identity matrix. This algebra is separable for any n. We verify this in the case $n = 2$, where the notational simplicity makes the proof very transparent. The general case is in principle the same. See Exercise 3. The transpose t is an antiautomorphism of $\mathrm{Mat}_2(R)$ and we use it to define $\mathrm{Mat}_2(R)^\circ$. Define $\theta : \mathrm{Mat}_2(R) \to \mathrm{Mat}_2(R) \otimes_R \mathrm{Mat}_2(R)^\circ = \mathrm{Mat}_2(R)^e$ by

$$\theta \begin{bmatrix} a & b \\ c & d \end{bmatrix} = \begin{bmatrix} a & 0 \\ c & 0 \end{bmatrix} \otimes \begin{bmatrix} 1 & 0 \\ 0 & 0 \end{bmatrix}^t + \begin{bmatrix} b & 0 \\ d & 0 \end{bmatrix} \otimes \begin{bmatrix} 0 & 1 \\ 0 & 0 \end{bmatrix}^t$$

for an arbitrary $\begin{bmatrix} a & b \\ c & d \end{bmatrix}$ in $\mathrm{Mat}_2(R)$. It is easy to check that θ is a homomorphism of $\mathrm{Mat}_2(R)^e$-modules. Since

$$(\phi_{\mathrm{Mat}_2(R)} \, \theta) \begin{bmatrix} a & b \\ c & d \end{bmatrix} = \begin{bmatrix} a & 0 \\ c & 0 \end{bmatrix} \begin{bmatrix} 1 & 0 \\ 0 & 0 \end{bmatrix} + \begin{bmatrix} b & 0 \\ d & 0 \end{bmatrix} \begin{bmatrix} 0 & 1 \\ 0 & 0 \end{bmatrix} = \begin{bmatrix} a & b \\ c & d \end{bmatrix},$$

θ fulfills the required splitting property. Therefore, $\mathrm{Mat}_2(R)$ is separable over R. It turns out that in fact $\mathrm{End}_R(P)$ is separable for any faithful finitely generated projective R-module P. See Exercise 5.

The current use of the concept separable is consistent with that from the theory of fields. Let E/F be a finite extension of fields. Then E is a finite-dimensional F-algebra in the obvious way. It turns out that E is a separable field extension of F if and only if E is a separable algebra over F. This fact will not be needed in this volume. In the case $[E : F] = 2$, it is verified in the

Example of Section C. In the general case, refer to the literature. See the Exercises of Chapter 9.

Continue to let F be a field and let A be a finite-dimensional algebra over F. Assume that A is central, i.e., that F = Cen A. An important fact asserts that A is separable over F if and only if A is simple (as a ring). According to a famous theorem of Wedderburn, an algebra satisfying this condition is isomorphic to the algebra of k × k matrices with coefficients in a finite-dimensional division algebra over F. Indeed, A is a finite-dimensional central simple algebra over F if and only if there exist a division ring D that is finite dimensional over its center F and a positive integer k such that $A \cong \text{Mat}_k(D)$ as algebras over F. These classical facts will be used later, but not until Chapter 9.

B. Separability Idempotents

Let A be any R-algebra. As already suggested, there are a number of equivalent characterizations of the property of separability. We single out one of them. Assume that A is separable. So $\phi : A^e \to A$ has a splitting map $\theta : A \to A^e$. Set $e = \theta(1) \in A^e$. Notice that $\phi(e) = 1$. Observe that

$$\theta(a) = \theta((1 \otimes a^\circ)1) = (1 \otimes a^\circ)\theta(1) = (1 \otimes a^\circ)e,$$

and in the same way, that $\theta(a) = (a \otimes 1^\circ)e$ for all a in A. Therefore, $(1 \otimes a^\circ)e = (a \otimes 1^\circ)e$ for all a in A. Assume conversely, that there is an element e in A^e such that

(∗) $\phi(e) = 1$ and $(1 \otimes a^\circ)e = (a \otimes 1^\circ)e$ for all a in A.

Define $\theta : A \to A^e$ by $\theta a = (1 \otimes a^\circ)e = (a \otimes 1^\circ)e$. Since θ is an A^e-homomorphism which splits ϕ, it follows that A is separable. Any element e in A^e that satisfies the conditions (∗) is called a *separability idempotent* for A. We have proved that

(2.1). A *is a separable R-algebra if and only if* A *has a separability idempotent.*

Supppose that e is a separability idempotent. So $e^2 - e = e(e - (1 \otimes 1^\circ))$. Note that $e - (1 \otimes 1^\circ) \in \ker \phi$. Define the splitting map θ as above. Since $\theta 1 = e$, it follows that $e^2 - e = (e - (1 \otimes 1^\circ))e \in (\ker \phi) \cap \theta A$. So $e^2 = e$, and hence e is an *idempotent* of the ring A^e. Put $e = \sum_i c_i \otimes d_i^o$, and note that

the conditions (∗) become $\sum_i c_i d_i = 1$ and $\sum_i ac_i \otimes d_i = \sum_i c_i \otimes (d_i a)^o$ for all

a in A.

Suppose that A is separable. The existence of a separability idempotent e

has implications for Cen A. Set $e = \sum_i c_i \otimes d_i^o$. Via the left A^e structure of

A, $ea = \sum_i c_i a d_i \in A$ for any a in A. Let a and b be any elements of A.

Then

$$(ea)b = (\sum_i c_i a d_i)b = (1 \otimes b^o)(ea) = (b \otimes 1^o)ea = b(\sum_i c_i a d_i) = b(ea).$$

So $ea \in$ Cen A. Conversely, if $c \in$ Cen A, then $ec = \sum_i c_i c d_i = c(\sum_i c_i d_i) = c$.

So $c \in eA$. We have proved that

$$\text{Cen } A = eA.$$

We illustrate this in the case $A = \text{Mat}_2(R)$. Here the separability idempotent

is

$$e = \theta\begin{bmatrix} 1 & 0 \\ 0 & 1 \end{bmatrix} = \begin{bmatrix} 1 & 0 \\ 0 & 0 \end{bmatrix} \otimes \begin{bmatrix} 1 & 0 \\ 0 & 0 \end{bmatrix}^t + \begin{bmatrix} 0 & 0 \\ 1 & 0 \end{bmatrix} \otimes \begin{bmatrix} 0 & 1 \\ 0 & 0 \end{bmatrix}^t.$$

Since

$$e\begin{bmatrix} a & b \\ c & d \end{bmatrix} = \begin{bmatrix} 1 & 0 \\ 0 & 0 \end{bmatrix}\begin{bmatrix} a & b \\ c & d \end{bmatrix}\begin{bmatrix} 1 & 0 \\ 0 & 0 \end{bmatrix} + \begin{bmatrix} 0 & 0 \\ 1 & 0 \end{bmatrix}\begin{bmatrix} a & b \\ c & d \end{bmatrix}\begin{bmatrix} 0 & 1 \\ 0 & 0 \end{bmatrix} = \begin{bmatrix} a & 0 \\ 0 & a \end{bmatrix},$$

we find that

$$\text{Cen Mat}_2(R) = \{rI \mid r \in R\}.$$

C. Separable Free Quadratic Algebras

We now consider the question as to when a free quadratic algebra S over R is

separable. Put $S = R[X]/(X^2 - aX - b)$. Since S is commutative, $S^o = S$ and

$S^e = S \otimes_R S$. Recall that $\{1_S, v\}$, where $v = X + (X^2 - aX - b)$, is a basis of

S. Of course, $v^2 = av + b$ and $\{1 \otimes 1, 1 \otimes v, v \otimes 1, v \otimes v\}$ is a basis of S^e.

Assume that $a^2 + 4b$ is in R^* and put $u = (a^2 + 4b)^{-1}$. Set

$$e = u((a^2 + 2b)(1 \otimes 1) - a(1 \otimes v) - a(v \otimes 1) + 2(v \otimes v)).$$

It is easy to verify that e is a separability idempotent for S. Note first that

$$\phi_S(e) = u(a^2 + 2b - av - av + 2v^2) = u(a^2 + 2b - 2av + 2(av + b))$$
$$= u(a^2 + 4b) = 1.$$

A routine expansion of terms shows that

$$u((a^2 + 2b)(1 \otimes v) - a \otimes v^2 - av \otimes v + 2v \otimes v^2)$$
$$= u(2v^2 \otimes v - av^2 \otimes 1 - av \otimes v + (a^2 + 2b)(v \otimes 1)),$$

and therefore that $(1 \otimes v)e = (v \otimes 1)e$. Since $\{1, v\}$ is a basis for S, e is a separability idempotent for S. We have proved that if $a^2 + 4b$ is in R^*, then the R-algebra $S = R[X]/(X^2 - aX - b)$ is separable.

Conversely, we will now see that if $S = R[X]/(X^2 - aX - b)$ is separable, then $a^2 + 4b$ is in R^*. So let e be a separability idempotent for S. Set

$$e = r_0(1 \otimes 1) + r_1(1 \otimes v) + r_2(v \otimes 1) + r_3(v \otimes v)$$

with r_0, r_1, r_2, r_3 in R. Since $\phi_S e = 1$, $r_1 + r_2 + ar_3 = 0$ and $r_0 + br_3 = 1$. Since $(1 \otimes v)e = (v \otimes 1)e$, it is easy to see that $br_1 = br_2$, $r_0 + ar_1 = br_3$, $r_0 + ar_2 = br_3$, and $r_2 + ar_3 = r_1 + ar_3$. Therefore,

i) $r_1 = r_2$, ii) $r_0 + ar_1 = br_3$, iii) $2r_1 + ar_3 = 0$, and iv) $r_0 + br_3 = 1$.

Observe that

$$(a^2 + 4b)r_1 = a(ar_1) + 2b(2r_1) = a(br_3 - r_0) - 2b(ar_3) = -ar_0 - bar_3 = -a$$

and

$$(a^2 + 4b)r_3 = a(ar_3) + 4(br_3) = a(-2r_1) + 4(r_0 + ar_1) = 4r_0 + 2ar_1$$

$$= 2r_0 + 2r_0 + 2ar_1 = 2r_0 + 2br_3 = 2.$$

Put $u = a^2 + 4b$. If u is not in R^*, then u is contained in some maximal ideal $\mathfrak{m}$ of R. It follows from the equalities that 2 and a are in $\mathfrak{m}$. By ii),

$r_0 - br_3 \in \mathfrak{m}$, so that by iv), $1 = r_0 + br_3$ is also in $\mathfrak{m}$. This contradiction shows that $u = a^2 + 4b \in R^*$. In summary, we have proved:

(2.2). *The* R-*algebra* $R[X]/(X^2 - aX - b)$ *is separable if and only if* $a^2 + 4b$ *is in* R^*.

An R-algebra that is isomorphic to some $R[X]/(X^2 - aX - b)$ with $a^2 + 4b$ in R^* is called a *free separable quadratic algebra* over R. By Exercise 3 in Chapter 1E, any R-algebra which is free of rank 2 and separable is free separable quadratic.

In the proof of (2.2), $e = r_0(1 \otimes 1) + r_1(1 \otimes v) + r_2(v \otimes 1) + r_3(v \otimes v)$, with the r_i in R, was taken to be an arbitrary separability idempotent of S and it was shown that: $u = a^2 + 4b$ is invertible, $r_1 = r_2 = -au^{-1}$, $r_3 = 2u^{-1}$, and $r_0 = -r_1 a + r_3 b = a^2 u^{-1} + 2bu^{-1} = (a^2 + 2b)u^{-1}$. A comparison of coefficients shows that e is the idempotent from the beginning of this section. In particular,

(2.3). *The separability idempotent of a free separable quadratic algebra* S *over* R *is unique.*

Example. Let F be a field and let $p(X) = X^2 - aX - b \in F[X]$. Then $p(X)$ is separable if and only if $a^2 + 4b \neq 0$. This is easily seen. Suppose $a^2 + 4b \neq 0$. Let E be any field extension of F such that $p(X) = (X - \gamma)(X - \delta)$ in $E[X]$. Note that $\delta = a - \gamma$. If $\delta = \gamma$, then $a = 2\gamma$. But this is not possible, since $0 = 4(\gamma^2 - a\gamma - b) = a^2 - 2a^2 - 4b = -(a^2 + 4b)$. So $p(X)$ is separable. Suppose conversely, that $p(X)$ is separable. Assume that $a^2 + 4b = 0$. If char $F \neq 2$, then $b = -(\frac{a}{2})^2$. So $p(X) = (X - \frac{a}{2})^2$ in $F[X]$, contradicting the separability of $p(X)$. If char $F = 2$, then $a = 0$. If γ is a root in a field extension E of F, then $p(X) = (X - \gamma)^2$ in $E[X]$, again a contradiction. Therefore, $a^2 + 4b \neq 0$ in either case. Assume now that $p(X)$ is irreducible and consider the extension field $E = F[X]/(p(X))$ of F. Since $a^2 + 4b \neq 0$ is obviously the same condition as $a^2 + 4b \in F^*$, we find by Proposition (2.2) that

E is a separable extension of F $\Leftrightarrow$ E is a separable algebra over F.

Notice that if char $F \neq 2$, then an irreducible $p(X)$ is automatically separable. This is clear since $a^2 + 4b = 0$ implies that $\frac{a}{2}$ is a root of $p(X)$ in F.

D. Properties of Conjugation

Let $S = R[X]/(X^2 - aX - b)$ be separable. So $a^2 + 4b$ is in R^*. Let $v = X + (X^2 - aX - b)$. Recall that the conjugation $\sigma : S \to S$ of S is defined by $(r + tv)^\sigma = (r + ta) - tv$. It is clear that $R \subseteq \{s \in S \mid s^\sigma = s\}$. Since

$$(r + tv)^\sigma = (r + tv) \Rightarrow ta = 0 \text{ and } 2t = 0$$
$$\Rightarrow t(a^2 + 4b) = 0 \Rightarrow t = 0 \Rightarrow r + tv = r,$$

we find that $R = \{s \in S \mid s^\sigma = s\}$. The conjugation σ is in fact the unique involution on S with this property. To see this, let τ be any involution on S. Put $v^\tau = r + tv$ with r and t in R. Since $\tau^2 = \mathrm{id}_S$, we find that $r + rt = 0$ and $t^2 = 1$. Since $(v^2)^\tau = (v^\tau)^2$, $ar = r^2$ and $2rt + a = at$. So $(a - r)r = 0$ and $(a - r)t = a - r$. By easy computations,

$$[(1 + t)v]^\tau = (1 + t)v \text{ and } [(a - r)v]^\tau = (a - r)v.$$

If R is the fixed point set of τ, then $r = a$ and $t = -1$, and indeed $\tau = \sigma$. We turn to the split exact sequence

$$(*) \qquad 0 \to \ker \phi_S \to S \otimes_R S \xrightarrow{\;\phi_S\;} S \to 0 .$$

Since S is commutative, ϕ_S is an R-algebra homomorphism. Set

$$T = \ker \phi_S.$$

The remainder of this section analyzes T and establishes its close connection with S and its conjugation.

Let $z = r_0(1 \otimes 1) + r_1(1 \otimes v) + r_2(v \otimes 1) + r_3(v \otimes v)$ be an arbitrary element of $S \otimes_R S$. By easy computation, z is in T if and only if

$$r_0 + r_1 v + r_2 v + r_3(av + b) = 0$$

if and only if $r_0 + r_3 b = 0$ and $r_1 + r_2 + r_3 a = 0$. It follows that the typical element of T has the form

$$z = -r_3 b(1 \otimes 1) + r_1(1 \otimes v) + (-r_1 - r_3 a)(v \otimes 1) + r_3(v \otimes v)$$

$$= r_1(1 \otimes v - v \otimes 1) - r_3(b(1 \otimes 1) + a(v \otimes 1) - v \otimes v).$$

Notice that $1 \otimes v - v \otimes 1$ and $b(1 \otimes 1) + a(v \otimes 1) - v \otimes v$ are both in T. Since $\{1 \otimes 1, 1 \otimes v, v \otimes 1, v \otimes v\}$ is a basis of $S \otimes_R S$, it follows that $1 \otimes v - v \otimes 1$ and $b(1 \otimes 1) + a(v \otimes 1) - v \otimes v$ together form a basis of T.

Let $u = (a^2 + 4b)^{-1}$. Recall from Section C that

$$e = u((a^2 + 2b)(1 \otimes 1) - a(1 \otimes v) - a(v \otimes 1) + 2(v \otimes v))$$

is the unique separability idempotent of S. Define the map $\varphi : S \rightarrow S \otimes_R S$ by $\varphi(s) = (s \otimes 1)(1 - e)$ for all s in S. Since $\phi_S((s \otimes 1)(1 - e)) = s(1 - 1) = 0$, φ maps into T. Clearly,

$$\varphi : S \rightarrow T$$

is a homomorphism of R-modules. We will show that it is in fact an isomorphism. Routine calculations show that

$1 - e$

$$= u(a^2 + 4b)(1 \otimes 1) - u((a^2 + 2b)(1 \otimes 1) - a(1 \otimes v) - a(v \otimes 1) + 2(v \otimes v))$$

$$= u(2b(1 \otimes 1) + a(1 \otimes v) + a(v \otimes 1) - 2(v \otimes v)).$$

This equality and long-winded multiplications in $S \otimes_R S$ give

(#)
$$\varphi(v) = uab(1 \otimes 1) - 2ub(1 \otimes v) + (1 - 2ub)(v \otimes 1) - ua(v \otimes v) \quad \text{and}$$

$$\varphi(r) = 2rub(1 \otimes 1) + rua(1 \otimes v) + rua(v \otimes 1) - 2ru(v \otimes v)$$

for any r in R. It follows that

$$\varphi(a - 2v) = 1 \otimes v - v \otimes 1 \quad \text{and} \quad \varphi(2b + av) = b(1 \otimes 1) + a(v \otimes 1) - v \otimes v.$$

The matrix $\begin{bmatrix} a & 2b \\ -2 & a \end{bmatrix}$ of coefficients that arises in expressing $a - 2v$ and $2b + av$ in the basis $\{1, v\}$ of S has determinant the unit $a^2 + 4b$ of R. So this matrix is invertible, and $\{a - 2v, 2b + av\}$ is a basis of S. Therefore φ takes a basis of S to a basis of T and it follows that $\varphi : S \rightarrow T$ is an isomorphism of R-modules.

Notice that T is closed under the product of $S \otimes_R S$. Note also that $1 - e$ is in T and that $(1 - e)(1 - e) = 1 - e$. Since any element of T has the form

$(s \otimes 1)(1 - e)$ for s in S, it follows that $1 - e$ is an identity element for T and that T is an R-algebra.

Observe that the assignment $S \times S \to S \otimes_R S$, given by $(s, s') \to s' \otimes s$ for all s and s' in S, induces an R-module map from $S \otimes_R S \to S \otimes_R S$ such that $s \otimes s' \to s' \otimes s$. This is an involution on the algebra $S \otimes_R S$, which we call "switch" and denote sw. A moment's reflection shows that switch takes the submodule $T = \ker \phi_S$ of $S \otimes_R S$ onto itself. Since e is the unique separability idempotent of S, sw $e = e$. Therefore $sw(1 - e) = 1 - e$. It follows that the restriction of sw to T is an involution on the algebra T.

(2.4). *The map* $\phi : S \to T$ *is an isomorphism of R-algebras and the diagram*

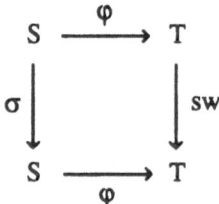

commutes.

Proof. Since $\phi(ss') = (ss' \otimes 1)(1 - e) = (s \otimes 1)(s' \otimes 1)(1 - e)^2 = \phi(s)\phi(s')$ for all s and s' in S, ϕ is an R-algebra map. Since we already know that ϕ is an isomorphism of R-modules, it is therefore an isomorphism of R-algebras. It remains to establish the commutativity of the diagram. Using the two equalities (#) we find that:

$$\phi(v) = uab(1 \otimes 1) - 2ub(1 \otimes v) + (1 - 2ub)(v \otimes 1) - ua(v \otimes v)$$

$$\phi(r) = 2rub(1 \otimes 1) + rua(1 \otimes v) + rua(v \otimes 1) - 2ru(v \otimes v)$$

$$\phi(v^\sigma) = \phi(a - v) = \phi(a) - \phi(v)$$

$$\begin{aligned}
&= 2aub(1 \otimes 1) + ua^2(1 \otimes v) + ua^2(v \otimes 1) - 2ua(v \otimes v) \\
&\quad - uab(1 \otimes 1) + 2ub(1 \otimes v) - (1 - 2ub)(v \otimes 1) + ua(v \otimes v)
\end{aligned}$$

$$\begin{aligned}
&= u(2ab(1 \otimes 1) + a^2(1 \otimes v) + a^2(v \otimes 1) - 2a(v \otimes v)) \\
&\quad + u(-ab(1 \otimes 1) + 2b(1 \otimes v) - (a^2 + 2b)(v \otimes 1) + a(v \otimes v))
\end{aligned}$$

$$= u(ab(1 \otimes 1) + (a^2 + 2b)(1 \otimes v) - 2b(v \otimes 1) - a(v \otimes v))$$

$$= sw(u(ab(1 \otimes 1) - 2b(1 \otimes v) + (a^2 + 2b)(v \otimes 1) - a(v \otimes v))$$

$$= sw((v \otimes 1)(1 - e)) = sw(\varphi v).$$

Since $\{1, v\}$ is a basis of S, it follows that the diagram commutes. QED.

E. Exercises

Consider the matrix ring $Mat_n(S)$ over a ring S. Several of the proofs of these exercises are facilitated by the use of the matrices $\{\varepsilon_{ij} \mid 1 \le i, j \le n\}$, where ε_{ij} is the $n \times n$ matrix which has a 1 in the (i,j) entry and 0 elsewhere. The set $\{\varepsilon_{ij} \mid 1 \le i, j \le n\}$ is a basis of $Mat_n(S)$ as S-module. Check that $\varepsilon_{ij}\varepsilon_{km} = \delta_{jk}\varepsilon_{im}$, where δ_{jk} is the Kronecker delta. As usual, R is a commutative ring.

1. Define $Mat_2(R) \times Mat_2(R) \to Mat_4(R)$ by

$$\left(\begin{bmatrix} r & s \\ t & u \end{bmatrix}, \begin{bmatrix} a & b \\ c & d \end{bmatrix} \right) \to \begin{bmatrix} r\begin{bmatrix} a & b \\ c & d \end{bmatrix} & s\begin{bmatrix} a & b \\ c & d \end{bmatrix} \\ t\begin{bmatrix} a & b \\ c & d \end{bmatrix} & u\begin{bmatrix} a & b \\ c & d \end{bmatrix} \end{bmatrix}.$$

Show that the induced map $Mat_2(R) \otimes_R Mat_2(R) \to Mat_4(R)$ is an algebra isomorphism.

2. Let S be an R-algebra. Show that the assignment from $S \times Mat_n(R)$ into $Mat_n(S)$ given by $(s, T) \to sT$ induces an isomorphism from $S \otimes_R Mat_n(R)$ onto $Mat_n(S)$. Deduce that $Mat_n(R) \otimes_R Mat_m(R)$ and $Mat_{nm}(R)$ are isomorphic for any n and m.

3. Consider the R-algebra $A = Mat_n(R)$. Show that A is central. Fix j with $1 \le j \le n$. Set $e = \sum_{i=1}^{n} \varepsilon_{ij} \otimes (\varepsilon_{ji})^t$ in $A \otimes_R A^o$. Show that $(\varepsilon_{km} \otimes I^t)e = (I \otimes (\varepsilon_{km})^t)e$ for any ε_{km}. Conclude that e is a separability idempotent for A and hence that A is a separable R-algebra.

4. Let A be a separable over R with separability idempotent e. Let M be left A^e-module. Set $M^A = \{x \in M \mid (1 \otimes a^\circ)x = (a \otimes 1^\circ)x \text{ for all } a \in A\}$ $= \{x \in M \mid xa = ax \text{ for all } a \in A\}$. Show that $M^A = eM$.

5. Let M be a faithful finitely generated projective module over R and consider the R-algebra $A = \operatorname{End}_R M$. Show that A is central and separable.

6. Consider the subrings $R = \mathbb{Z}[\sqrt{-3}]$ and $A = R[\omega]$ of $\mathbb{C}$, where ω is a primitive cube root of 1. Show that A is a separable R-algebra.

7. Let S be a free quadratic algebra over R. For any $s \in S$, let $\operatorname{Tr}(s)$ be the trace of the module homomorphism $m_s : S \to S$ defined by $m_s(x) = sx$. Show that the map σ defined by $s^\sigma = \operatorname{Tr}(s) - s$ agrees with the conjugation of S on a basis, and that it is therefore equal to the conjugation of S. Let $\operatorname{Nr}(s) = s(s)^\sigma$ and show that $\operatorname{Nr}(s)$ is the determinant of m_s.

Hints:

5. See page 19 of Orzech and Small.

6. Since $\omega = e^{(2/3)\pi i} = \cos\left(\frac{2}{3}\pi\right) + i\sin\left(\frac{2}{3}\pi\right) = -\frac{1}{2} + \frac{\sqrt{-3}}{2}$, $2\omega \in R$. Therefore, $2\omega \otimes 1 = 1 \otimes 2\omega$ in $A \otimes_R A$. Since $\omega^2 + \omega + 1 = 0$, it is not very difficult to check that $1 \otimes 1 + 1 \otimes \omega - \omega \otimes 1$ is a separability idempotent for A.

3
Groups of Free Quadratic Algebras

Overview

In this chapter we provide the set of isomorphism classes of free separable quadratic algebras over a commutative ring R with a group structure and focus on the properties of this group as well as those of its graded analogue. These will be important in Chapter 7 in the analysis of the Clifford algebra of a quadratic module. Certain "projective" versions of these groups will have crucial impact on the structure of the Brauer and Witt groups over R. See Chapters 13 and 14.

A. The Group $Qu_f(R)$

Shorten notation and denote the free quadratic R-algebra $R[X]/(X^2 - aX - b)$ by (a, b). Let

$$Q = \{ (a, b) \mid a^2 + 4b \in R^* \}.$$

By (2.2), any separable free quadratic R-algebra is isomorphic to one in Q. If (a, b) and (c, d) in Q are isomorphic, we write $(a, b) \cong (c, d)$.

(3.1). *Let* (a, b) *and* (c, d) *be in* Q. *The following statements are equivalent.*

(1) $(a, b) \cong (c, d)$

(2) *There exist elements* r *in* R *and* u *in* R^* *such that*
  (i) $c = ua + 2r$ *and* (ii) $d = u^2b - rua - r^2$.

(3) $X^2 - acX - (a^2d + c^2b + 4bd)$ *has a root in* R.

Proof. For the equivalence of (1) and (2) see (1.1). We prove that (3) implies (2). Let γ be a root of $X^2 - acX - (a^2d + c^2b + 4bd)$. Set $t = a^2 + 4b$ and put $r = t^{-1}(\gamma a + 2cb)$ and $u = t^{-1}(ac - 2\gamma)$. We show that r and u satisfy the required conditions. Since

$$tc = c(a^2 + 4b) = ca^2 - 2\gamma a + 2\gamma a + 4cb = (ac - 2\gamma)a + 2(\gamma a + 2cb),$$

we find that $c = ua + 2r$, and (i) is satisfied. As to (ii), proceed as follows:

$$d = u^2 b - rua - r^2 \Leftrightarrow d + rc = r^2 + u^2 b \Leftrightarrow t^2 rc + t^2 d = t^2 r^2 + t^2 u^2 b$$

$$\Leftrightarrow tc(\gamma a + 2cb) + t^2 d = (\gamma a + 2cb)^2 + (ac - 2\gamma)^2 b$$

$$\Leftrightarrow t[c(\gamma a + 2cb) + td] = \gamma^2(a^2 + 4b) + bc^2(a^2 + 4b)$$

$$\Leftrightarrow c\gamma a + 2c^2 b + (a^2 + 4b)d = \gamma^2 + bc^2$$

$$\Leftrightarrow \gamma^2 - ac\gamma - a^2 d - c^2 b - 4bd = 0.$$

The proof that (3) implies (2) is complete. That (2) implies (3) is now easy. Let r and u satisfy (i) and (ii) and put $\gamma = ra - 2bu$. Using (i), we find again with $t = a^2 + 4b$ that $r = t^{-1}(\gamma a + 2cb)$ and $u = t^{-1}(ac - 2\gamma)$. The chain of equivalences above shows that γ is a root of $X^2 - acX - (a^2 d + c^2 b + 4bd)$.

QED.

Let (a, b) and (c, d) be in Q, and define the element $(a, b) * (c, d)$ by

$$(a, b) * (c, d) = (ac, a^2 d + c^2 b + 4bd).$$

Since $(ac)^2 + 4(a^2 d + c^2 b + 4bd) = (a^2 + 4b)(c^2 + 4d) \in R^*$, $(a, b) * (c, d)$ is in Q.

(3.2). *Let* (a, b), (a', b'), (c, d), *and* (c', d') *be in* Q. *Suppose that* $(a, b) \cong (a', b')$ *and* $(c, d) \cong (c', d')$. *Then* $(a, b) * (c, d) \cong (a', b') * (c', d')$.

Proof. By (3.1), there exist scalars r_1 and r_2 in R and u_1 and u_2 in R^*, such that

$$(\#) \quad \begin{cases} a' = u_1 a + 2r_1 & \text{and} \quad b' = u_1^2 b - r_1 u_1 a - r_1^2 \\ c' = u_2 c + 2r_2 & \text{and} \quad d' = u_2^2 d - r_2 u_2 c - r_2^2 . \end{cases}$$

We now put $r = 2r_1 r_2 + r_1 u_2 c + r_2 u_1 a$ and $u = u_1 u_2$. It suffices to show that r and u satisfy the conditions (i) and (ii) of part (2) of (3.1) relative to the algebras $(a, b) * (c, d) = (ac, a^2 d + c^2 b + 4bd)$ and $(a', b') * (c', d') =$

(a'c', $(a')^2d' + (c')^2b' + 4b'd'$). Since, a'c' = $(u_1a + 2r_1)(u_2a + 2r_2)$ = uac + 2r, we have (i). As to (ii), set s = $a^2d + c^2b + 4bd$ and s' = $(a')^2d' + (c')^2b' + 4b'd'$. We must show that s' = $u^2s - ruac - r^2$. To verify this equality, expand s' by using the equations (#), then expand $u^2s - ruac - r^2$ and compare the two sides. This is exhausting, but routine.   QED.

Denote the R-algebra isomorphism class of an element (a, b) of Q by [a, b] and let $Qu_f(R)$ = {[a, b] | (a, b) ∈ Q} be the set of all isomorphism classes. Define a product on $Qu_f(R)$ by

$$[a, b][c, d] = [(a, b) * (c, d)] .$$

This operation is well defined by (3.2), and it follows easily from the definitions that it is associative and commutative. The class [1, 0] is an identity element. Also, [a, b][a, b] = $[a^2, 2a^2b + 4b^2]$ = [1, 0], since (1, 0) ≅ $(a^2, 2a^2b + 4b^2)$ by taking r = −2b and u = $a^2 + 4b$ in (2) of (3.1). So every element is its own inverse. Therefore,

$$Qu_f(R)$$

is a group − the *free quadratic group of* R. Observe that it is an elementary Abelian 2-group.

(3.3). *Let* [a, b] ∈ $Qu_f(R)$. *Then* [a, b] = [1, 0] *if and only if* $X^2 − aX − b$ *has a root in* R.

Proof. By the equivalence of (1) and (3) of (3.1), (a, b) is isomorphic to (1, 0) if and only if $X^2 − aX − b$ has a root in R.   QED.

Example 1.   Refer to Examples 1 and 3 of Chapter 1A. Observe that $\mathbb{C}[X]/(X^2 − X)$ is up to isomorphism the only free quadratic separable algebra over $\mathbb{C}$ and that $\mathbb{R}[X]/(X^2 − X)$ and $\mathbb{R}[X]/(X^2 + 1)$ are the only ones over $\mathbb{R}$. It follows that

$$Qu_f(\mathbb{C}) = 1 \quad \text{and} \quad Qu_f(\mathbb{R}) \cong \mathbb{Z}_2.$$

Observe that $Qu_f(\mathbb{C}) \cong \mathbb{C}^*/(\mathbb{C}^*)^2$ and $Qu_f(\mathbb{R}) \cong \mathbb{R}^*/(\mathbb{R}^*)^2$. We will see shortly that this is no coincidence.

B. The Discriminant δ

Define the *discriminant*

$$\delta \ : Qu_f(R) \ \rightarrow \ R^*/(R^*)^2$$

by $[a, b] \rightarrow (a^2 + 4b)(R^*)^2$. That δ is well defined is an easy consequence of the equivalence of (1) and (2) of (3.1). That it is a homomorphism follows from the equality $(ac)^2 + 4(a^2d + c^2b + 4bd) = (a^2 + 4b)(c^2 + 4d)$.

If the polynomial $X^2 - aX - b$ has a root, say γ, in R, then $a^2 + 4b = a^2 + 4(\gamma^2 - a\gamma) = a^2 - 4a\gamma + 4\gamma^2 = (a - 2\gamma)^2$, so that $a^2 + 4b \in (R^*)^2$. A moment's reflection and (3.3) show that ker δ is a measure of the failure of the converse of this statement.

(3.4). *The homomorphism* $\delta : Qu_f(R) \rightarrow R^*/(R^*)^2$ *is injective under each of the following assumptions:*

(1) $2 \in R^*$,

(2) *R is an integrally closed domain with* char R $\neq 2$, *or*

(3) 2 *is not a zero divisor of R and* 2R *is a prime ideal* .

Proof. Suppose that $(a, b) = R[X]/(X^2 - aX - b)$ satisfies $a^2 + 4b = u^2$ with u in R^*. Then by (1.3), $R[X]/(X^2 - aX - b) \cong R[X]/(X^2 - X) = (1, 0)$. So $[a, b] = [1, 0]$, and we are done. QED.

(3.5). *If* $2 \in R^*$, *then* $\delta : Qu_f(R) \rightarrow R^*/(R^*)^2$ *is an isomorphism.*
Proof. It must be shown that δ is onto. Let u in R^*. Take any a in R and set $b = 4^{-1}(u - a^2)$. So $u = a^2 + 4b$ and $\delta[a, b] = u(R^*)^2$.   QED.

Example 2. By Exercise 4 of Chapter 1E, $Qu_f(\mathbb{Q}) \cong \mathbb{Q}^*/(\mathbb{Q}^*)^2$ is a countably infinite group.

The natural ring homomorphism $R \rightarrow R/4R$ restricts to a group homomorphism $R^* \rightarrow (R/4R)^*$, which in turn induces a homomorphism $R^*/(R^*)^2 \rightarrow (R/4R)^*/((R/4R)^*)^2$. This map determines the image of δ.

(3.6). *The sequence* $Qu_f(R) \xrightarrow{\delta} R^*/(R^*)^2 \rightarrow (R/4R)^*/((R/4R)^*)^2$ *is exact.*
Proof. Suppose $a^2 + 4b \in R^*$. Since

$$(a + 4R)^2 = a^2 + 4R = (a^2 + 4b) + 4R,\ a + 4R \in (R/4R)^*,$$

and hence $(a^2 + 4b) + 4R \in ((R/4R)^*)^2$. So the image of $(a^2 + 4b)(R^*)^2$ in $(R/4R)^*/((R/4R)^*)^2$ is trivial. It remains to show that δ maps $Qu_f(R)$ onto the kernel of $R^*/(R^*)^2 \to (R/4R)^*/((R/4R)^*)^2$. Let $u(R^*)^2 \in R^*/(R^*)^2$ be any element such that $u + 4R = (a + 4R)^2$ for some $a + 4R$ in $(R/4R)^*$. So $u = a^2 + 4b$ for some b in R. Since $\delta[a, b] = u(R^*)^2$, we are done.

QED.

Example 3. Consider the case $R = \mathbb{Z}$. Put $\mathbb{Z}/4\mathbb{Z} = \{0, 1, 2, 3\}$. Since $(\mathbb{Z}/4\mathbb{Z})^* = \{1, 3\}$, $((\mathbb{Z}/4\mathbb{Z})^*)^2 = \{1\}$. So $\mathbb{Z}^*/(\mathbb{Z}^*)^2 \to (\mathbb{Z}/4\mathbb{Z})^*/((\mathbb{Z}/4\mathbb{Z})^*)^2$ is an isomorphism. Therefore the image of $\delta : Qu_f(\mathbb{Z}) \to \mathbb{Z}^*/(\mathbb{Z}^*)^2$ is trivial. It follows that $Qu_f(\mathbb{Z}) = 1$.

Remarks. 1) Suppose that $2 \in R^*$. Then $4 \in R^*$, and hence $R' = R/4R = \{0\}$. As a consequence, (3.5) is a special case of (3.6). Observe also that if $R = \{0\}$, then $R[X] = \{0\} = R[X]/(X^2 - X)$ is the only R-algebra. So $Qu_f(R) = 1$.

2) Suppose char $R = 2$. Since $4R = \{0\}$, $R^*/(R^*)^2 \to (R/4R)^*/((R/4R)^*)^2$ is an isomorphism. So by (3.6), δ is the trivial map and gives no information about $Qu_f(R)$. The approach to the structure of $Qu_f(R)$ in this case involves the subgroup $\wp(R) = \{r + r^2 \mid r \in R\}$ of R and the map $R/\wp(R) \to Qu_f(R)$ given by $r + \wp(R) \to [1, r]$. This turns out to be an isomorphism if R is a field or a local ring. These matters are left to Exercises 6 – 11.

3) The structure of the group $Qu_f(R)$ seems to be very delicate in general. See Exercises 3 – 5 for the case where R is the ring of integers in a quadratic number field. For general number theoretic rings, it is closely related to certain aspects of global class field theory. See Exercise 5 in this chapter and Exercise 15 in Chapter 14F.

C. The Group $QU_f(R)$

We will now construct a graded version of the group $Qu_f(R)$. In order to rule out possible confusion, the elements of the ring $\mathbb{Z}_2$ are written as $\overline{0}$ and $\overline{1}$. Let $(a, b) = R[X]/(X^2 - aX - b)$ be a free separable quadratic algebra over R. For $\varepsilon \in \mathbb{Z}_2$, the graded algebra $(a, b)^\varepsilon$ was defined in Chapter 1C: $(a, b)^{\overline{0}}$ is the algebra (a, b) supplied with the trivial grading; and $(a, b)^{\bar{1}}$ is the algebra (a, b) with $a = 0$ and the grading $R \oplus Rv$, where $v = X + (X^2 - aX - b)$. Since $a^2 + 4b \in R^*$, 2 and b are both in R^* in the latter case.

Denote by GQ the set of all graded quadratic algebras of the form $(a, b)^\varepsilon$ with $a^2 + 4b \in R^*$ and $\varepsilon \in \mathbb{Z}_2$.

(3.7). *Let $(a, b)^\varepsilon$ and $(c, d)^\eta$ be in GQ. Then $(a, b)^\varepsilon \cong (c, d)^\eta$ (as graded algebras) $\Leftrightarrow (a, b) \cong (c, d)$ and $\varepsilon = \eta$.*

Proof. If $(a, b)^\varepsilon \cong (c, d)^\eta$, then clearly $(a, b) \cong (c, d)$ and $\varepsilon = \eta$. We assume that $(a, b) \cong (c, d)$ and $\varepsilon = \eta$ and verify the converse. If $\varepsilon = \eta = \bar{0}$, this is clear. Suppose $\varepsilon = \eta = \bar{1}$. Now $a = c = 0$. Let $\varphi : (c, d) \to (a, b)$ be an isomorphism and refer to Proposition (1.1) and its proof. So $\varphi w = r + uv$, with $r \in R, u \in R^*$ and $2r = 0$. Since $2 \in R^*$ in the present case, $r = 0$ and hence $\varphi w = uv$. It follows that $\varphi : (c, d)^\eta \to (a, b)^\varepsilon$ is graded.   QED.

For $-1 \in \mathbb{Z}$ and ε in $\mathbb{Z}_2$, let $(-1)^\varepsilon = 1$ or -1 (both in $\mathbb{Z}$) according to whether $\varepsilon = \bar{0}$ or $\bar{1}$. Let $(a, b)^\varepsilon$ and $(c, d)^\eta$ be elements of GQ. If either ε or η equals $\bar{0}$, then $(ac)^2 + 4(a^2d + c^2b + (-1)^{\varepsilon\eta}4bd) = (a^2 + 4b)(c^2 + 4d)$ and if $\varepsilon = \eta = \bar{1}$, then $a = c = 0$ and $(ac)^2 + 4(a^2d + c^2b + (-1)^{\varepsilon\eta}4bd) = -16bd$. Observe that $(ac)^2 + 4(a^2d + c^2b + (-1)^{\varepsilon\eta}4bd) \in R^*$ in either case, and that therefore,

$$(a, b)^\varepsilon * (c, d)^\eta = (ac, a^2d + c^2b + (-1)^{\varepsilon\eta} 4bd)^{\varepsilon+\eta}$$

defines an element in GQ.

Denote the graded isomorphism class of $(a, b)^\varepsilon$ by $[a, b]^\varepsilon$ and let $QU_f(R) = \{[a, b]^\varepsilon \mid (a, b)^\varepsilon \in GQ\}$ be the set of all isomorphism classes. In view of (3.7), $[a, b]^\varepsilon = [c, d]^\eta$ in $QU_f(R)$ if and only if $[a, b] = [c, d]$ in $Qu_f(R)$ and $\varepsilon = \eta$. Define a product on $QU_f(R)$ by

$$[a, b]^\varepsilon[c, d]^\delta = [(a, b)^\varepsilon * (c, d)^\delta].$$

It follows from (3.2) and (3.7) that this operation is well defined. It is easy to check that it is associative and commutative and that $[1, 0]^{\bar{0}}$ is an identity element. Consider the element $([a, b]^{\varepsilon})^2 = [a^2, 2a^2b + (-1)^{\varepsilon\varepsilon}4b^2]^{\bar{0}}$ of $QU_f(R)$.

If $\varepsilon = \bar{0}$, then $[a^2, 2a^2b + (-1)^{\varepsilon\varepsilon}4b^2]^{\bar{0}} = [1, 0]^{\bar{0}}$, since $[a^2, 2a^2b + 4b^2] = [1, 0]$ in $Qu_f(R)$. It follows that $([a, b]^{\varepsilon})^4 = [1, 0]^{\bar{0}}$ for all $[a, b]^{\varepsilon}$ in $QU_f(R)$. Therefore,

$$QU_f(R)$$

is a group – the *graded free quadratic group* of R. It is Abelian of exponent 4.

It is easy to see that the assignment $[a, b] \to [a, b]^{\bar{0}}$ defines an injective homomorphism

$$Qu_f(R) \to QU_f(R) .$$

If $2 \notin R^*$, then there are no algebras of the form $(a, b)^{\bar{1}}$ and this map is an isomorphism. So we assume for the remainder of this section that $2 \in R^*$.

Define $QU_f(R) \to \mathbb{Z}_2$ by $[a, b]^{\varepsilon} \to \varepsilon$. This is a homomorphism which is onto since $(0, 1)^{\bar{1}}$ is in GQ. It is clear that it extends the map above to an exact sequence

$$1 \to Qu_f(R) \to QU_f(R) \to \mathbb{Z}_2 \to 0.$$

If $-1 \in (R^*)^2$, then the assignment $\bar{0} \to [1, 0]^{\bar{0}}$ and $\bar{1} \to [0, 1]^{\bar{1}}$ defines a homomorphism $\mathbb{Z}_2 \to QU_f(R)$ by (3.3) and (3.7). This map splits the sequence. Conversely, if the sequence splits, then $-1 \in (R^*)^2$. See Exercise 2.

To describe the structure of $QU_f(R)$, start with the set $\mathbb{Z}_2 \times R^*/(R^*)^2$. Define a product by $(\varepsilon, x)(\eta, y) = (\varepsilon + \eta, (-1)^{\varepsilon\eta}xy)$ for any (ε, x) and (η, y) in $\mathbb{Z}_2 \times R^*/(R^*)^2$. Here $(-1)^{\bar{0}}x = x$, and with $x = rR^*$, $(-1)^{\bar{1}}x = (-r)R^*$. Denote the resulting Abelian group by

$$\mathbb{Z}_2 \times R^*/(R^*)^2.$$

Recall the discriminant homomorphism $\delta : Qu_f(R) \to R^*/(R^*)^2$ and define

$$QU_f(R) \to \mathbb{Z}_2 \times R^*/(R^*)^2$$

by $[a, b]^\varepsilon \to (\varepsilon, \delta[a, b]) = (\varepsilon, (a^2 + 4b)(R^*)^2)$. An easy comparison of the multiplications shows that this is a homomorphism. By (3.5), its proof, and the fact that $a = 0$ if $\varepsilon = \bar{1}$, it is an isomorphism. In summary,

(3.8). *Let* R *be a commutative ring. Then,*

$$QU_f(R) \cong \begin{cases} Qu_f(R) & \text{if } 2 \notin R^* \\ \\ \mathbb{Z}_2 \times R^*/(R^*)^2 & \text{if } 2 \in R^* . \end{cases}$$

It follows from this description that $QU_f(R)$ is an elementary Abelian 2-group, except when both $2 \in R^*$ and $-1 \notin (R^*)^2$ (where it is not). See Exercise 2.

Example 4. $QU_f(\mathbb{C}) \cong \mathbb{Z}_2$, $QU_f(\mathbb{R}) \cong \mathbb{Z}_4$, $QU_f(\mathbb{Z}) = 1$.

D. Another Look at $(a, b)^\varepsilon * (c, d)^\eta$

An alternative description of the algebra $(a, b)^\varepsilon * (c, d)^\eta$ follows. This development will be of importance in Chapter 7 in the analysis of "special elements" of Clifford algebras.

Let S^ε and T^η be two graded quadratic algebras in GQ. So $S = (a_1, b_1) = R[X]/(X^2 - a_1 X - b_1)$ and $T = (a_2, b_2) = R[X]/(X^2 - a_2 X - b_2)$, where $a_i^2 + 4b_i \in R^*$ for i equal to 1 or 2. Set $v = X + (X^2 - a_1 X - b_1)$ and $w = X + (X^2 - a_2 X - b_2)$. Let σ and τ be the respective conjugations of S and T and recall that $v^\sigma = a_1 - v$ and $w^\tau = a_2 - w$.

We now turn to the graded algebra $S^\varepsilon \hat{\otimes}_R T^\eta$ and the structural properties that were developed at the end of Chapter 1D. Since $\sigma \otimes \tau$ is an automorphism of $S^\varepsilon \hat{\otimes}_R T^\eta$,

$$P = \{p \in S \otimes_R T \mid p^{\sigma \otimes \tau} = p\}$$

is a subalgebra. This subalgebra is the focus of this section.

We begin by showing that P is free of rank 2 over R. Let

$$p = r_0(1 \otimes 1) + r_1(v \otimes 1) + r_2(1 \otimes w) + r_3(v \otimes w)$$

with r_0, r_1, r_2, r_3 in R be an arbitrary element in $S^{\mathcal{E}} \hat{\otimes}_R T^{\eta}$. By an easy computation,

$$p^{\sigma \otimes \tau} = (r_0 + r_1 a_1 + r_2 a_2 + r_3 a_1 a_2)(1 \otimes 1) - (r_1 + r_3 a_2)(v \otimes 1)$$
$$- (r_2 + r_3 a_1)(1 \otimes w) + r_3(v \otimes w).$$

Therefore, $p^{\sigma \otimes \tau} = p$ if and only if $r_1 a_1 + r_2 a_2 + r_3 a_1 a_2 = 0$, $2r_1 + r_3 a_2 = 0$, and $2r_2 + r_3 a_1 = 0$. Observe that this condition is the same as $r_1 a_1 = r_2 a_2$, $r_3 a_2 = -2r_1$, and $r_3 a_1 = -2r_2$. Consider the element

$$z = (v \otimes w) + (v^{\sigma} \otimes w^{\tau}) = a_1 a_2(1 \otimes 1) - a_2(v \otimes 1) - a_1(1 \otimes w) + 2(v \otimes w)$$

of $S^{\mathcal{E}} \hat{\otimes}_R T^{\eta}$. By the criterion above, z is in P.

We will prove that $\{1_P = 1 \otimes 1, z\}$ is a basis of P. Independence is easy. For suppose that $r1 + r'z = 0$ for some r and r' in R. Then $r + r'a_1 a_2 = 0$, $r'a_1 = 0$, and $2r' = 0$. Thus $r = 0$ and $r'(a_1^2 + 4b_1) = 0$. So $r' = 0$, and $\{1_P, z\}$ is independent. Now let $p = r_0(1 \otimes 1) + r_1(v \otimes 1) + r_2(1 \otimes w) + r_3(v \otimes w)$, with $r_1 a_1 = r_2 a_2$, $r_3 a_2 = -2r_1$ and $r_3 a_1 = -2r_2$, be an arbitrary element in P. Set $u_i = r'(a_i^2 + 4b_i)$. By trivial computation,

$$r_3 = 2u_1^{-1}(2r_3 b_1 - r_2 a_1) = 2u_2^{-1}(2r_3 b_2 - r_1 a_2).$$

Put $t_1 = u_1^{-1}(2r_3 b_1 - r_2 a_1)$ and $t_2 = u_2^{-1}(2r_3 b_2 - r_1 a_2)$. By another easy computation,

$$u_2 u_1 t_1 = (a_2^2 + 4b_2)(2r_3 b_1 - r_2 a_1) = -4r_1 a_2 b_1 - r_1 a_1^2 a_2 + 8r_3 b_1 b_2 + 2r_3 a_1^2 b_2$$
$$= (a_1^2 + 4b_1)(2r_3 b_2 - r_1 a_2) = u_1 u_2 t_2.$$

It follows that $t_1 = t_2$. Set $t_1 = t_2 = t$. Note that $r_3 = 2t$. We show next that $r_1 = -t a_2$ and $r_2 = -t a_1$. First observe that

$$a_2 t = a_2 u_2^{-1}(2r_3 b_2 - r_1 a_2) = u_2^{-1}(4t_2 b_2 a_2 + t_2 a_2^3 - t_2 a_2^3 - r_1 a_2^2)$$

$$= u_2^{-1}(t_2 a_2(a_2^2 + 4b_2) - a_2^2(r_1 + t_2 a_2))$$

$$= u_2^{-1}(t_2 a_2 u_2) - u_2^{-1} a_2^2(r_1 + t_2 a_2).$$

It follows that $a_2^2(r_1 + t_2 a_2) = 0$. Also, $2(r_1 + t_2 a_2) = 2r_1 + r_3 a_2 = 0$. So, $u_2(r_1 + t_2 a_2) = (a_2^2 + 4b_2)(r_1 + t_2 a_2) = 0$. Therefore, $r_1 = -ta_2$, as required. In a completely similar way, $r_2 = -ta_1$. It is now clear that $p = (r_0 - ta_1 a_2)1 + tz$, and the proof of the fact that $\{1_p, z\}$ is a basis of P is complete.

The study of the multiplicative structure of P is next. Set $a = a_1 a_2$ and $b = a_1^2 b_2 + a_2^2 b_1 + (-1)^{\varepsilon \eta} 4b_1 b_2$. We assert that

$$z^2 = b + az.$$

This equality is easy to check if $\varepsilon = \delta = \bar{1}$. For then, $a_1 = a_2 = 0$, $z = 2(v \otimes w)$, so that $z^2 = -4(v^2 \otimes w^2) = -4b_1 b_2 = b$. In all other cases, $b = a_1^2 b_2 + a_2^2 b_1 + 4b_1 b_2$, and the verification of the equality proceeds by a lengthy expansion of the product $z \cdot z$ and use of the equations $v^2 = a_1 v + b_1$ and $w^2 = a_2 w + b_2$. The details are routine and left to the reader.

Consider the involutions $\sigma \otimes id_T$ and $id_S \otimes \tau$ of $S^\varepsilon \overset{\wedge}{\otimes}_R T^\delta$. Let $p \in P$ be arbitrary. Since $p^{(\sigma \otimes id_T)} = p^{(\sigma \otimes \tau)(\sigma \otimes id_T)} = p^{(id_S \otimes \tau)}$, we see that $p^{(\sigma \otimes id_T)(\sigma \otimes \tau)} = p^{(id_S \otimes \tau)} = p^{(\sigma \otimes id_T)}$. Therefore, $p^{(\sigma \otimes id_T)}$ is in P. So the involution $\sigma \otimes id_T = id_S \otimes \tau$ is an involution on P. We call $\sigma \otimes id_T = id_S \otimes \tau$ the *conjugation* of P. Observe that

$$z^{(\sigma \otimes id_T)} = ((v \otimes w) + (v^\sigma \otimes w^\tau))^{(\sigma \otimes id_T)} = (v^\sigma \otimes w) + (v \otimes w^\tau)$$

$$= (a_1 - v) \otimes w + v \otimes (a_2 - w) = a_1(1 \otimes w) + a_2(v \otimes 1) - 2(v \otimes w)$$

$$= a_1 a_2 1_p - (a_1 a_2 1_p - a_2(v \otimes 1) - a_1(1 \otimes w) + 2(v \otimes w)) = a1_p - z,$$

so that $z^{(\sigma \otimes id_T)} = a1_p - z$.

Finally, we supply P with a grading. If ε and η are both equal to $\bar{0}$ or both equal to $\bar{1}$, put the trivial grading on P. In the cases where either ε or η is $\bar{1}$ and the other is $\bar{0}$, either $a_1 = 0$ or $a_2 = 0$. So $a = 0$, and we supply P with the grading $P = R1_P \oplus Rz$. Refer to the discussion of the grading of $S^{\varepsilon} \hat{\otimes}_R T^{\eta}$ in Chapter 1D. If $\varepsilon = \eta = \bar{0}$, then this grading is trivial, and z is obviously in the 0-component of the grading. If $\varepsilon = \eta = \bar{1}$, then $a_1 = a_2 = 0$ and $z = 2(v \otimes w)$ is again in the 0-component. Therefore when $\varepsilon = \delta, z$ is in the 0-component of the grading of $S^{\varepsilon} \hat{\otimes}_R T^{\eta}$. Suppose that $\varepsilon \neq \eta$. If $\varepsilon = \bar{1}$ and $\eta = \bar{0}$, then $z = - a_2(v \otimes 1) + 2(v \otimes w)$. So z is in the 1-component of the grading of $S^{\varepsilon} \hat{\otimes}_R T^{\eta}$. The situation is the same if $\varepsilon = \bar{0}$ and $\eta = \bar{1}$. We have now verified in all cases that the grading of P is obtained by restricting that of $S^{\varepsilon} \hat{\otimes}_R T^{\eta}$.

We have now provided the subalgebra P of $S^{\varepsilon} \hat{\otimes}_R T^{\eta}$ with the structure of a free graded quadratic algebra with involution. The goal of developing an alternative description of the product $(a_1, b_1)^{\varepsilon} * (a_2, b_2)^{\eta}$ is now reached. Refer to the definitions of a and b and observe that $(a_1, b_1)^{\varepsilon} * (a_2, b_2)^{\eta}$ is equal to $(a, b)^{\varepsilon + \eta}$. Define

$$\theta : (a, b)^{\varepsilon + \eta} \to P$$

by $\theta(1) = 1_P, \theta(X + (X^2 - aX - b)) = z$ and by extending linearly.

(3.9). *The map* $\theta : (a, b)^{\varepsilon + \eta} \to P$ *is an isomorphism of graded* R*-algebras with conjugation.*

Proof. Put $X + (X^2 - aX - b) = y$. Since it takes a basis to a basis, θ is bijective. Since

$$\theta(y^2) = \theta(b + ay) = b + az = z^2 = \theta(y)^2,$$

θ is an algebra isomorphism. Let ρ be the conjugation of (a, b). The fact that

$$\theta(y^P) = \theta(a1 - y) = a1_p - z = (z)^{(\sigma \otimes idT)} = (\theta y)^{(\sigma \otimes idT)},$$

implies that θ preserves involutions. The grading of P was defined to be trivial if $\varepsilon = \eta$ and given by $P = R1_p \oplus Rz$ if $\varepsilon \neq \eta$. A moment's reflection shows that therefore θ preserves the gradings.   QED.

Remark. Under the assumption that the gradings on $S = (a_1, b_1) = R[X]/(X^2 - a_1 X - b_1)$ and $T = (a_2, b_2) = R[X]/(X^2 - a_2 X - b_2)$ are both trivial, (3.9) asserts that the subalgebra $P = \{p \in S \otimes_R T \mid p^{\sigma \otimes \tau} = p\}$ of $S \otimes_R T$ is isomorphic to $(a_1, b_1) * (a_2, b_2) = (a_1 a_2, a_1^2 b_2 + a_2^2 b_1 + 4b_1 b_2)$.

E. Exercises

Unless additional assumptions are made, R is an arbitrary commutative ring.

1. Let F be a finite field with char $F \neq 2$. Show that $Qu_f(F) \cong \mathbb{Z}_2$. (This also holds if char $F = 2$. See Exercise 11.)

2. Let $2 \in R^*$. Show that $([0, 1]^{\overline{1}})^2 = [1, 0]^{\overline{0}}$ if and only if $-1 \in (R^*)^2$. Deduce that $1 \to Qu_f(R) \to QU_f(R) \to \mathbb{Z}_2 \to 0$ splits if and only if $-1 \in (R^*)^2$ if and only if $QU_f(R)$ is an elementary Abelian 2-group.

For the next three problems, let F be the quadratic number field $\mathbb{Q}(\sqrt{d})$, where d is square free, and let R be the ring of integers of F. Refer to Samuel or Ribenboim for the properties of F used here.

3. Set $T = \{r \in R^* \mid r = a^2 + 4b$ with a and b in $R\}$. Show that $-1 \in T$ if and only if $d \equiv 3 \pmod 4$.

4. Use the description of R^* and conclude for $d < 0$, that

$$Qu_f(R) \cong \begin{cases} \mathbb{Z}_2, & \text{if } d \equiv 3 \pmod 4 \text{ and } d \neq -1 \\ 1 & \text{in all other cases} . \end{cases}$$

5. Suppose $d > 0$. By (3.4, 2), $Qu_f(R) \cong T/(R^*)^2 \cong \mathbb{Z}_2^k$, with $0 \leq k \leq 2$. All three possibilities for k actually occur. By the Dirichlet Unit Theorem, R^*

$= \{\pm 1\} \times U$, where U is infinite cyclic. It can be shown that the generator u can be chosen uniquely with the property that $u > 1$. This u is the *fundamental unit* of $\mathbb{Q}(\sqrt{d})$. Since u is an algebraic integer in $\mathbb{Q}(\sqrt{d})$, $u = (\frac{n}{2} + \frac{m}{2}\sqrt{d})$ with n and m in $\mathbb{Z}$ either both even or both odd (the latter case is possible only if $d \equiv 1 \bmod 4$). Show that the following is a reformulation of Theorem 1 of Therond [1975]:

$$Qu_f(R) \cong \begin{cases} 1, & \text{if } d \equiv 1,2 \pmod 4 \text{ and } \frac{m}{2} \text{ is an odd integer} \\ \mathbb{Z}_2 \times \mathbb{Z}_2, & \text{if } d \equiv 3 \pmod 4 \text{ and } \frac{m}{2} \text{ is an even integer} \\ \mathbb{Z}_2, & \text{in all other cases}. \end{cases}$$

6. Consider the subset $\{x \in R \mid 1 + 4x \in R^*\}$ of R. Call x and y in this set equivalent if $y = x + (r + r^2)(1 + 4x)$ for some r in R. Show this to be an equivalence relation. Let $[x]$ be the equivalence class of x. Show that the set $V(R)$ of equivalence classes is an Abelian group with operation $[x][y] = [x.y]$, where $x.y = x + y + 4xy$. Note that $[0]$ is the identity. Observe that if char $R = 2$, then $V(R) = R/\wp(R)$, where $\wp(R)$ is the subgroup $\wp(R) = \{r + r^2 \mid r \in R\}$ of R.

7. Define $\rho : V(R) \to Qu_f(R)$ by $\rho([x]) = [1, x]$. Show that ρ is an injective homomorphism.

8. Consider the subset $\{a \in R \mid a^2 + 4b \in R^* \text{ for } b \text{ in } R\}$ of R. Call a and a' in this set equivalent if there exist r in R and u in R^* such that $a' = ua + 2r$. Show that this is an equivalence relation. Denote the class of a by $[a]$. Show that the set $S(R)$ of equivalence classes is an Abelian group under the operation $[a][c] = [ac]$.

9. Define $\omega : Qu_f(R) \to S(R)$ by $\omega([a, b]) = [a]$. Show that ω is a homomorphism and that there is an exact sequence
$$0 \to V(R) \xrightarrow{\rho} Qu_f(R) \xrightarrow{\omega} S(R) \to 1.$$

10. If R is a local ring, i.e., one which has a unique maximal ideal, or if $4 \in \text{Rad}(R)$ − this is the intersection of all maximal ideals of R − then ρ is surjective, so that $Qu_f(R) \cong V(R)$. If $2 = 0$, then $Qu_f(R) \cong R/\wp(R)$.

11. Let F be a finite field with char $F = 2$. Then $Qu_f(F) \cong \mathbb{Z}_2$.

A *monoid* is a set M together with a (multiplicative) binary operation which is associative and commutative and has an identity element denoted 1. Submonoids and homomorphisms between monoids will have the obvious meaning. Let M be a monoid and let k, n and m denote elements of M. An element k is invertible if $kn = 1$ for some n. Note that the set of invertible elements of M form an Abelian group. If, for any k, m, and n in M, $kn = mn$ implies that $k = m$, then M is a *cancellation monoid*.

12. Let M be a monoid. Let 0 be any symbol. Consider $N = M \cup \{0\}$ and extend the multiplication of M to N by defining $n0 = 0$ for all n in N. More generally, let P be any set with distinguished element e and define a product on P by $pq = e$ for all p and q in P. Extend the products on M and P to the disjoint union $N = M \cup P$ by setting $mp = pm = p$ for all $m \in M$ and $p \in P$. Show that N is a monoid and that the inclusion $M \rightarrow N$ is a monoid homomorphism. The monoid N is never a cancellation monoid.

13. Let R be a commutative ring, but regard only its product. Then R is a commutative monoid. Consider the set $R/(R^*)^2 = \{r(R^*)^2 \mid r \in R\}$. Show that the product of R provides $R/(R^*)^2$ with the structure of a monoid. Suppose that R is a domain and let $R^\bullet = R - \{0\}$. Then $R^\bullet$ and $R^\bullet/(R^*)^2 = \{r(R^*)^2 \mid r \in R^\bullet\}$ are both cancellation monoids. Show that R and $R/(R^*)^2$ are isomorphic to $R^\bullet \cup \{0\}$ and $R^\bullet/(R^*)^2 \cup \{0\}$. respectively.

14. Consider the set $\text{Qum}_f(R)$ of isomorphism classes of free quadratic algebras (separable or not). Denote the isomorphism class of $R[X]/(X^2 - aX - b)$ by [a, b]. Show that $[a, b][c, d] = [ac, a^2d + c^2b + 4bd]$ provides $\text{Qum}_f(R)$ with the structure of a monoid.

15. The group $\text{Qu}_f(R)$ is a submonoid of $\text{Qum}_f(R)$. Let $[a, b] \in \text{Qum}_f(R)$ and show that [a, b] is invertible if and only if $[a, b] \in \text{Qu}_f(R)$. So $\text{Qu}_f(R)$ is the group of invertible elements of $\text{Qum}_f(R)$.

16. Define $\delta : \text{Qum}_f(R) \rightarrow R/(R^*)^2$ by $[a, b] \rightarrow (a^2 + 4b)(R^*)^2$ and show that δ is a homomorphism of monoids.

17. Show that the homomorphism $\delta : \text{Qum}_f(R) \rightarrow R/(R^*)^2$ is injective under any of the assumptions: i) $2 \in R^*$, ii) R is an integrally closed domain with char $R \neq 2$, or iii) 2R is a prime ideal of R.

18. If F is a field with char $F \neq 2$, then $\text{Qum}_f(F) \cong F^*/(F^*)^2 \cup \{0\}$.

19. Suppose that F is a field with char $F = 2$. Consider the vector space F over F^2. For any $b \in F$, let $\langle b \rangle = bF^2$ be the line spanned by b. Let P be the set consisting of $\langle 1 \rangle$ and all the planes of F containing $\langle 1 \rangle$. Define a product on P by setting the product of any two elements equal to $\langle 1 \rangle$ and consider the monoid $F/\wp(F) \cup P$. Extend the isomorphism $\text{Qu}_f(F) \cong F/\wp(F)$ to a map $\text{Qum}_f(F) \rightarrow F/\wp(F) \cup P$ by sending $[0, b]$ to $\langle 1 \rangle$ if $b \in F^2$ and to the plane $\langle 1 \rangle \oplus \langle b \rangle$ if $b \notin F^2$. Show that this map is an isomorphism of monoids.

20. Let F be a finite field. Then $\text{Qum}_f(F) \cong \text{Qu}_f(F) \cup \{0\} \cong \mathbb{Z}_2 \cup \{0\}$.

21. Formulate monoid analogues of the group $\text{QU}_f(R)$ and the constructions of the last section of the chapter.

Hints:

1. The map $F^* \rightarrow (F^*)^2$ given by squaring has kernel $\{\pm 1\}$. It follows that $F^*/(F^*)^2 \cong \mathbb{Z}_2$.

3. See Lemma 3 of Therond [1975].

4. If $d < 0$, then by (3.4) and Exercise 3, $\text{Qu}_f(R) \cong T/(R^*)^2 \cong \mathbb{Z}_2^k$ with k 0 or 1. If $d \neq -1, -3$, then $R^* = \{\pm 1\}$ and the description of $\text{Qu}_f(R)$ follows. Check that $\text{Qu}_f(R) = 1$ in the two remaining cases. Use Exercises 14, 15, and 16 of Chapter 1E.

7. To show that ρ is well defined proceed as follows. If $[x] = [y]$, then $y = x + (z + z^2)(1 + 4x)$ with z in R. Let $u = 1 + 2z$. Since
$$u^2 = 1 + 4z + 4z^2 = 1 + 4(y - x)(1 + 4x)^{-1}$$
$$= (1 + 4x)(1 + 4x)^{-1} + (4y - 1 + 1 - 4x)(1 + 4x)^{-1}$$
$$= (1 + 4y)(1 + 4x)^{-1} \in R^*,$$
$u \in R^*$. Now let $r = -z$. So $1 = u + 2r$ and by an easy computation, $y = u^2x - ru - r^2$. By (3.1), $[1, x] = [1, y]$.

9. That ω is well defined follows by Proposition (3.1). By Exercise 7, ρ is injective; that ω is onto is easy. Clearly, im $\rho \subseteq \ker \omega$. It remains to check that $\ker \omega \subseteq \text{im } \rho$. Suppose that $\omega([c, d]) = 1$. So $c = u + 2r$ for

some r in R and u in R^*. Put $b = u^{-2}(d + ru + r^2)$. Hence $d = u^2 b - ru - r^2$. Since $c^2 + 4d \in R^*$, it follows by an easy computation that $1 + 4b \in R^*$. So by (3.1), $[c, d] = [1, b] = \rho([b])$.

10. Observe that the second statement follows from Exercise 6 and the first statement. To prove the first, we show that $S(R) = 1$ in either case and apply Exercise 9. Suppose $4 \in Rad(R)$. Let $a^2 + 4b = u \in R^*$. So $u^{-1}a^2 = 1 + 4u^{-1}b$. By Nakayama's Lemma, $1 + Rad(R)$ contains only elements of R^*. So $a \in R^*$ and $S(R) = 1$. Let R be a local ring. In view of the case already considered, we can assume that $2 \in R^*$. Let $a^2 + 4b \in R^*$ be arbitrary. If $a \in R^*$ we are done as before, so assume that a is in the unique maximal ideal of R. Since $1 + a \in R^*$ and $a = 2(-\frac{1}{2}) + 1(1 + a)$, $[a] = [1]$ in $S(R)$. So $S(R) = 1$ as required.

11. Consider the additive map $F \to \wp(F)$ given by $z \to z + z^2$.

19. Use Exercise 7 of Chapter 1E.

4
Bilinear and Quadratic Forms

Overview

It is the purpose of this paragraph to introduce additional important concepts and their basic properties. This will include bilinear and quadratic forms, discriminant modules, and the group Dis(R). Proof by localization, i.e., by reduction to the case of a local ring, is introduced here. For the entire chapter, we fix a commutative ring R and a right R-module M.

A. Localization

We refer to Chapter 7 of Jacobson [1985, II] for the basic properties of localization. For purposes of notation and illustration, we recall some of these.

Let $\mathfrak{p}$ be a prime ideal of R. So $R/\mathfrak{p}$ is a domain. Since $\mathfrak{p} \neq R$, $R = \{0\}$ is ruled out. The localization $R_{\mathfrak{p}} = \{r/s \mid r \in R, s \in R - \mathfrak{p}\}$ of R is a local ring with unique maximal ideal $\mathfrak{p}R_{\mathfrak{p}}$. The element $r/1$ of $R_{\mathfrak{p}}$ is denoted $r_{\mathfrak{p}}$. If $r \in R - \mathfrak{p}$, then $r_{\mathfrak{p}} \in (R_{\mathfrak{p}})^*$, since $(r/1)(1/r) = 1$. The ring homomorphism $R \to R_{\mathfrak{p}}$ given by $r \to r_{\mathfrak{p}}$ gives $R_{\mathfrak{p}}$ the structure of a commutative algebra over R. Recall that if R is a domain, then $\{0\}$ is a prime ideal and $F = R_{\{0\}}$ is the field of fractions of R. In this case, the localization $R_{\mathfrak{p}}$ can be taken to be the subring $\{r/s \in F \mid r \in R$ and $s \in R - \mathfrak{p}\}$ of F.

(4.1). *If* $r_{\mathfrak{p}} = 0$ *for all prime ideals* $\mathfrak{p}$ *of* R, *then* $r = 0$.

Proof. Let r in R be arbitrary. Note that $\mathfrak{a} = \{t \in R \mid tr = 0\}$ is an ideal of R. Assume that $r \neq 0$. Then $1 \notin \mathfrak{a}$. So $\mathfrak{a} \neq R$ and $\mathfrak{a}$ is contained in a maximal ideal and hence a prime ideal $\mathfrak{q}$ of R. We claim that $r_{\mathfrak{q}} \neq 0$. If $r_{\mathfrak{q}} = 0$, then $tr = 0$ for some $t \in R - \mathfrak{q}$. But this is a contradiction, since $\mathfrak{a}$ is contained in $\mathfrak{q}$. QED.

Let $\mathfrak{p}$ be a prime ideal of R and consider the right $R_\mathfrak{p}$-module $M \otimes_R R_\mathfrak{p}$. There is an isomorphism of $R_\mathfrak{p}$-modules from $M \otimes_R R_\mathfrak{p}$ onto the $R_\mathfrak{p}$-module obtained by localizing M at $\mathfrak{p}$ which satisfies $x \otimes (1/s) \to x/s$ for any x in M and s in $R - \mathfrak{p}$. We therefore often denote $M \otimes_R R_\mathfrak{p}$ by $M_\mathfrak{p}$. Observe that the map $M \to M_\mathfrak{p}$ given by $x \to x \otimes 1$ is an additive map. If M is finitely generated and projective, then $M_\mathfrak{p}$ is free of finite rank over $R_\mathfrak{p}$.

(4.2). *The kernel of* $M \to M_\mathfrak{p}$ *is* $\{x \in M \mid xs = 0$ *for some* $s \in R - \mathfrak{p}\}$. *Let* M *be finitely generated. Then* $M_\mathfrak{p} = \{0\}$ *if and only if there exists* $r \in R - \mathfrak{p}$ *such that* $Mr = \{0\}$.

Proof. The first statement is an easy consequence of the isomorphism between $M_\mathfrak{p}$ and the $R_\mathfrak{p}$-module obtained by localizing M at $\mathfrak{p}$. The proof of the second statement is next. Observe that $M_\mathfrak{p} = \{0\}$ if and only if $x \otimes (s/t) = 0$ for all x in M and s/t in $R_\mathfrak{p}$. Let $r \in R - \mathfrak{p}$ such that $Mr = \{0\}$. Then $(x \otimes (s/t))(r/1) = x \otimes (r/1)(s/t) = xr \otimes (s/t) = 0$, for all x in M and (s/t) in $R_\mathfrak{p}$. Since $r/1 = r_\mathfrak{p}$ is in $(R_\mathfrak{p})^*$, it follows that $M_\mathfrak{p} = \{0\}$. Suppose conversely, that $M_\mathfrak{p} = \{0\}$ and let $\{x_1, ..., x_n\}$ be a set of generators of M. Since $x_i \otimes 1 = 0$, $x_i r_i = 0$ for some $r_i \in R - \mathfrak{p}$. Let $r = r_1 \cdots r_n$. Since $x_i r = 0$ for all i, $Mr = \{0\}$.   QED.

(4.3). *Let* M *be finitely generated and projective. Then* M *is faithful if and only if* rank $M_\mathfrak{p} \geq 1$ *for all prime ideals* $\mathfrak{p}$ *of R.*

Proof. Suppose that rank $M_\mathfrak{p} = 0$ for some $\mathfrak{p}$. Then $M_\mathfrak{p} = \{0\}$ and by (4.2), $Mr = \{0\}$ with $r \in R - \mathfrak{p}$. So M cannot be faithful. Conversely, suppose that rank $M_\mathfrak{p} \geq 1$ for all $\mathfrak{p}$. Assume that $Mr = \{0\}$ for some r. Let $\mathfrak{p}$ be arbitrary. Then for any x in M, the image $x \otimes r_\mathfrak{p}$ of xr under $M \to M_\mathfrak{p}$ is zero. It follows that $(M_\mathfrak{p})r_\mathfrak{p} = \{0\}$, and therefore that $r_\mathfrak{p} = 0$. So $r = 0$ by (4.1), and M is faithful. QED.

Let Spec(R) be the set of all prime ideals of R. The *rank* of a finitely generated projective M is defined to be the map

$$\text{rank } M : \text{Spec}(R) \to \mathbb{Z}$$

given by $(\text{rank } M)(\mathfrak{p}) = m(\mathfrak{p})$, where $m(\mathfrak{p}) = \text{rank } M_\mathfrak{p}$. For any subset X of R, let V(X) be the set of all prime ideals $\mathfrak{p}$ of R that contain X. Define a topology on Spec(R) by letting the closed sets of Spec(R) be precisely the sets of the form V(X). This is the Zariski topology. If $\mathbb{Z}$ is supplied with the discrete topology (all subsets are open and closed), then rank M is continuous. We say that rank M is *odd* if $m(\mathfrak{p})$ is odd for all $\mathfrak{p}$, and that rank M is *even* if $m(\mathfrak{p})$ is even (possibly zero) for all $\mathfrak{p}$. In the first case, M is faithful by (4.3). If rank M is constant, then clearly rank M is either odd or even. Evidently, rank M is constant if M is free.

(4.4). *Suppose that* R *is a domain with field of fractions* F. *Let* M *be finitely generated projective. Then* $\text{rank } M_{\mathfrak{p}} = \dim_F (M \otimes_R F)$ *for any prime ideal* $\mathfrak{p}$ *of* R. *In particular,* rank M *is constant.*

Proof. We may assume that $M \neq \{0\}$. Since M is a submodule of a free module, it is faithful. So by (4.2), the map $M \to M \otimes_R F$ is injective. Let $\mathfrak{p}$ be any prime ideal of R. We consider $R \subseteq R_{\mathfrak{p}} \subseteq F$. Observe that the composite

$$M \to M_{\mathfrak{p}} = M \otimes_R R_{\mathfrak{p}} \to M \otimes_R F$$

is the injective map just referred to. We show that $M \otimes_R R_{\mathfrak{p}} \to M \otimes_R F$ is also injective. Let $y = \sum_i (x_i \otimes s_i)$, with $x_i \in M$ and $s_i \in R_{\mathfrak{p}}$, be in the kernel. Since $s_i \in F$, it follows that there is a nonzero r in R such that yr is in the image of M in $M \otimes_R R_{\mathfrak{p}}$. By the injectivity of the composite, yr = 0. Since $M \otimes_R R_{\mathfrak{p}}$ is free, y = 0. So $M \otimes_R R_{\mathfrak{p}} \to M \otimes_R F$ is injective. Suppose that $\{y_1,..., y_n\}$ is a basis of $M_{\mathfrak{p}}$. An argument similar to that just used shows that the image of $\{y_1,..., y_n\}$ under $M \otimes_R R_{\mathfrak{p}} \to M \otimes_R F$ is a basis of $M \otimes_R F$. Therefore, $\text{rank } M_{\mathfrak{p}} = \dim_F(M \otimes_R F)$ as required.   QED.

Let M and N be two R-modules and let $\varphi : M \to N$ be an R-homomorphism. The $R_{\mathfrak{p}}$-homomorphism $\varphi \otimes 1$ from $M_{\mathfrak{p}} = M \otimes_R R_{\mathfrak{p}}$ to $N_{\mathfrak{p}} = N \otimes_R R_{\mathfrak{p}}$ is denoted by $\varphi_{\mathfrak{p}}$.

(4.5). *Let* M *and* N *be finitely generated projective and assume that* rank M = rank N. *If* $\varphi : M \to N$ *is a surjective R-homomorphism, then* φ *is an isomorphism.*

Proof. Let $\mathfrak{p}$ be any prime ideal of R. Since $M_{\mathfrak{p}}$ and $N_{\mathfrak{p}}$ have the same finite rank, and since $\varphi_{\mathfrak{p}}$ is surjective, it follows by §3.4 of Jacobson [1985, I] that $\varphi_{\mathfrak{p}}$ is an isomorphism. Therefore by the basic properties of localization, φ is an isomorphism.   QED.

If A is an R-algebra, then $A \otimes_R R_{\mathfrak{p}}$ is an R-algebra; in fact via the map $s \to 1 \otimes s$ for $s \in R_{\mathfrak{p}}$, it is an $R_{\mathfrak{p}}$-algebra.

B. Bilinear Forms

Let A be a commutative R-algebra. Let $f : M \times M \to A$ be an R-*bilinear form*, i.e., f is a map which is R-linear in each of the variables. By the defining property of the tensor product, f induces an R-module map $M \otimes_R M \to A$, which sends $x \otimes y$ to $f(x, y)$ for all x and y in M. This map will also be denoted by f.

Assume for a moment that M is a free R-module with finite basis $\mathcal{X} = \{x_1, ..., x_n\}$. Then $C = (f(x_i, x_j))$ is a matrix with coefficients in A. It is *the matrix of f in the basis $\mathcal{X}$*. Let $\mathcal{X}' = \{x_1', ..., x_n'\}$ be another basis of M. Let $C' = (f(x_i', x_j'))$ be the matrix of f in the basis $\mathcal{X}'$ and let $T = (t_{ij})$ be the change-of-basis matrix given by $x_j' = \sum_i x_i t_{ij}$. By routine computations, $C' = (T^t)CT$, where t is the transpose. Consider $\det_{\mathcal{X}} f = \det (f(x_i, x_j)) \in A$, and observe that the image $(\det_{\mathcal{X}} f)(A^*)^2$ in the monoid $A/(A^*)^2$ is independent of the choice of the basis $\mathcal{X}$. If (a_{ij}) is any $n \times n$ matrix with coefficients in A, then the map $M \times M \to A$ defined by $(x_i, x_j) \to a_{ij}$ and R-linearity is an R-bilinear form on M. Clearly, this form has the matrix (a_{ij}) in the basis $\mathcal{X}$. Note therefore that any R-bilinear form $M \times M \to A$ can be constructed by specifying an $n \times n$ matrix over A.

Let M and N be two R-modules equipped with R-bilinear forms $f : M \times M \to A$ and $g : N \times N \to A$ respectively. Consider the following maps: The map

(i) $(M \otimes_R N) \times (M \otimes_R N) \to (M \otimes_R N) \otimes_R (M \otimes_R N)$.

The "switching" isomorphism

(ii) $(M \otimes_R N) \otimes_R (M \otimes_R N) \to (M \otimes_R M) \otimes_R (N \otimes_R N)$

of R-modules given by $(x \otimes x') \otimes (y \otimes y') \to (x \otimes y) \otimes (x' \otimes y')$ for all x and y in M and x' and y' in N. This map is established in a manner similar to the associativity property of the tensor product. The tensor product map

(iii) $(M \otimes_R M) \otimes_R (N \otimes_R N) \xrightarrow{\ f \otimes g\ } A \otimes_R A$

provided by $f : M \otimes_R M \to A$ and $g : N \otimes_R N \to A$. And finally,

(iv) $$A \otimes_R A \xrightarrow{\text{product}} A.$$

The composite of (i) – (iv) will also be denoted by $f \otimes g$. A check of the definitions and linearity properties of the maps above shows that

$$f \otimes g : (M \otimes_R N) \times (M \otimes_R N) \to A$$

is additive and that it satisfies $(f \otimes g)\big((x \otimes x'), (y \otimes y')\big) = f(x, y)g(x', y')$ for all x and y in M and x' and y' in N. It follows that it is a R-bilinear form.

The first application of this development is the change-of-scalars construction. This follows next.

An R-bilinear form $f : M \times M \to A$ will be called a *bilinear form on* M if we are in the special case $A = R$. In this case, the pair (M, f) or briefly M is called a *bilinear module over* R. This is the special case, which will receive most of our attention.

Let $f : M \times M \to R$ be a bilinear form on M. Consider f to be an R-bilinear form $f : M \times M \to A$ by composing f with $R \to A$. Take $N = A$ and let $g : A \times A \to A$ be the product of A. Apply the preceding construction to this situation. The bilinear form $f \otimes g$ is denoted by f_A. Observe that

$$f_A : (M \otimes_R A) \times (M \otimes_R A) \to A$$

satisfies $f_A((x \otimes a), (y \otimes b)) = f(x, y)ab$, for all x and y in M and a and b in A. Notice that f_A is an A-bilinear form, i.e., f_A is a bilinear form on the A-module $M \otimes_R A$. So f has been extended from M to a form f_A on $M \otimes_R A$. In the special case, where $A = R_{\mathfrak{p}}$ for a prime ideal $\mathfrak{p}$ of R, f_A will be denoted by $f_{\mathfrak{p}}$. So $M_{\mathfrak{p}}$ with $f_{\mathfrak{p}}$ is a bilinear module over $R_{\mathfrak{p}}$. This is the *localization of* M *at* $\mathfrak{p}$.

Let M and N be bilinear modules over R with underlying forms f and g. An R-module homomorphism $\varphi : M \to N$ is called a *representation* if $g(\varphi x, \varphi y) = f(x, y)$ for all x and y in M. If, in addition, φ is bijective, then it is called an *isometry*. If an isometry $\varphi : M \to N$ exists, then M and N are *isometric* and we write $M \cong N$.

(4.6). *Let* M *and* N *be finitely generated projective bilinear modules. Let* $\varphi : M \to N$ *be an* R-*module map with the property that* $\varphi_{\mathfrak{p}} : M_{\mathfrak{p}} \to N_{\mathfrak{p}}$ *is an isometry for all prime ideals* $\mathfrak{p}$ *of* R. *Then* $\varphi : M \to N$ *is an isometry.*

Proof. Let f and g be the underlying forms of M and N. Let x and y in M be arbitrary and consider the elements $g(\varphi x, \varphi y)$ and $f(x, y)$. Fix a prime ideal $\mathfrak{p}$ of R. Since $\varphi_{\mathfrak{p}} : M_{\mathfrak{p}} \to N_{\mathfrak{p}}$ is an isometry,

$$g_{\mathfrak{p}}(\varphi_{\mathfrak{p}}(x \otimes 1), \varphi_{\mathfrak{p}}(y \otimes 1)) = f_{\mathfrak{p}}(x \otimes 1, y \otimes 1).$$

So $g_{\mathfrak{p}}(\varphi x \otimes 1, \varphi y \otimes 1) = f_{\mathfrak{p}}(x \otimes 1, y \otimes 1)$. This implies that the elements $g(\varphi x, \varphi y)$ and $f(x, y)$ have the same image under $R \to R_{\mathfrak{p}}$. So $g(\varphi x, \varphi y) - f(x, y)$ is in the kernel of $R \to R_{\mathfrak{p}}$. Since this is true for any $\mathfrak{p}$, $g(\varphi x, \varphi y)$ and $f(x, y)$ are equal by Proposition (4.1).   QED.

Let $f : M \times M \to R$ be a bilinear form on M. Suppose that f is *symmetric*, i.e., that $f(x, y) = f(y, x)$ for all x and y in M. The bilinear module M will now be called a *symmetric bilinear module*. Let $M^* = \text{Hom}_R(M, R)$ be the dual of M (the context will prevent ambiguity with the notation R^* for the units of R) and consider the R-homomorphism $\bar{f} : M \to M^*$ given by $x \to f(x, \ )$. We call M or the underlying form f *nonsingular* if $\bar{f}$ is an isomorphism. If M is finitely generated projective of constant rank, then by (4.5), M is nonsingular if and only if $\bar{f}$ is surjective. Assume for a moment that $M \neq \{0\}$ is free with finite basis $\mathcal{X}$. It is a routine exercise in linear algebra (see Exercise 8) to show that M is nonsingular if and only if the element $\det_{\mathcal{X}} f = \det (f(x_i, x_j))$ is in R^*. It follows that f is nonsingular if and only if the element $(\det_{\mathcal{X}} f)(R^*)^2$ in the monoid $R/(R^*)^2$ is in the group $R^*/(R^*)^2$.

(4.7). *Let M be a finitely generated projective symmetric bilinear module. Then M is nonsingular if and only if $M_{\mathfrak{p}}$ is nonsingular for all prime ideals $\mathfrak{p}$ of R.*

Proof: By the properties of localization, $\bar{f} : M \to M^*$ is an isomorphism if and only if $(\bar{f})_{\mathfrak{p}} : M_{\mathfrak{p}} \to (M^*)_{\mathfrak{p}}$ is an isomorphism for all prime ideals $\mathfrak{p}$ of R. But for any $\mathfrak{p}$, $(\bar{f})_{\mathfrak{p}} : M_{\mathfrak{p}} \to (M^*)_{\mathfrak{p}}$ is, up to isomorphism, the same map as $\overline{(f_{\mathfrak{p}})} : M_{\mathfrak{p}} \to (M_{\mathfrak{p}})^*$. In other words, there is a commutative diagram of isomorphisms which links $(\bar{f})_{\mathfrak{p}}$ and $\overline{(f_{\mathfrak{p}})}$. See Exercise 7.   QED.

(4.8). *Let* M *be a finitely generated projective symmetric bilinear module with form* f. *If* M *is faithful and nonsingular, then the ideal generated by* $\{f(x, y) \mid x$ *and* y *in* M$\}$ *equals* R.

Proof. If not, then this ideal is contained in a maximal ideal $\mathfrak{p}$ of R. It follows that the image of the localized form $f_{\mathfrak{p}} : M_{\mathfrak{p}} \times M_{\mathfrak{p}} \to R_{\mathfrak{p}}$ is contained in the maximal ideal $\mathfrak{p} R_{\mathfrak{p}}$. Let $\mathcal{X}$ be a basis of $M_{\mathfrak{p}}$ and observe that the element $\det_{\mathcal{X}} f_{\mathfrak{p}}$ is in $\mathfrak{p} R_{\mathfrak{p}}$. But this is impossible, since $f_{\mathfrak{p}}$ is nonsingular by (4.7). QED.

Let M be a symmetric bilinear module over R with form f. For a submodule N of M, define the *orthogonal complement* $N^{\perp}$ of N by

$$N^{\perp} = \{y \in M \mid f(x, y) = f(y, x) = 0 \text{ for all } x \in N\}.$$

(4.9). *Let* M *be a symmetric bilinear module with form* f. *Let* N *be a submodule of* M. *If the restriction of* f *to* $N \times N$ *is nonsingular, then* $M = N \oplus N^{\perp}$.

Proof. Fix $z \in M$. Let x in N be arbitrary. The assignment $x \to f(x, z)$ defines an element in $N^* = \text{Hom}_R(N, R)$. Since N with the restricted f is nonsingular, there is a y in N such that $x \to f(x, y)$ is the same map as $x \to f(x, z)$. So $f(x, z) = f(x, y)$ for all x in N. So $z - y \in N^{\perp}$. Therefore, $M = N + N^{\perp}$. Suppose that $z \in N \cap N^{\perp}$. So $x \to f(x, z)$ is the zero map on N. The nonsingularity of the restriction of f implies that $z = 0$. Therefore, M $= N \oplus N^{\perp}$. QED.

If $M_1, \ldots, M_k$ are submodules of M such that $M = M_1 \oplus \ldots \oplus M_k$ and $f(M_i, M_j) = 0$, for $i \neq j$, we write $M = M_1 \perp \ldots \perp M_k$. In particular, if N is nonsingular, then $N \perp N^{\perp}$. There is an "external" version of this construction. Namely, let $M_1, M_2, \ldots, M_k$ be R-modules equipped with symmetric bilinear forms $f_1, f_2, \ldots, f_k$, and supply the Cartesian sum $M = M_1 \oplus \ldots \oplus M_k$ with the form

$$f(x_1 + \ldots + x_k, y_1 + \ldots + y_k) = \sum_{1 \leq i \leq k} f_i(x_i, y_i).$$

It is clear that M with f is a symmetric bilinear module. Denote it by $M_1 \perp ... \perp M_k$. Identify the modules M_i with the obvious submodules of M, and observe that $M = M_1 \perp ... \perp M_k$ in the earlier sense. If $M = M_1 \perp ... \perp M_k$ (in either situation), then it is not hard to show that M is nonsingular if and only if all the M_i are nonsingular. See Exercise 9.

C. The Group Dis(R)

We now illustrate the preceding concepts by focusing on bilinear modules which are finitely generated projective of rank 1. The following development is due to Bass [1974]. We present it in expanded form.

Begin with the free case, in particular with the R-module R. Let u be an element of R. Then $f_u : R \times R \to R$ defined by $f_u(r, s) = rus$ is a bilinear form on R. Notice that f_u is symmetric. The bilinear module R with f_u is denoted by (R, f_u). Let u and v be in R. Check that if $u = vt^2$ for some $t \in R^*$, then $r \to rt$ provides an isometry $(R, f_v) \to (R, f_u)$. Conversely, if $(R, f_v) \to (R, f_u)$ is an isometry, then $u = vt^2$ for some $t \in R^*$. Since $[u]$ is the matrix of f_u in the basis $\{1\}$, f_u is nonsingular if and only if $u \in R^*$. Let M with f be any bilinear module which is free of rank 1. Let x be a basis and set $f(x, x) = u$. Check that $M \cong (R, f_u)$. Note as a consequence, that if M is free of rank 1, then any bilinear form on M is symmetric.

Let P be any R-module which is finitely generated projective of rank 1, and let $f : P \times P \to R$ be a bilinear form on P. We claim that f is symmetric. Let $\mathfrak{p}$ be any prime ideal of R. Since $P_{\mathfrak{p}}$ is free of rank 1 over $R_{\mathfrak{p}}$, $f_{\mathfrak{p}}$ is symmetric. So for any p and q in P, $f(p, q)$ and $f(q, p)$ have the same image under $R \to R_{\mathfrak{p}}$. It follows that $f(p, q) = f(q, p)$ for any p and q in P.

Consider the map $f : P \otimes_R P \to R$. We claim that this map is an isomorphism if and only if the map $\bar{f} : P \to P^*$ is an isomorphism, i.e., if and only if P is nonsingular. Since $\bar{f}$ is an isomorphism if and only if $(\bar{f})_{\mathfrak{p}}$ is an isomorphism for all prime ideals $\mathfrak{p}$ of R, we find by (4.7) that the question is reduced to the case where P is free of rank 1, i.e., to the case $P = R$ and $f = f_u$. Suppose f_u is nonsingular. So $u \in R^*$. For s in R arbitrary, put $y = u^{-1}s$. Since $1 \otimes y \to f_u(1, y) = 1uu^{-1}s = s$, $f_u : R \otimes_R R \to R$ is surjective, and hence an isomorphism by (4.5). If, on the other hand, this map is an isomorphism, then 1 can be expressed as a sum of elements of the form rus, with r and s in R. So 1 is in the ideal Ru. It follows that $u \in R^*$. So f_u is nonsingular and the assertion is verified.

A nonsingular symmetric bilinear module P which is finitely generated projective of rank 1 is called a *discriminant module*. If the underlying form f needs emphasis, P will be denoted by (P, f).

(4.10). *Let* (P, f) *be a discriminant module. Then* f *is symmetric, and for all* $p, p', q,$ *and* q' *in* P, $f(p, q)q' = f(q', p)q$ *and* $f(p, q)f(p', q') = f(p, p')f(q, q')$.

Proof. That f is symmetric was already verified. The equalities $f(p, q)q' = f(q', p)q$ and $f(p, q)f(p', q') = f(p, p')f(q, q')$ are established in a similar way. We do the first. Fix $p, q,$ and q' and let $\mathfrak{p}$ be any prime ideal of R. It suffices to show that the image of the element $f(p, q)q' - f(q', p)q$ under $P \to P_{\mathfrak{p}}$ is 0. Since $P_{\mathfrak{p}} = P \otimes_R R_{\mathfrak{p}}$ is free of rank 1 over $R_{\mathfrak{p}}$, $(P_{\mathfrak{p}}, f_{\mathfrak{p}}) \cong (R_{\mathfrak{p}}, f_u)$ where u is a unit in $R_{\mathfrak{p}}$. So replace $(P_{\mathfrak{p}}, f_{\mathfrak{p}})$ by $(R_{\mathfrak{p}}, f_u)$. Since $f_u(p, q)q' - f_u(q', p)q = (puq)q' - (q'up)q = 0$ in $R_{\mathfrak{p}}$ for any $p, q,$ and q' in $R_{\mathfrak{p}}$, we are done.   QED.

Let (P, f) and (Q, g) be two discriminant modules. The construction in Section B provides the bilinear form

$$f \otimes g : (P \otimes_R Q) \times (P \otimes_R Q) \to R.$$

Recall that $(f \otimes g)(p \otimes q, p' \otimes q') = f(p, p')g(q, q')$. Since both f and g are symmetric, so is $f \otimes g$. Consider the special case $(P, f) = (R, f_u)$ and $(Q, g) = (R, f_v)$. Both are nonsingular; thus u and v are in R^*. Observe that the R-module map $R \otimes_R R \to R$ given by multiplication is an isometry $(R \otimes_R R, f_u \otimes f_v) \to (R, f_{uv})$ of bilinear modules. It follows that $(R \otimes_R R, f_u \otimes f_v)$ is nonsingular. An application of the nonsingularity criterion (4.7), shows that $(P \otimes_R Q, f \otimes g)$ is nonsingular in general.

Denote the isometry class of a discriminant module (P, f) by $[P, f]$ and let

$$\text{Dis}(R)$$

be the set of isometry classes of all discriminant modules. It follows from basic properties of the tensor product that the product law

$$[P, f] [Q, g] = [P \otimes_R Q, f \otimes g]$$

makes Dis(R) into an Abelian group with identity element $[R, f_1]$.

(4.11). Dis(R) *is an elementary Abelian 2-group.*

Proof. We must show that $[P, f]^2 = 1$ for any $[P, f]$ in Dis(R). For this, it suffices to show that the isomorphism $f : P \otimes_R P \to R$ is an isometry from $(P \otimes_R P, f \otimes f)$ onto (R, f_1). Observe that for any $p, p', q,$ and q' in P,

$$f_1(f(p \otimes q), f(p' \otimes q')) = f(p, q)f(p', q') \quad \text{and}$$

$$(f \otimes f)(p \otimes q, p' \otimes q') = f(p, p')f(q, q').$$

By (4.10), these elements are equal, so f is an isometry as required.   QED.

For an arbitrary $u \in R^*$ we will denote the element $[R, f_u]$ of Dis(R) by $[R, u]$. Define a map $R^*/(R^*)^2 \to$ Dis(R) by $u(R^*)^2 \to [R, u]$.

(4.12). *The map* $R^*/(R^*)^2 \to$ Dis(R) *is an injective homomorphism.*

Proof. Let u and v be elements in R^*. If $u = vt^2$ for some $t \in R^*$, then $(R, f_v) \cong (R, f_u)$, so that the map is well defined. That it is a homomorphism follows, since it was already observed that $(R \otimes_R R, f_u \otimes f_v) \cong (R, f_{uv})$. Suppose that $u(R^*)^2$ is in the kernel. So (R, f_u) and (R, f_1) are isometric. So $u \in (R^*)^2$. Therefore, $R^*/(R^*)^2 \to$ Dis(R) is injective.   QED.

Let A be a commutative R-algebra. Refer to the change-of-scalars construction and verify that the assignment $(P, f) \to (P \otimes_R A, f_A)$ induces a homomorphism

$$\text{Dis}(R) \to \text{Dis}(A).$$

For the remainder of this section, R will be a domain with field of fractions F. The goal is a more explicit description of Dis(R) in this case. For the proofs of the basic facts used later, refer, for example, to Chapter 10.1 of Jacobson [1985, II].

A nonzero R-submodule $\mathfrak{a}$ of F is called a *fractional ideal* if $r\mathfrak{a} \subseteq R$ for some nonzero r in R. If $\mathfrak{a}$ is an R-submodule of F which is finitely generated, then it is easy to see that $\mathfrak{a}$ is a fractional ideal. For any $c \in F^*$, Rc is a fractional ideal. A fractional ideal of this form is *principal*. For fractional ideals $\mathfrak{a}$ and $\mathfrak{b}$,

$$\mathfrak{ab} = \{\sum_{\text{fin}} ab \mid a \in \mathfrak{a}, b \in \mathfrak{b}\}$$

is also a fractional ideal called the *product* of $\mathfrak{a}$ and $\mathfrak{b}$. For a fractional ideal $\mathfrak{a}$ define $\mathfrak{a}^{-1}$ by $\mathfrak{a}^{-1} = \{d \in F \mid d\mathfrak{a} \subseteq R\}$. It is easy to see that $\mathfrak{a}^{-1}$ is also a fractional ideal. The fractional ideal $\mathfrak{a}$ is *invertible* if the product $\mathfrak{a}\mathfrak{a}^{-1}$ is equal to R. Any invertible fractional ideal is a finitely generated R-module, and a fractional ideal is invertible if and only if it is projective. An invertible fractional ideal has (constant) rank 1.

Let (P, f) be a discriminant module. Let $(P \otimes_R F, f_F)$ be the bilinear module over F obtained by change of scalars. Fix a nonzero x in P. Since P is projective, $P \to P \otimes_R F$ is injective, and therefore $x \otimes 1$ is a basis of $P \otimes_R F$. Set $f_F(x \otimes 1, x \otimes 1) = c \in F^*$. The assignment $x \otimes 1 \to 1$ provides an isometry $(P \otimes_R F, f_F) \to (F, f_c)$. Let $\varphi : P \to F$ be the composite $P \to P \otimes_R F \to F$ and denote the image of φ by $\mathfrak{a}$. Since $\mathfrak{a}$ is a finitely generated R-module, we find by "clearing" denominators of a finite set of generators that $\mathfrak{a}$ is a fractional ideal of F. It is invertible, since P is projective. Since f induces an isomorphism $f : P \otimes_R P \to R$, we obtain the commutative diagram of isomorphisms

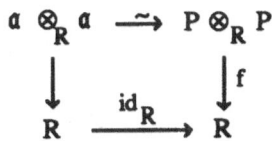

with the map on the left given by $a \otimes b \to acb$. It follows that $R = c\mathfrak{a}^2$ and hence that $c^{-1}R = \mathfrak{a}^2$. We have shown that any discriminant module over R is isometric to one of the form $(\mathfrak{a}, f_c)$, where $\mathfrak{a}$ is an invertible fractional ideal of F, $c \in F^*$, $\mathfrak{a}^2 = c^{-1}R$, and the form $f_c : \mathfrak{a} \times \mathfrak{a} \to R$ is given by $f_c(a, b) = acb$. If, conversely, a pair $(\mathfrak{a}, f_c)$ satisfies these conditions, then it is a discriminant module. See Exercise 12.

Suppose that the discriminant modules $(\mathfrak{a}, f_c)$ and $(\mathfrak{b}, f_d)$ are isometric and let $\varphi : \mathfrak{a} \to \mathfrak{b}$ be an isometry. Choose a nonzero v in R such that $v\mathfrak{a} \subseteq R$. Fix a nonzero $c \in \mathfrak{a}$. So $vc \in R$. Now, $vc\varphi(a) = \varphi(vac) = va\varphi(c)$. So $\varphi(a) = (\varphi(c)c^{-1})a$ for all a in $\mathfrak{a}$. Put $\varphi(c)c^{-1} = z$. We have shown that φ is given by multiplication by a nonzero element z of F. In particular, $z\mathfrak{a} = \mathfrak{b}$. Since φ is an isometry, $zadzb = f_d(za, zb) = f_c(a, b) = acb$ for any a and b in $\mathfrak{a}$. So $c = z^2 d$. Conversely, if $z\mathfrak{a} = \mathfrak{b}$ and $c = z^2 d$ for some nonzero z in F, then $(\mathfrak{a}, f_c)$ and $(\mathfrak{b}, f_d)$ are isometric. Denoting the isometry class of $(\mathfrak{a}, f_c)$ by $[\mathfrak{a}, c]$, we can set

$$\text{Dis}(R) = \{[\mathfrak{a}, c] \mid \mathfrak{a}^2 = c^{-1}R\}.$$

Fix a prime ideal $\mathfrak{p}$ of R and consider the change-of-scalars homomorphism

$$\text{Dis}(R) \to \text{Dis}(R_{\mathfrak{p}}).$$

Refer to Exercise 13 and show that this map is given by $[\mathfrak{a}, c] \to [\mathfrak{a}R_{\mathfrak{p}}, c]$, where $\mathfrak{a}R_{\mathfrak{p}} = \{ \underset{\text{fin}}{\sum} ax \in F \mid a \in \mathfrak{a} \text{ and } x \in R_{\mathfrak{p}} \}$. Denote by $\text{Max}(R)$ the set of maximal ideals of R.

(4.13). *If R is a domain, then* $\text{Dis}(R) \to \underset{\mathfrak{m} \in \text{Max}(R)}{\prod} \text{Dis}(R_{\mathfrak{m}})$ *is injective.*

Proof. Let $[\mathfrak{a}, c]$ be in the kernel. Let $\mathfrak{m} \in \text{Max}(R)$ be arbitrary. Since $(\mathfrak{a}R_{\mathfrak{m}}, f_c) \cong (R_{\mathfrak{m}}, f_1)$, $z(\mathfrak{a}R_{\mathfrak{m}}) = R_{\mathfrak{m}}$ and $c = z^2$ for some $z \in F^*$. Put $\mathfrak{b} = (zR)\mathfrak{a} = z\mathfrak{a}$. Note that $[\mathfrak{a}, c] = [\mathfrak{b}, 1]$ and $\mathfrak{b}^2 = R$. Since $\mathfrak{b}R_{\mathfrak{m}} = R_{\mathfrak{m}}$, $\mathfrak{b} \subseteq R_{\mathfrak{m}}$. Since this holds for any $\mathfrak{m}$ in $\text{Max}(R)$, $\mathfrak{b} \subseteq R$. So $R = \mathfrak{b}^2 \subseteq \mathfrak{b} \subseteq R$, and hence $\mathfrak{b} = R$.   QED.

D. Quadratic Forms

Let M be an R-module. A *quadratic form* on M is a map $q : M \to R$ such that

$$q(xr) = r^2 q(x)$$

for all x in M and r in R and such that the map $h : M \times M \to R$ defined by the equation

$$h(x, y) = q(x + y) - q(x) - q(y)$$

is a bilinear form. This h is the *bilinear form associated* to q. Clearly, h is symmetric and $h(x, x) = 2q(x)$ for all x in M. By induction we find that for any $x_1, ..., x_k$ in M and $r_1, ..., r_k$ in R,

$$q(x_1 r_1 + ... + x_k r_k) = \sum_{1 \le i \le n} r_i^2 q_i(x_i) + \sum_{1 \le i < j \le n} r_i r_j h_i(x_i, x_j).$$

An R-module M equipped with a quadratic form is called a *quadratic module* over R. A quadratic module is *nonsingular* if it is nonsingular relative to the associated symmetric bilinear form. A quadratic module is automatically a symmetric bilinear module relative to the associated bilinear form. If $2 \in R^*$, then the converse is true. Namely, let M be a symmetric bilinear module with form f and define $q(x) = \frac{1}{2} f(x, x)$. Then q is a quadratic form with associated bilinear form f. So if $2 \in R^*$, then the concepts symmetric bilinear module and quadratic module are interchangeable.

More generally, there is the following connection between symmetric bilinear and quadratic forms.

(4.14). *Let f be a symmetric bilinear form on a module M over a commutative ring R. Assume that 2 is not a zero divisor in R. Then f is the symmetric bilinear form associated to a quadratic form q on M if and only if $f(x, x) \in 2R$ for all x in M.*

Proof. Assume $f(x, x) \in 2R$ for all x in M. For x in M, $f(x, x) = 2r_x$ for a unique r_x in R. Define $q : M \to R$ by $q(x) = r_x$. Since $f(xr, xr) = 2r^2 r_x$, $q(xr) = r^2 r_x = r^2 q(x)$. Let x and y be in M. So $q(x) = r_x$, $q(y) = r_y$ and $q(x + y) = r_{x+y}$, where $f(x, x) = 2r_x$, $f(y, y) = 2r_y$ and $f(x + y, x + y) = 2r_{x+y}$. Since $f(x + y, x + y) = f(x, x) + 2f(x, y) + f(y, y)$, $r_{x+y} = r_x + r_y + f(x, y)$. So $f(x, y) = q(x + y) - q(x) - q(y)$. We have proved that q is a quadratic form on M with associated symmetric bilinear form f. The other direction is clear.
QED.

Let M and M' be two quadratic modules over R. An R-homomorphism from M into M' which preserves the underlying quadratic forms is a *representation* from M to M', and a bijective representation from M to M' is an *isometry* from M onto M'. If such a map exists, then M and M' are *isometric* and we write $M \cong M'$.

Example. Let $S = R[X]/(X^2 - aX - b)$ be a free quadratic algebra. Recall that the conjugation σ of S is given by $(r + tv)^\sigma = (r + ta) - tv$, where $v = X + (X^2 - aX - b)$. Define $q : S \to R$ by $q(r + tv) = (r + tv)(r + tv)^\sigma = r^2 + rta - t^2 b$. Check that this is a quadratic form on S. The associated symmetric bilinear form h is routinely shown to be given by

$$h(r + tv, r' + t'v) = 2rr' + rt'a + tr'a - 2tt'b.$$

The matrix of h in the basis $\{1, v\}$ is $\begin{bmatrix} 2 & a \\ a & -2b \end{bmatrix}$. In this way, the free quadratic algebra S has been made into a quadratic module over R. The form q is called the *norm form* of S. Note that $\det \begin{bmatrix} 2 & a \\ a & -2b \end{bmatrix} = -(a^2 + 4b)$. Therefore by Proposition (2.2), S is nonsingular as a quadratic module if and only if S is separable as an algebra.

(4.15). *Let S and T be free quadratic algebras. Then S and T are isomorphic as algebras over R if and only if they are isometric as quadratic modules over R.*

Proof. Let $S = R[X]/(X^2 - aX - b)$ and $T = R[X]/(X^2 - cX - d)$. Let $v = X + (X^2 - aX - b)$ and $w = X + (X^2 - cX - d)$. Let $q : S \to R$ be the norm form of S and let $q' : T \to R$ be the norm form of T. The associated symmetric bilinear forms are given by h and h'.

Let $\varphi : T \to S$ be an isomorphism of algebras. So $\varphi(1) = 1$ and $-$ see the proof of Proposition (1.1) $- \varphi(w) = r + uv$, with $r \in R$ and $u \in R^*$, where $c = ua + 2r$ and $d = u^2 b - rua - r^2$. Since

$$q(\varphi(r_1 + r_2 w)) = q(r_1 + r_2(r + uv)) = (r_1 + r_2 r)^2 + (r_1 + r_2 r)r_2 ua - (r_2 u)^2 b$$
$$= r_1^2 + r_1 r_2 c - r_2^2 d = q'(r_1 + r_2 w),$$

$\varphi : T \to S$ is an isometry. Conversely, let $\varphi : T \to S$ be an isometry. Put $\varphi(1) = r_1 + r_2 v$, and $\varphi(w) = r_3 + r_4 v$. Since $\{r_1 + r_2 v, r_3 + r_4 v\}$ is a basis of S, $\begin{bmatrix} r_1 & r_3 \\ r_2 & r_4 \end{bmatrix}$ is invertible and hence $u = r_1 r_4 - r_3 r_2 \in R^*$. Note that

$$1 = q'(1) = q(\varphi(1)) = q(r_1 + r_2 v) = r_1^2 + r_1 r_2 a - r_2^2 b,$$

$$-d = q'(w) = q(\varphi(w)) = q(r_3 + r_4 v) = r_3^2 + r_3 r_4 a - r_4^2 b, \text{ and}$$

$$c = h'(1, w) = h(\varphi(1), \varphi(w)) = h(r_1 + r_2 v, r_3 + r_4 v)$$

$$= 2r_1 r_3 + r_1 r_4 a + r_2 r_3 a - 2r_2 r_4 b.$$

Putting $r = r_1 r_3 + r_2 r_3 a - r_2 r_4 b$, we see that $c = 2r + au$. To show that $S \cong T$ as R-algebras, we prove that $d = u^2 b - rua - r^2$ and apply Proposition (1.1). This is a somewhat lengthy, but completely routine, bookkeeping operation. Plug the expressions for r and u into $u^2 b - rua - r^2$, collect and simplify terms, and note that nine terms remain: three have r_3^2 as a factor, three have $r_3 r_4 a$ as a factor, and three contain the factor $r_3^2 b$. Factor these terms out, use the equalities, and note that the expression now equals d.   QED.

Any submodule N of M is a quadratic module with the restricted quadratic form $q|_N$. If $M_1, \ldots, M_k$ are submodules of M and $M = M_1 \perp \ldots \perp M_k$ as bilinear modules, we say that the quadratic module M is the *orthogonal direct sum* of the M_i. As in the symmetric bilinear case, there is an "external" version of orthogonal sum. Namely, if quadratic modules $M_1, \ldots, M_k$ over R with quadratic forms $q_1, \ldots, q_k$ are given, then $q(x_1 + \ldots + x_k) = \sum_{1 \leq i \leq k} q_i(x_i)$ defines a quadratic form on the Cartesian sum $M_1 \oplus \ldots \oplus M_k$. We also call this quadratic module the *orthogonal direct sum* of the M_i and denote it by $M_1 \perp \ldots \perp M_k$. If $h_1, \ldots, h_k$ are the symmetric bilinear forms associated to the q_i, then the bilinear form associated to q is given by

$$h(x_1 + \ldots + x_k, y_1 + \ldots + y_k) = \sum_{1 \leq i \leq k} h_i(x_i, y_i).$$

So the notation $M_1 \perp \ldots \perp M_k$ is completely consistent with that used in the symmetric bilinear case. As a consequence, M is nonsingular if and only if all the M_i are nonsingular.

(4.16). *Let* R *be a local ring and let* M *be nonzero, free of finite rank, and nonsingular. Then there exists a splitting* $M = M_1 \perp \ldots \perp M_k$ *such that each* M_i *is free of rank either* 1 *or* 2 *and nonsingular.*

Proof. Let the rank of M be n and let h be the underlying symmetric bilinear form. Suppose first that there is an element x in M such that $h(x, x) \in R^*$. Then xR is a nonsingular submodule of M. So $M = xR \perp (xR)^\perp$ where $(xR)^\perp$ is nonsingular of rank $n - 1$. If no such x exists, then $h(x, x)$ is in the maximal ideal $\mathfrak{m}$ of R for all x in M. Since the matrix of h in a basis of M is invertible, it follows that $h(x, y) \in R^*$ for some x and y in M. It is easy to see that x and y are independent and that the determinant of the matrix of the restriction of h to $xR \oplus yR$ is in R^*. In particular, $xR \oplus yR$ is a nonsingular submodule of M. Therefore, $M = (xR \oplus yR) \perp (xR \oplus yR)^\perp$ with $(xR \perp yR)^\perp$ nonsingular of rank $n - 2$. A repetition of these constructions completes the proof.   QED.

(4.17). *Let* M *be a finitely generated projective quadratic module over* R *with quadratic form* q *and associated bilinear form* h. *Then there exists a bilinear form* $f : M \times M \to R$, *such that* $q(x) = f(x, x)$ *and* $h(x, y) = f(x, y) + f(y, x)$ *for all* x *and* y *in* M.

Proof. Suppose first that M is free with finite basis $\{x_1, \ldots, x_n\}$. Set $f(x_i, x_j) = h(x_i, x_j)$, $q(x_i)$, or 0, according to whether $i < j$, $i = j$, or $i > j$. Extend this to a bilinear form f on M by linearity. Let $x = x_1 r_1 + \ldots + x_n r_n$ and $y = x_1 s_1 + \ldots + x_n s_n$ in M be arbitrary. Then

$$q(x) = q(x_1 r_1 + \ldots + x_n r_n) = \sum_{1 \le i \le n} r_i^2 q(x_i) + \sum_{1 \le i < j \le n} r_i r_j h(x_i, x_j)$$

$$= \sum_{1 \le i \le n} r_i^2 f(x_i, x_i) + \sum_{1 \le i, j \le n} r_i r_j f(x_i, x_j) = f(x, x),$$

and since $q(x_i) = 2h(x_i, x_i)$,

$$h(x, y) = \sum_{1 \le i, j \le n} r_i s_j h_i(x_i, x_j) = 2 \sum_{1 \le i \le n} r_i s_i h_i(x_i, x_i) + 2 \sum_{1 \le i < j \le n} r_i s_j h_i(x_i, x_j)$$

$$= \sum_{1 \le i \le n} r_i s_i f(x_i, x_i) + \sum_{1 \le i \ne j \le n} r_i s_j f(x_i, x_j) + \sum_{1 \le i \ne j \le n} s_i r_j f(x_j, x_i)$$

$$= f(x, y) + f(y, x).$$

So f satisfies the requirements. In the projective situation, choose N such that $M \oplus N$ is free of finite rank. Let q' be the quadratic form on N given by $q'(z) = 0$ for all z in N. The associated bilinear form h' is also identically zero. Denote the quadratic form and symmetric bilinear form of the quadratic module $M \perp N$ by q'' and h''. Let g be the bilinear form provided by applying the special case of the proposition already proved to $M \perp N$. So $q''(x + z) = g(x + z, x + z)$ and $h''(x + z, y + z) = g(x + z, y + z) + g(y + z, x + z)$ for all x and y in M and z in N. Define $f : M \times M \to R$, by $f(x, y) = g(x + 0, y + 0)$. For all x and y in M, and z in N,

$$q(x) = q''(x + 0) = g(x + 0, x + 0) = f(x, x) \text{ and}$$

$$h(x, y) = h''(x + 0, y + 0) = g(x + 0, y + 0) + g(y + 0, x + 0) = f(x, y) + f(y, x).$$

Therefore, f fills the bill and the proof is complete.   QED.

Let A be a commutative R-algebra. Assume that M is finitely generated projective. Then the A-module $M \otimes_R A$ can be supplied with a quadratic form $q_A : M \otimes_R A \to A$ such that q_A and the associated bilinear form h_A satisfy

$$q_A(x \otimes a) = a^2(q(x)) \text{ and } h_A(x \otimes a, y \otimes b) = ab(h(x, y))$$

for all x, y in M and a, b in A. This is a consequence of (4.17): Choose an f for the pair (h, q). Refer to Section B and consider the corresponding f_A on $M \otimes_R A$. Now define q_A and h_A by

$$q_A(u) = f_A(u, u) \quad \text{and} \quad h_A(u, v) = f_A(u, v) + f_A(v, u),$$

respectively, for all u and v in $M \otimes_R A$. We find that if q is nonsingular, then q_A is also. This construction applied to $A = R_{\mathfrak{p}}$ supplies the module $M_{\mathfrak{p}}$ over $R_{\mathfrak{p}}$ with a quadratic form $q_{\mathfrak{p}}$ – the *localization of* M *at* $\mathfrak{p}$.

(4.18). *Let* M *be finitely generated projective and nonsingular. If* rank M *is odd, then* $2 \in R^*$.

Proof. We first prove the proposition in the case where M is free with basis $\{x_1, \ldots, x_n\}$ where n is odd. Consider the expansion

$$\det (h(x_i, x_j)) = \sum_{\pi \in S_n} (-1)^{\text{sign } \pi} h(x_1, x_{\pi 1}) \cdots h(x_n, x_{\pi n}) .$$

Suppose $\pi \in S_n$ satisfies $\pi = \pi^{-1}$. Consider the composition of π into disjoint cycles. Since $\pi^2 = 1$, all cycles must be transpositions. Since n is odd, it follows that $\pi i = i$ for some i. So $h(x_i, x_{\pi^{-1}i}) = h(x_i, x_i) = 2q(x_i)$.

Suppose $\pi \neq \pi^{-1}$. Consider the term $(-1)^{\text{sign}(\pi^{-1})} h(x_1, x_{\pi^{-1}1}) \cdots h(x_n, x_{\pi^{-1}n})$. Each factor $h(x_i, x_{\pi^{-1}i})$ equals $h(x_{\pi j}, x_j)$ for some j. Since h is symmetric, $h(x_i, x_{\pi^{-1}i}) = h(x_j, x_{\pi j})$. Therefore,

$$(-1)^{\text{sign}(\pi^{-1})} h(x_1, x_{\pi^{-1}1}) \cdots h(x_n, x_{\pi^{-1}n}) = (-1)^{\text{sign } \pi} h(x_1, x_{\pi 1}) \cdots h(x_n, x_{\pi n}).$$

We have now proved that $\det (h(x_i, x_j)) = 2r$ for some r in R. Since q is nonsingular, $2r \in R^*$, so $2 \in R^*$. The case of a finitely generated projective M is now easy. If $2 \notin R^*$, then 2 is contained in some prime ideal $\mathfrak{p}$ of R. So $2 \notin (R_{\mathfrak{p}})^*$. Since the localized quadratic module $M_{\mathfrak{p}}$ over $R_{\mathfrak{p}}$ is free of odd rank, this is a contradiction.   QED.

E. Exercises

1. Refer, for example, to Chapter 7 of Jacobson [1985, II] and study the basic constructions of the localization theory of rings and modules.

2. Let R be a local ring with maximal ideal $\mathfrak{m}$. So $R = \mathfrak{m} \cup R^*$. Prove that $2 \in R^*$ if and only if char $R/\mathfrak{m} \neq 2$.

3. Consider the ring $\mathbb{Z}$. Let (p) be the principal prime ideal determined by a prime p and describe the localization $\mathbb{Z}_{(p)}$ as a subring of $\mathbb{Q}$.

4. Let R be a domain with field of fractions F and let $\mathfrak{p}$ be a prime ideal of R. Show that $R_{\mathfrak{p}}$ is isomorphic to the subring $\{(r/s) \in F \mid s \in R - \mathfrak{p}\}$ of F and that F is also the field of fractions of $R_{\mathfrak{p}}$.

5. Describe the Zariski toplogy of $\mathrm{Spec}(\mathbb{Z})$.

6. Let M and N be R-modules. Let $\mathfrak{p}$ be a prime ideal of R. Show that $(M \otimes_R N)_{\mathfrak{p}} \cong (M_{\mathfrak{p}} \otimes_{R_{\mathfrak{p}}} N_{\mathfrak{p}})$.

7. Let M be an R-module. Let $\mathfrak{p}$ be a prime ideal of R. Show that there is a natural isomorphism $(M^*)_{\mathfrak{p}} \cong (M_{\mathfrak{p}})^*$.

8. Let M be a free R-module of finite rank with bilinear form f. Let $\mathcal{X} = \{x_1,..., x_n\}$ be a basis of M and let $\mathcal{X}^* = \{\rho_1,..., \rho_n\}$ be the dual basis.

   Show that $C = (f(x_i, x_j))$ is the matrix of $\bar{f} : M \to M^*$ in the bases $\mathcal{X}$ and $\mathcal{X}^*$. So M is nonsingular, if and only if C is invertible, if and only if $\det_{\mathcal{X}} f = \det C$ is in R^*.

9. Let M be an R-module. Suppose that $M = M_1 \oplus ... \oplus M_k$ is a direct sum of submodules M_i. Show that

   $$M^* = \mathrm{Hom}_R(M, R) \cong \mathrm{Hom}_R(M_1, R) \oplus ... \oplus \mathrm{Hom}_R(M_k, R)$$
   $$= M_1^* \oplus ... \oplus M_k^*$$

   in a natural way. Deduce that if $M = M_1 \perp ... \perp M_k$ is a symmetric bilinear module over R, then M is nonsingular if and only if all the M_i are nonsingular.

10. Consider all pairs (P, f) where P is a finitely generated projective R-module of rank 1 and f is a bilinear form on P which is not necessarily nonsingular. Let $[P, f]$ denote the isometry class of (P, f). Show that the set of all isometry classes $[P, f]$ form a monoid with operation induced by tensor product and $[R, f_1]$ as identity. Denote it by $\mathrm{Dim}(R)$. Is $[P, f]$ invertible in $\mathrm{Dim}(R)$ if and only if f is nonsingular, i.e., is $\mathrm{Dis}(R)$ the submonoid of invertible elements of $\mathrm{Dim}(R)$? Is the map $R/(R^*)^2 \to \mathrm{Dim}(R)$ a homomorphism?

11. Let R be a domain with field of fractions F and let $\mathfrak{a}$ be an invertible fractional ideal. Let $\mathfrak{p}$ be a prime ideal of R. Consider $R_{\mathfrak{p}} \subseteq F$. Show that $\mathfrak{a} \otimes_R R_{\mathfrak{p}} \to \mathfrak{a} R_{\mathfrak{p}} = \{\sum_{\text{fin}} ax \in F \mid a \in \mathfrak{a} \text{ and } x \in R_{\mathfrak{p}}\}$, defined by $(a, r') \to ar'$ for $a \in \mathfrak{a}$ and $r' \in R_{\mathfrak{p}}$, is an isomorphism of $R_{\mathfrak{p}}$-modules.

12. Let R be a domain with field of fractions F. Consider a pair $(\mathfrak{a}, f_c)$ where $\mathfrak{a}$ is a fractional ideal and c is a nonzero element of F such that $\mathfrak{a}^2 = c^{-1}R$. Since $\mathfrak{a}(c\mathfrak{a}) = R$, $\mathfrak{a}$ is invertible and hence finitely generated projective of rank 1. Show that $f_c : \mathfrak{a} \times \mathfrak{a} \to R$ induces an isomorphism $\mathfrak{a} \otimes_R \mathfrak{a} \cong R$. So $(\mathfrak{a}, f_c)$ is a discriminant module.

13. Let R be a domain with field of fractions F and let $(\mathfrak{a}, f_c)$ be a discriminant module. Let $\mathfrak{p}$ be a prime ideal of R. Show that the localization $(\mathfrak{a} \otimes_R R_{\mathfrak{p}}, (f_c)_{\mathfrak{p}})$ is a discriminant module over $R_{\mathfrak{p}}$, that $(\mathfrak{a} R_{\mathfrak{p}}, f_c)$ is a discriminant module over $R_{\mathfrak{p}}$, and that $\mathfrak{a} \otimes_R R_{\mathfrak{p}} \to \mathfrak{a} R_{\mathfrak{p}}$ is an isometry $(\mathfrak{a} \otimes_R R_{\mathfrak{p}}, (f_c)_{\mathfrak{p}}) \to (\mathfrak{a} R_{\mathfrak{p}}, f_c)$.

14. Is the product $\prod_{\mathfrak{m} \in \text{Max}(R)} \text{Dis}(R_{\mathfrak{m}})$ of (4.13) a *restricted product* in general? In other words, is it always the case for $[\mathfrak{a}, c]$ in $\text{Dis}(R)$, that $[\mathfrak{a} R_{\mathfrak{m}}, c] = [R_{\mathfrak{m}}, 1]$ for all but finitely many $\mathfrak{m}$ in $\text{Max}(R)$?

15. Let M be a free nonsingular quadratic module of rank 2 over R. Let q be the quadratic form on M and assume that $q(x) = 1$ for some x in M. Show that x can be completed to a basis of M. Deduce that there is a free separable quadratic algebra S over R such that $M \cong S$ as quadratic modules.

16. Let M be a free nonsingular quadratic module of finite rank over a local ring R. Let q be the quadratic form on M. If $M \neq \{0\}$, then there exists $x \in M$ such that $q(x) \in R^*$.

Hints:

14. No. Consider $[\mathfrak{a}, c]$ with $c \notin (F^*)^2$.

15. Let $\{x_1, x_2\}$ be a basis of M and put $x = x_1 r_1 + x_2 r_2$. Let h be the bilinear form associated to q. Note that the determinant of

$$\begin{bmatrix} r_1 & -(q(x_2)r_2 + r_1 h(x_1, x_2)) \\ r_2 & q(x_1)r_1 \end{bmatrix}$$ is 1. Let $s_1 = -(q(x_2)r_2 + r_1 h(x_1, x_2))$ and $s_2 = q(x_1)r_1$, and $y = x_1 s_1 + x_2 s_2$. Note that $\{x, y\}$ is a basis of M. Put $h(x, y) = a$ and $-q(y) = b$ and check that $a^2 + 4b \in R^*$.

16. Let X be a basis of M. Since $\det_X h \in R^*$, there are basis elements x and y such that $q(x)$ or $q(x + y)$ is in R^*.

5
Clifford Algebras: The Basics

Overview

This chapter introduces the Clifford algebra of a quadratic module over a commutative ring. This theory is well established in the literature. Refer, for example, to Baeza [1978], Hahn-O'Meara, or Knus [1991], for the theory over rings, and to Lam, O'Meara [1971], or Scharlau [1985] for the theory over fields. This chapter recalls the basic concepts, constructions, and facts. Only a few proofs are provided. Throughout, R is a commutative ring and M is a quadratic module over R with underlying quadratic form q and bilinear form h.

A. Definition and Existence

A pair (A, α) consisting of an R-algebra A and a homomorphism of R-modules $\alpha : M \to A$ is *compatible with* M if

$$\alpha(x)^2 = q(x)1_A$$

for all x in M. This equation applied to $x + y$ implies that

$$\alpha(x)\alpha(y) + \alpha(y)\alpha(x) = h(x, y)1_A$$

for all x and y in M. A *Clifford algebra* of M is a universal such pair. Therefore, a Clifford algebra of M is a pair $C(M) = (C(M), \gamma)$ such that

    (i) $C(M)$ is an R-algebra,

    (ii) $\gamma : M \to C(M)$ is an R-module map that satisfies

$$\gamma(x)^2 = q(x)1 \quad \text{and} \quad \gamma(x)\gamma(y) + \gamma(y)\gamma(x) = h(x, y)1$$

        for all x and y in M, and

(iii)   If (A, α) is any pair compatible with M, then there is a unique R-algebra homomorphism $\varphi : C(M) \to A$ such that the diagram

$$M \xrightarrow{\ \alpha\ } A$$
$$\gamma \searrow \quad \nearrow \varphi$$
$$C(M)$$

commutes.

Example 0. Suppose that $M = \{0\}$. So $q = 0$ and the pair consisting of R and the zero map from M to R is easily seen to be a Clifford algebra of M.

Example 1. Suppose that M is free of rank 1 and that $\{x\}$ is a basis of M. Let $C(M)$ be the R-algebra $R[X]/(X^2 - q(x))$ and let $\gamma : M \to C(M)$ be defined by $\gamma(xr) = rX + (X^2 - q(x))$ for all r in R. Then the pair $(C(M), \gamma)$ is a Clifford algebra of M. To see this, let (A, α) be any pair compatible with M and define $\varphi : C(M) \to A$ by

$$\varphi ((r + sX) + (X^2 - q(x))) = r1 + \alpha(xs)$$

for all r and s in R.

(5.1). *Theorem. Let* M *be a quadratic module over* R. *Then* M *has a Clifford algebra* $((C(M), \gamma)$. *If* (C', γ) *is another Clifford algebra of* M, *then there is a unique* R-algebra isomorphism $\varphi : C(M) \to C'$ *such that the diagram*

$$C(M) \xrightarrow{\ \varphi\ } C'$$
$$\gamma \nwarrow \quad \nearrow \gamma'$$
$$M$$

commutes.

B. Generation, Grading, and Involutions

According to Theorem (5.1), the Clifford algebra $(C(M), \gamma)$ of M exists in an essentially unique way. A discussion of the very basic properties of $C(M)$ follows.

(5.2). *If* $\{x_i\}_{i \in I}$, *where* I *is some index set, spans* M *as* R-*module, then* 1_C, *together with the elements* $\gamma(x_i)$, $i \in I$, *generates* C(M) *as an* R-*algebra. If* I *is linearly ordered, then the elements* 1_C *and all*

$$\gamma(x_{i_1})\gamma(x_{i_2}) \cdots \gamma(x_{i_k}), \quad i_1 < \dots < i_k$$

span C(M) *as* R-*module.*

(5.3). *Suppose* M *is nonzero and free with finite basis* $\{x_1, \dots, x_n\}$. *Then* C(M) *is a free* R-*module with basis*

$$\{1_C, \gamma(x_{i_1})\gamma(x_{i_2}) \cdots \gamma(x_{i_k}) \mid 1 \leq k \leq n, \ i_1 < \dots < i_k\}.$$

In particular, $\text{rank}_R C(M) = 2^n$.

If M is free with basis $\{x_1, \dots, x_n\}$, then as a consequence of this theorem, the ring homomorphism $R \to C(M)$ given by $r \to r1_C$ is injective, and the module homomorphism $\gamma : M \to C(M)$ is also injective.

Return to a general M and let $\{x_i\}_{i \in I}$ span M as an R-module. Let

$$C_0(M)$$

be the submodule of C(M) spanned by 1_C and all the $\gamma(x_{i_1})\gamma(x_{i_2}) \cdots \gamma(x_{i_k})$ with k even, and let

$$C_1(M)$$

be the submodule spanned by all the $\gamma(x_{i_1})\gamma(x_{i_2}) \cdots \gamma(x_{i_k})$ with k odd. These definitions are easily seen to be independent of the spanning set $\{x_i\}_{i \in I}$ of M. The submodules $C_0(M)$ and $C_1(M)$ are the *even* and *odd* parts of C(M).

(5.4). *The decomposition* $C(M) = C_0(M) \oplus C_1(M)$ *is a grading of* C(M). *In particular*, $C_0(M)$ *is a subalgebra of* C(M) *and* $C_1(M)$ *is a* $C_0(M)$-*module.*

If M is free with basis $\{x_1, \dots, x_n\}$, then 1_C and all the $\gamma(x_{i_1})\gamma(x_{i_2}) \cdots \gamma(x_{i_k})$, with $1 \leq k \leq n$, $i_1 < \dots < i_k$ and k even, are a basis of $C_0(M)$. Similarly, all the $\gamma(x_{i_1})\gamma(x_{i_2}) \cdots \gamma(x_{i_k})$, with $1 \leq k \leq n$, $i_1 < ... < i_k$, and k odd, are a basis for $C_1(M)$. In particular, $C_0(M)$ and $C_1(M)$ are both free of rank 2^{n-1}.

Example 2. Consider first the trivial case $M = \{0\}$. Since by Example 0 $C(M) \cong R$, it follows that $C_0(M) = C(M) = R1_C$ and $C_1(M) = \{0\}$. Suppose that M is free of rank 1 and that $\{x\}$ is a basis of M. Then $C_0(M) = R1_C$ and $C_1(M) = R\gamma x$.

Let $M' = (M, q')$ be another quadratic module over R and let $(C(M'), \gamma)$ be a Clifford algebra of M. Let $\sigma : M \rightarrow M'$ be a representation and consider the diagram

$$
\begin{array}{ccc}
M & \xrightarrow{\gamma} & C(M) \\
\sigma \downarrow & \searrow{\alpha} & \\
M' & \xrightarrow{\gamma'} & C(M')
\end{array}
$$

where $\alpha = \gamma'\sigma$. Since

$$\alpha(x)^2 = (\gamma'(\sigma x))^2 = q'(\sigma x)1_{C'} = q(x)1_{C'}$$

for all x in M, $(C(M'), \alpha)$ is compatible with M. So there exists a unique algebra homomorphism

$$C(\sigma) : C(M) \rightarrow C(M'),$$

which completes the diagram to a commutative diagram. Observe that it is graded. Let M'' be a third quadratic module over R and let $(C(M''), \gamma'')$ be a Clifford algebra for M''. For a representation σ' from $M' \rightarrow M''$, $\sigma'\sigma$ is a representation from M into M'', and there are the corresponding algebra homomorphisms $C(\sigma')$ from $C(M')$ into $C(M'')$ and $C(\sigma'\sigma)$ from $C(M)$ into $C(M'')$. It follows by the uniqueness property of these homomorphisms that

$$C(\sigma'\sigma) = C(\sigma')C(\sigma) .$$

Note in particular that if σ is an isometry, then $C(\sigma)$ is a graded algebra isomorphism with inverse $C(\sigma^{-1})$. If σ is an isometry from M onto M, then $C(\sigma)$ is an algebra automorphism of $C(M)$.

Consider the isometry -1_M on M defined by $(-1_M)(x) = -x$. The algebra automorphism $\beta = C(-1_M)$ of $C(M)$ satisfies $(\gamma x)^\beta = -\gamma x$ for all x in M. It follows that

$$\beta = 1_{C_0(M)} \oplus -1_{C_1(M)} .$$

So β is an involution which preserves the grading of $C(M)$.

Let $C(M)^\circ$ be the opposite algebra of $C(M)$. By the defining property of $(C(M), \gamma)$ there is a unique algebra homomorphism $\sim : C(M) \to C(M)^\circ$ such that

$$
\begin{array}{ccc}
M & \xrightarrow{\;\gamma\;} & C(M)^\circ \\
{\scriptstyle\gamma}\big\downarrow & \nearrow_{\scriptstyle\sim} & \\
C(M) & &
\end{array}
$$

commutes. We will consider $\sim$ as a map $\sim : C(M) \to C(M)$. Note that $\widetilde{cc_1}$ $= \tilde{c}_1 \tilde{c}$ for all c and c_1 in $C(M)$. Since $\tilde{\gamma}(x) = \gamma(x)$ for all x in M, it follows from (5.2) that $\tilde{\tilde{c}} = c$. So $\sim$ is an involution on $C(M)$. Define $\bar{} : C(M) \to C(M)$ to be the composite

$$C(M) \xrightarrow{\;\beta\;} C(M) \xrightarrow{\;\sim\;} C(M) .$$

Observe that $\bar{}$ is an antiautomorphism of $C(M)$. Since $\overline{\gamma(x)} = -\gamma(x)$ for all x in M, it follows from (5.2) that $\bar{}$ is also an involution and that it preserves the grading of $C(M)$. Note, again by (5.2), that $\bar{}$ is the unique R-linear anti-automorphism of $C(M)$ with the property that $\overline{\gamma(x)} = -\gamma(x)$ for all x in M. It is called the *canonical involution* of $C(M)$.

C. Graded Tensor Product

Fix a Clifford algebra $(C(M), \gamma)$ of M. Suppose that $M = N \perp N'$ is an orthogonal splitting of M. Let $(C(N), \delta)$ and $(C(N'), \delta')$ be Clifford algebras of the quadratic modules N and N'. Since the restrictions $q|_N$ and $q|_{N'}$

determine q, one would expect that the algebra structures of $C(N)$ and $C(N')$ determine that of $C(M)$. That this is indeed the case is an important application of the graded tensor product of algebras.

Observe that the pairs $(C(M), \gamma_N)$ and $(C(M), \gamma_{N'})$ are compatible with the quadratic modules N and N', respectively. Therefore, there exist unique algebra homomorphisms $j : C(N) \to C(M)$ and $j' : C(N') \to C(M)$ such that the diagrams

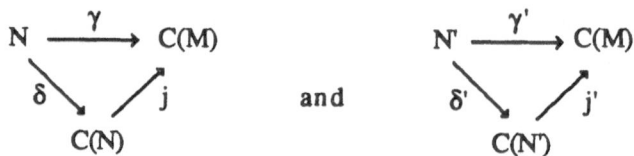

commute. The assignment $(c, c') \to (jc)(j'c')$ induces a module homomorphism $\varphi : C(N) \otimes_R C(N') \to C(M)$ which satisfies $\varphi(c \otimes c') = (jc)(j'c')$ for all c in $C(N)$ and c' in $C(N')$.

(5.5). *Theorem. The map* $\varphi : C(N) \hat{\otimes}_R C(N') \to C(M)$ *is a graded algebra isomorphism.*

We assume for the rest of this section that M is finitely generated projective and conclude our summary of basic facts about the Clifford algebra of M with the change-of-scalars construction and some of its consequences.

Let A be a commutative algebra over R. Recall from Chapter 4D, that the quadratic module $(M \otimes_R A, q_A)$ over A, where $q_A : M \otimes_R A \to A$ satisfies $q_A(x \otimes a) = a^2 q(x)$ for all x in M and a in A. Consider the A-algebra $C(M) \otimes_R A$ and the homomorphism of A-modules

$$\gamma \otimes 1 : M \otimes_R A \to C(M) \otimes_R A .$$

It is a routine matter to check that the pair $(C(M) \otimes_R A, \gamma \otimes 1)$ is compatible with $M \otimes_R A$. Therefore, there exists a unique homomorphism

$$\psi : C(M \otimes_R A) \to C(M) \otimes_R A$$

of A-algebras such that the diagram

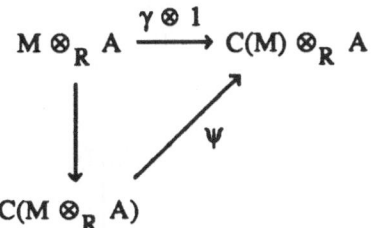

commutes. It is not difficult to prove that ψ is an isomorphism which takes $C_0(M \otimes_R A)$ onto $C_0(M) \otimes_R A$. Supply $C(M) \otimes_R A$ with the grading which that of $C(M)$ naturally provides and observe that ψ is a graded isomorphism.

As an initial application of the change of scalars construction, we show that the maps $R \to C(M)$ and $\gamma : M \to C(M)$ are both injective. This was already observed in the special case where M is free. Let $\mathfrak{p}$ be any prime ideal of R and let $M_{\mathfrak{p}} = M \otimes_R R_{\mathfrak{p}}$ be the quadratic module over $R_{\mathfrak{p}}$ obtained by change of scalars. Let $C(M) \to C(M) \otimes_R R_{\mathfrak{p}}$ be the map into the first component and let $C(M) \otimes_R R_{\mathfrak{p}} \to C(M \otimes_R R_{\mathfrak{p}})$ be the inverse of the isomorphism ψ (as specialized to the present situation). Consider the diagram

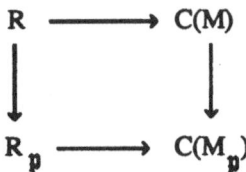

where $C(M) \to C(M_{\mathfrak{p}})$ is the composite of the maps just referred to and $R_{\mathfrak{p}} \to C(M_{\mathfrak{p}})$ is the analogue of $R \to C(M)$. Check that the diagram commutes, and observe that $R_{\mathfrak{p}} \to C(M_{\mathfrak{p}})$ is injective, since $M_{\mathfrak{p}}$ is free of finite rank. If r is in the kernel of $R \to C(M)$, then r must be in the kernel $R \to R_{\mathfrak{p}}$ for any $\mathfrak{p}$. But this implies by (4.1), that $r = 0$. Therefore, $R \to C(M)$ is injective. A similar argument shows that $\gamma : M \to C(M)$ is injective. We will therefore identify r with $r1_C$ for any r in R and x with $\gamma(x)$ for any x in M, and consider

$$R \subseteq C(M) \quad \text{and} \quad M \subseteq C(M).$$

Suppose that $M = N \perp N'$. Note that N is projective, and since $N \cong M/N'$, that N is finitely generated. Similarly, N' is finitely generated and projective. Let $(C(N), \delta)$ be a Clifford algebra of N. We will now verify that the algebra homomorphism $j : C(N) \to C(M)$ induced by the inclusion $N \to M$ is injective. In view of the identification $\gamma x = x$, note that $jx = x$ for all x in N. Observe that the diagram

$$
\begin{array}{ccc}
N & \xrightarrow{\ \delta\ } & C(N) \\
\big\uparrow & & \big\downarrow j \\
M & \xrightarrow{\ \gamma\ } & C(M)
\end{array}
$$

commutes. Suppose first that both M and N are free. Let $c \in C(N)$ be arbitrary and assume that $jc = 0$. By (5.3), $C(N)$ is a free module with finite basis, say $\{c_1,..., c_n\}$. Put $c = \sum_i c_i r_i$. Recall from above that the isomorphism

$\varphi : C(N) \otimes_R C(N') \to C(M)$ satisfies $\varphi(c \otimes 1) = (jc)(j'1)$. So $c \otimes 1 = 0$.

Therefore, $\sum_i (c_i \otimes 0) = 0 = \left(\sum_i c_i r_i\right) \otimes 1 = \sum_i (c_i \otimes r_i 1)$. Since $\{c_1,..., c_n\}$ is

a basis, we find by a basic property of tensor products (see Hungerford, for example) that $r_i 1 = 0$ for all i. So $r_i = 0$ for all i, and $c = 0$. Therefore, j is injective as asserted. Now turn to the general case. For any prime ideal $\mathfrak{p}$ of R, consider the diagram

$$
\begin{array}{ccc}
C(N) \otimes_R R_{\mathfrak{p}} & \xrightarrow{\ j \otimes 1\ } & C(M) \otimes_R R_{\mathfrak{p}} \\
\psi_N \big\uparrow & & \big\uparrow \psi \\
C(N \otimes_R R_{\mathfrak{p}}) & \xrightarrow{\ j'\ } & C(M \otimes_R R_{\mathfrak{p}})
\end{array}
$$

where ψ and its analogue ψ_N are the change-of-scalars isomorphisms already discussed and j' is the analogue of j. Since $M \otimes_R R_{\mathfrak{p}}$ and $N \otimes_R R_{\mathfrak{p}}$ are both free of finite rank, j' is injective by the case already dealt with. It follows that $j \otimes 1$ is injective. Since this is true for any prime ideal $\mathfrak{p}$ of R, j is injective and we are done.

In the situation of a finitely generated projective M, where $M = N \perp N'$, we will therefore regard $C(N)$ and $C(N')$ to be contained in $C(M)$.

D. Exterior Algebras

Assume that $q(x) = 0$ for all x in M. In this important special case, the Clifford algebra of M is called the *exterior algebra* of M and is denoted $\Lambda(M)$. We will not go into a detailed analysis of the exterior algebra (for this we refer the interested reader to Bourbaki [1972, Algebra I] or McDonald) — instead, some of the basic features of the theory are pointed to and some facts are singled out that will be important later. Assume throughout that M is finitely generated projective.

The product of two elements c and d in $\Lambda(M)$ is denoted by $c \wedge d$. Since $q(x) = 0 = h(x, y)$ for all x and y in M, $x^2 = x \wedge x = 0$ and $x \wedge y = -y \wedge x$ for any x and y in M. Let $\{x_i\}_{i \in I}$, where I is a finite index set of cardinality m, span M as R-module. Define $\Lambda^0(M) = R$, and for $k \geq 1$ let $\Lambda^k(M)$ be the R-submodule of $\Lambda(M)$ spanned by all elements of the form $x_{i_1} \wedge x_{i_2} \wedge \ldots \wedge x_{i_k}$. Observe that $\Lambda^k(M)$ is independent of the spanning set. Let $k > m$. Then a typical generator $x_{i_1} \wedge x_{i_2} \wedge \ldots \wedge x_{i_k}$ of $\Lambda^k(M)$ must contain some x_i at least twice. It follows that this typical generator is equal to 0. So $\Lambda^k(M) = 0$. In view of (5.2), $\Lambda(M) = \Lambda^0(M) + \ldots + \Lambda^m(M)$. In fact

$$\Lambda(M) = \Lambda^0(M) \oplus \ldots \oplus \Lambda^m(M).$$

If R is a local ring, then M is free of finite rank and this is a consequence of (5.3). For a general commutative R, this follows by a localization argument: Let $\mathfrak{p}$ be a prime ideal of R. By the change-of-scalars construction for the Clifford algebra,

$$\Lambda(M)_{\mathfrak{p}} = \Lambda(M) \otimes_R R_{\mathfrak{p}} \cong \Lambda(M \otimes_R R_{\mathfrak{p}}) = \Lambda(M_{\mathfrak{p}}).$$

By restriction, $\Lambda^k(M)_{\mathfrak{p}} \cong \Lambda^k(M_{\mathfrak{p}})$. Consider the inclusion

$$(\Lambda^k(M) \cap \sum_{i \neq k} \Lambda^k(M)) \to \Lambda^k(M).$$

Its localization $(\Lambda^k(M) \cap \sum_{i \neq k} \Lambda^k(M))_{\mathfrak{p}} \to \Lambda^k(M)_{\mathfrak{p}} \xrightarrow{\sim} \Lambda^k(M_{\mathfrak{p}})$ is injective and its image is contained in $\Lambda^k(M_{\mathfrak{p}}) \cap \sum_{i \neq k} \Lambda^k(M_{\mathfrak{p}})$. But by the local case, this intersection is $\{0\}$. Therefore, $(\Lambda^k(M) \cap \sum_{i \neq k} \Lambda^k(M))_{\mathfrak{p}} = \{0\}$. Since this is true for any $\mathfrak{p}$, $\Lambda^k(M) \cap \sum_{i \neq k} \Lambda^k(M) = \{0\}$. So the sum is direct, as asserted. Since $\Lambda^0(M) = R$, R is a direct summand of $\Lambda(M)$. This fact has the following consequence for the Clifford algebra:

(5.6). *Let* M *be any finitely generated projective quadratic module over* R. *Then the Clifford algebra* C(M) *and its subalgebra* $C_0(M)$ *are both finitely generated projective over* R.

Proof. Since M is finitely generated projective, there is a finitely generated projective module N such that $M \oplus N$ is free of finite rank. Supply N with the form $q' = 0$ and consider the orthogonal sum $M \perp N$. By Theorem (5.5), $C(M \perp N) \cong C(M) \otimes_R \Lambda(N)$ as R-modules. By (5.3), $C(M \perp N)$ is free. Since $\Lambda(N)$ contains R as direct summand, it follows that C(M) is isomorphic to a direct summand of $C(M \perp N)$. Therefore, C(M) is projective. It is finitely generated by (5.2). Since $C_0(M)$ is a direct summand of C(M), the same is true for $C_0(M)$.   QED.

Return to the "exterior" case where $q = 0$ and assume that M has constant rank n. From above, $\Lambda^k(M) = 0$ for $k > n$ and

$$\Lambda(M) = \Lambda^0(M) \oplus \ldots \oplus \Lambda^n(M).$$

If M is free, then by (5.3),

$$\Lambda^1(M) = M \quad \text{and} \quad \Lambda^n(M) \cong R.$$

By localizing, $\Lambda^1(M) = M$ in general and $\Lambda^n(M)$ is finitely generated projective of rank 1.

Suppose that $M = N \oplus N'$. Since $M = N \perp N'$, $\Lambda(M) \cong \Lambda(N) \otimes_R \Lambda(N')$ as R-modules, by Theorem (5.5). A moment's reflection shows that

$$\Lambda^k(M) \cong \bigoplus_{i+j=k} (\Lambda^i(N) \otimes_R \Lambda^j(N')).$$

Assume now that N has rank 1 and that N' is free of rank $n - 1$. From above, $\Lambda^n(N') = 0$ and $\Lambda^j(N) = 0$ for any $j \geq 2$. It follows that

$$\Lambda^n(M) \cong \Lambda^1(N) \otimes_R \Lambda^{n-1}(N') \cong N \otimes_R R \cong N.$$

We conclude this section with the following direct consequence of this fact:

(5.7). *Suppose that* M *is finitely generated projective of constant rank* n. *If* M $= N \oplus P = N' \oplus P'$ *where* N *and* N' *are free of rank* $n - 1$ *and both* P *and* P' *have rank* 1, *then* $P \cong P'$.

E. Exercises

Unless additional assumptions are made, M is an arbitrary quadratic module over a commutative ring R.

1. Suppose that M is a *hyperbolic plane*. This means that M is a quadratic module over R with a *hyperbolic basis* $\{x_1, x_2\}$, i.e., one with $q(x_1) = q(x_2) = 0$ and $h(x_1, x_2) = 1$. Show that the R-algebra $\text{Mat}_2(R)$, together with the map $M \to \text{Mat}_2(R)$ defined by $(x_1 r + x_2 s) \to \begin{bmatrix} 0 & r \\ s & 0 \end{bmatrix}$ for all r and s in R, is compatible with M. Show that there exists a unique algebra isomorphism $C(M) \to \text{Mat}_2(R)$ extending this map. What grading on $\text{Mat}_2(R)$ corresponds to $C(M) = C_0(M) \oplus C_1(M)$?

2. The prededing example has the following generalization. Let M be finitely generated projective. Suppose $M = H \perp N$, where H is a hyperbolic plane with hyperbolic basis $\{x_1, x_2\}$. Show that $\text{Mat}_2(C(N))$, together with $M \to \text{Mat}_2(C(N))$, given by $(x_1 r + x_2 s + y) \to \begin{bmatrix} y & r \\ s & -y \end{bmatrix}$ for all r and s in R and y in N, is compatible with M. Show that the corresponding $C(M) \to \text{Mat}_2(C(N))$ is an algebra isomorphism. What grading on $\text{Mat}_2(C(N))$ corresponds to $C(M) = C_0(M) \oplus C_1(M)$?

3. Let $M = N \perp N'$ and assume that both N and N' are free of rank 1. Combine Theorem (5.5), Example 2, and the description of $S^\varepsilon \hat{\otimes}_R T^\eta$ in Chapter 1D to analyze $C(M)$.

4. Let $M = N \perp N'$. Let $\mathfrak{p}$ be a prime ideal of R. Show that the algebra maps $C(N) \to C(N) \hat{\otimes}_R C(N')$ and $C(N') \to C(N) \hat{\otimes}_R C(N')$ are injective. The maps $C(N)_\mathfrak{p} \to (C(N) \hat{\otimes}_R C(N'))_\mathfrak{p}$ and $C(N')_\mathfrak{p} \to (C(N) \hat{\otimes}_R C(N'))_\mathfrak{p}$ give rise to $C(N)_\mathfrak{p} \times C(N')_\mathfrak{p} \to (C(N) \hat{\otimes}_R C(N'))_\mathfrak{p}$. Show that this defines an isomorphism $C(N)_\mathfrak{p} \hat{\otimes}_{R_\mathfrak{p}} C(N')_\mathfrak{p} \to (C(N) \hat{\otimes}_R C(N'))_\mathfrak{p}$ of graded algebras, where the last grading is induced from $C(N) \hat{\otimes}_R C(N')$.

5. Let $M = N \perp N'$. Let $\mathfrak{p}$ be any prime ideal of R. Consider: the isomorphism of graded algebras $C(N)_{\mathfrak{p}} \hat{\otimes}_{R_{\mathfrak{p}}} C(N')_{\mathfrak{p}} \to (C(N) \hat{\otimes}_R C(N'))_{\mathfrak{p}}$ defined in Exercise 4; the localization $(C(N) \hat{\otimes}_R C(N'))_{\mathfrak{p}} \to C(M)_{\mathfrak{p}}$ of the isomorphism of Theorem (5.5); $C(N_{\mathfrak{p}}) \hat{\otimes}_{R_{\mathfrak{p}}} C(N'_{\mathfrak{p}}) \to C(M_{\mathfrak{p}})$ obtained by applying Theorem (5.5) to the quadratic module $M_{\mathfrak{p}} = N_{\mathfrak{p}} \perp N'_{\mathfrak{p}}$ over $R_{\mathfrak{p}}$; and finally the change-of-scalars isomorphisms. Check that the resulting diagram

$$
\begin{array}{ccc}
(C(N) \hat{\otimes}_R C(N'))_{\mathfrak{p}} & \longrightarrow & C(M)_{\mathfrak{p}} \\
\uparrow & & \uparrow \\
C(N_{\mathfrak{p}}) \hat{\otimes}_{R_{\mathfrak{p}}} C(N'_{\mathfrak{p}}) & \longrightarrow & C(M_{\mathfrak{p}})
\end{array}
$$

commutes.

6. Fill in the details in the discussion of Section D.

Hints:

1. To show that it is an isomorphism, compute the images of the basis vectors $1, x_1, x_2$, and $x_1 x_2$.

2. Use (5.6) and (4.5) to show that the map is an isomorphism.

5. Check the commutativity on generators.

6
Algebras with Standard Involution

Overview

Again, R is any commutative ring. The goal of this chapter is the analysis of the structure of an algebra over R with "standard" involution. The Clifford algebra $C(M)$ of a quadratic module M over R which is nonsingular and free of rank 2 is the most prominent example and will receive particular attention. A number of the concepts and constructions of the previous chapter are illustrated in the process. In addition, we will see that $C(M)$ is separable over R, that the center of $C(M)$ is R, and that $\text{Cen } C_0(M) = C_0(M)$ is a free separable quadratic algebra over R. These matters will be taken up for a general non-singular M in Chapter 9. The special case of rank 2 is a cornerstone for this investigation.

A. Standard Involutions

Let A be an R-algebra and assume throughout that A is faithful. Since the assignment $r \to r1_A$ injects R into A, we consider $R \subseteq A$. Let σ be an anti-involution on A. Define the *norm* and *trace* of A, respectively, by

$$nr(a) = a(a^\sigma) \quad \text{and} \quad tr(a) = a + a^\sigma$$

for $a \in A$. We call σ a *standard involution* if $nr(a) \in R$ for all $a \in A$. If σ is standard, A is *an algebra with standard involution*. Since

$$nr(1 + a) = (1 + a)(1 + a^\sigma) = 1 + a + a^\sigma + aa^\sigma,$$

it follows that if σ is standard, then $tr(a) = a + a^\sigma \in R$ for all a in A.

Example 1. Let $A = R$ and $\sigma = \text{id}_R$. This is the *trivial* algebra with standard involution.

Example 2. Let $A = R[X]/(X^2 - cX - d)$ be a free quadratic algebra. The conjugation of A is a standard involution. See Chapter 1B.

Example 3. Let $A = \text{Mat}_2(R)$ and define σ by $\begin{bmatrix} r & s \\ t & u \end{bmatrix} \rightarrow \begin{bmatrix} u & -s \\ -t & r \end{bmatrix}$. Then A with σ is an algebra with standard involution. The norm and trace are respectively the determinant and the matrix trace. Consider the algebra $R[X]/(X^2 - cX - d)$ and put $v = X + (X^2 - cX - d)$. Embed $R[X]/(X^2 - cX - d)$ into A by $1 \rightarrow I$ and $v \rightarrow \begin{bmatrix} 0 & d \\ 1 & c \end{bmatrix}$. This is an injective homomorphism of R-algebras. The restriction of σ to the image corresponds to the conjugation of $R[X]/(X^2 - cX - d)$.

Return to an arbitrary algebra A with standard involution σ. Let $a \in A$. Since $a + a^\sigma \in R$, $a(a + a^\sigma) = (a + a^\sigma)a$. So

(i) $aa^\sigma = a^\sigma a$ and $a^2 - tr(a)a + nr(a) = 0$.

For a and b in A, $(ab)(ab)^\sigma = a(bb^\sigma)a^\sigma$, so

(ii) $nr(ab) = nr(a)nr(b)$.

Deduce that $a \in A^*$ if and only if $nr(a) \in R^*$.

Verify next that $f(a, b) = tr(ab^\sigma) = ab^\sigma + ba^\sigma$ defines a symmetric bilinear form

$$f : A \times A \rightarrow R.$$

This is the *symmetric bilinear form defined by* σ. Since $(a + b)(a + b)^\sigma = aa^\sigma + ab^\sigma + ba^\sigma + bb^\sigma$, it follows that

(iii) $nr(a + b) = nr(a) + nr(b) + f(a, b)$.

Therefore, $nr : A \rightarrow R$ is a quadratic form on A with associated symmetric bilinear form f.

These formulas, when "linearized," provide a number of additional connections between nr, f, and the multiplication of A:

Let a, b, c, and d be arbitrary elements of A. Replacing a by $a + c$ in (ii) and using (iii), we get the chain of equalities

$$nr(a + c)nr(b) = nr(ab + cb)$$

$$(nr(a) + nr(c) + f(a, c))nr(b) = nr(ab) + nr(cb) + f(ab, cb)$$

$$nr(a)n(b) + nr(c)nr(b) + nr(b)f(a, c) = nr(a)nr(b) + nr(c)nr(b) + f(ab, cb),$$

so that

(iv) $\qquad\qquad f(ab, cb) = nr(b)f(a, c).$

Replacing b by $b + d$ in (ii) and repeating this computation provides the identity

(v) $\qquad\qquad f(ab, ad) = nr(a)f(b, d).$

Replacing b by $b + d$ in (iv) and computing as above, shows that

(vi) $\qquad\qquad f(a, c)f(b, d) = f(ab, cd) + f(ad, cb).$

Since $f(1, b^\sigma) = b^\sigma + b$, $f(1, b^\sigma)a = ab^\sigma + ab$. Therefore, using (vi),

$f(ab^\sigma, c)$

$\qquad = f(f(1, b^\sigma)a - ab, c) = f(1, b^\sigma)f(a, c) - f(ab, c) = f(a, c)f(b^\sigma, 1) - f(ab, c)$

$\qquad = f(ab^\sigma, c) + f(a, cb^\sigma) - f(ab, c).$

This and a similar calculation verify the equalities:

(vii) $\qquad f(ab, c) = f(a, cb^\sigma) \quad \text{and} \quad f(ab, c) = f(b, a^\sigma c).$

As a consequence of (i) we find for any c in A that $R + Rc$ is a subalgebra of A and that this subalgebra is a free quadratic algebra if 1 and c are independent over R.

(6.1). *Let A be an algebra with standard involution σ. The following are equivalent for an element c in A.*

(1) $c - c^\sigma \in A^*$

(2) $(c - c^\sigma)^2 \in R^*$

(3) $R + Rc = R \oplus Rc$ *is a separable free quadratic algebra over* R.

Proof. Let c in A be arbitrary. Since $c + c^\sigma$ and cc^σ are in R, $(c + c^\sigma)^2 = c^2 + 2cc^\sigma + (c^\sigma)^2$ and $-4cc^\sigma$ are in R. Therefore, $(c - c^\sigma)^2 \in R$. Put $a = c - c^\sigma$. Since $a^\sigma = -a$, $nr(a) = -a^2$. So $a \in A^*$ if and only if $a^2 \in R^*$, and

(1) and (2) are equivalent. We now show that (2) and (3) are equivalent. Assume that $R + Rc = R \oplus Rc$. So $R \oplus Rc$ is a free quadratic algebra in A. With $\alpha = tr(c)$ and $\beta = -nr(c)$, we find that $R \oplus Rc \cong R[X]/(X^2 - \alpha X - \beta)$ where c corresponds to $v = X + (X^2 - \alpha X - \beta)$. Note that $\alpha^2 + 4\beta = (c + c^\sigma)^2 - 4cc^\sigma = c^2 - 2cc^\sigma + (c^\sigma)^2 = (c - c^\sigma)^2$. So, by the Example in Chapter 4D, $R \oplus Rc$ is separable if and only if $(c - c^\sigma)^2 \in R^*$. For the equivalence of (2) and (3) it remains to check that if $(c - c^\sigma)^2 \in R^*$, then $\{1, c\}$ is a basis of $R + Rc$. Suppose that $r + sc = 0$ with r and s in R. So $s(c - c^\sigma) = sc - (sc)^\sigma = 0$. If $(c - c^\sigma)^2 \in R^*$, then $s = 0$ and $r = 0$.   QED.

Example 4. Let c in A satisfy the equivalent conditions. Let $\alpha = tr(c)$ and $\beta = -nr(c)$. From the preceding proof the separable free quadratic algebra $S = R \oplus Rc$ is isomorphic to $R[X]/(X^2 - \alpha X - \beta)$, where c corresponds to the element $v = X + (X^2 - \alpha X - \beta)$. Recall from Chapter 1B and the beginning of Chapter 2D that the conjugation of $R[X]/(X^2 - \alpha X - \beta)$, given by $v \to \alpha - v$, is the unique involution with fixed point set R. Since $c^\sigma = \alpha - c$, the conjugation of S is the restriction of σ to S. By the Example in Chapter 4D, the restriction of the bilinear form f of A to S is nonsingular.

The existence of an element c satisfying the conditions of (6.1), i.e., the existence of a single separable free quadratic algebra in A, has considerable impact on the structure of A.

(6.2). *Theorem. Let A be an algebra with standard involution σ. Suppose there is an element c in A such that $c - c^\sigma \in A^*$. Then*

(1)   *A has a grading*

$$A = A_0 \oplus A_1,$$

where $A_0 = R \oplus Rc$ is a separable free quadratic algebra and A_1 is the orthogonal complement of A_0 in A relative to f.

(2)   *$\sigma = \tau \oplus -id_{A_1}$, where τ is the conjugation of A_0, and*

(3)   *$Cen_A A_0 = A_0$ and $Cen A = \{r + r'c \in A_0 \mid r'A_1 = 0\}$. In particular, $Cen A = R$ if and only if A_1 is a faithful R-module.*

Proof. By (6.1), $A_0 = R + Rc = R \oplus Rc$ is a separable free quadratic algebra. By Example 4, the restriction of σ is the conjugation τ of A_0, and A_0 is a nonsingular submodule of A relative to f. Thus by (4.9), $A = A_0 \perp (A_0)^\perp$.

Let $A_1 = (A_0)^\perp$, and note that $A = A_0 \oplus A_1$ as R-modules.

We prove (2) and (3). Let $a \in A_1$. So $a^\sigma + a = f(1, a) = 0$. Therefore, $a^\sigma = -a$, and $\sigma = \tau \oplus -\mathrm{id}_{A_1}$ as required. Now let $a = a_0 + a_1$ with $a_0 \in A_0$ and $a_1 \in A_1$ be any element such that $ac = ca$. Since A_0 is commutative, $a_1 c = ca_1$. Now, $0 = f(c, a_1) = ca_1^\sigma + a_1 c^\sigma = -ca_1 + a_1 c^\sigma$. So $a_1 c = ca_1 = a_1 c^\sigma$. Thus $a_1 (c - c^\sigma) = 0$, and hence $a_1 = 0$. We have shown that $\mathrm{Cen}_A A_0 \subseteq A_0$, and therefore, that $\mathrm{Cen}_A A_0 = A_0$. The fact about $\mathrm{Cen}\, A$ is now easy.

To prove that $A = A_0 \oplus A_1$ is a grading of A, we must check the inclusions $A_0 A_1 \subseteq A_1$, $A_1 A_0 \subseteq A_1$, and $A_1 A_1 \subseteq A_0$. To verify that $A_0 A_1 \subseteq A_1$, it suffices to show that $cd \in A_1$ for any $d \in A_1$. Let $d \in A_1$ be arbitrary. Using formula (vi) with $a = b = 1$ and then formula (v) with $a = c$ and $b = 1$, we see that

$$f(1, c)f(1, d) = f(1, cd) + f(d, c) \quad \text{and} \quad f(c, cd) = nr(c)f(1, d).$$

Therefore, $f(1, cd) = 0$ and $f(c, cd) = 0$. As a consequence, $cd \in (A_0)^\perp = A_1$. The proof of $A_1 A_0 \subseteq A_1$ is parallel. So $A_1 A_1 \subseteq A_0$ remains. Put $c + c^\sigma = r$. Let $a \in A_1$ be arbitary. Since $0 = f(c, a) = ca^\sigma + ac^\sigma = -ca + a(r - c)$, we get $ca = a(r - c)$. So for a and b in A_1,

$$c(ab) = a(r - c)b = arb - a(cb) = arb - ab(r - c) = (ab)c.$$

Therefore, $ab \in \mathrm{Cen}_A A_0 = A_0$.   QED.

B. Free Quaternion Algebras

A *free quaternion algebra* is defined to be the Clifford algebra of a quadratic module which is free of rank 2. This algebra will be the focus of the next two sections of this chapter.

Let M be a quadratic module over R with basis $\{x_1, x_2\}$. Let $q : M \to R$ be the quadratic form and h the associated symmetric bilinear form. Put

$$q(x_1) = s_1, \quad q(x_2) = -s_2, \quad \text{and} \quad h(x_1, x_2) = t.$$

The matrix of h in the basis $\{x_1, x_2\}$ is $\begin{bmatrix} 2s_1 & t \\ t & -2s_2 \end{bmatrix}$. By (5.3), the Clifford algebra $C(M)$ has basis $\{1, x_1, x_2, x_1x_2\}$. The basic equations in $C(M)$ say that

$$x_1^2 = s_1, \quad x_2^2 = -s_2, \quad \text{and} \quad x_2x_1 = h(x_1, x_2)1 - x_1x_2 = t1 - x_1x_2.$$

Recall the canonical involution $\bar{}$ of $C(M)$ from Chapter 5B. Observe that it satisfies

$$\bar{1} = 1, \quad \bar{x}_1 = -x_1, \quad \bar{x}_2 = -x_2, \quad \text{and} \quad \overline{x_1x_2} = \bar{x}_2\bar{x}_1 = x_2x_1 = t1 - x_1x_2.$$

Let $c = r_0 1 + r_1 x_1 + r_2 x_2 + r_3 x_1 x_2$, where r_0, r_1, r_2, r_3 are in R, be any element of $C(M)$. Routine computations (the first is easy, the second lengthy) show that

$$\bar{c} = (r_0 + r_3 t) - r_1 x_1 - r_2 x_2 - r_3 x_1 x_2 \quad \text{and}$$

$$c\bar{c} = r_0^2 + r_0 r_3 t - r_3^2 s_1 s_2 - (r_1^2 s_1 + r_1 r_2 t - r_2^2 s_2).$$

Observe in particular that

(6.3). *The canonical involution $\bar{}$ of $C(M)$ is a standard involution.*

The norm form corresponding to $\bar{}$ is, of course, given by $nr(c) = c\bar{c}$ and the corresponding symmetric bilinear form by $f(c, d) = c\bar{d} + d\bar{c}$. This supplies $C(M)$ with the structure of a quadratic module over R. Routine computations show that the matrix of f in the basis $\{1, x_1x_2, x_1, x_2\}$ of $C(M)$ is

$$\begin{bmatrix} 2 & t & 0 & 0 \\ t & -2s_1 s_2 & 0 & 0 \\ 0 & 0 & -2s_1 & -t \\ 0 & 0 & -t & 2s_2 \end{bmatrix}.$$

For example,

$$f(x_1x_2, x_1x_2) = 2(x_1x_2)(\overline{x_1x_2}) = 2x_1x_2\bar{x}_2\bar{x}_1 = -2s_1 s_2 \quad \text{and}$$

$$f(x_1x_2, x_1) = (x_1x_2)\bar{x}_1 + x_1(\overline{x_1x_2}) = -x_1(x_2x_1) + x_1(x_2x_1) = 0.$$

Consider the grading $C(M) = C_0(M) \oplus C_1(M)$. By the remark that follows (5.4), $C_0(M)$ has basis $\{1, x_1x_2\}$ and $C_1(M)$ has basis $\{x_1, x_2\}$. So $C_1(M) = M$, and relative to f,

$$C(M) = C_0(M) \perp M.$$

(6.4). *The following statements are equivalent.*

(1) M *is nonsingular relative to* q *and* h.

(2) $C(M)$ *is nonsingular relative to* nr *and* f.

(3) $C_0(M)$ *is a separable free quadratic algebra.*

(4) $x_1x_2 - \overline{x_1x_2}$ *is an invertible element of* $C(M)$.

Proof. We have already computed the matrix of the form h in the basis $\{x_1, x_2\}$ as well as that of f in the basis $\{1, x_1, x_2, x_1x_2\}$. Their determinants are $-4s_1s_2 - t^2$ and $(-4s_1s_2 - t^2)^2$, respectively. Set $c = x_1x_2$. Since $c - \bar{c} = x_1x_2 - \overline{x_1x_2} = 2x_1x_2 - t$,

$$(c - \bar{c})^2 = 4(x_1x_2)^2 - 4tx_1x_2 + t^2 = 4x_1(t - x_1x_2)x_2 - 4tx_1x_2 + t^2$$
$$= 4s_1s_2 + t^2.$$

An application of (6.1) now provides the equivalence of (1) – (4).   QED.

(6.5). *Suppose* M *is nonsingular. Then*

$$\mathrm{Cen}_{C(M)}C_0(M) = \mathrm{Cen}\, C_0(M) = C_0(M) \text{ and } \mathrm{Cen}\, C(M) = R.$$

Proof. Combine (6.3), (6.4), and (6.2).   QED.

We now consider an important special case. Suppose that M is nonsingular and that the basis $\{x_1, x_2\}$ is orthogonal, i.e., $h(x_1, x_2) = t = 0$. Since the determinant of the matrix of h in the basis $\{x_1, x_2\}$ is $-4s_1s_2, 2 \in R^*$. Setting $x_1x_2 = x_3$, we find that the free quaternion algebra $C(M)$ has the multiplication table

	1	x_1	x_2	x_3
1	1	x_1	x_2	x_3
x_1	x_1	$s_1 1$	x_3	$s_1 x_2$
x_2	x_2	$-x_3$	$-s_2 1$	$s_2 x_1$
x_3	x_3	$-s_1 x_2$	$-s_2 x_1$	$s_1 s_2 1$

This free quaternion algebra is denoted $\left(\frac{s_1, -s_2}{R}\right)$. The special case $\left(\frac{-1, -1}{\mathbb{R}}\right)$ is Hamilton's classical quaternion division algebra. Compare this with the construction that concludes Chapter 1D.

C. Separability of Free Quaternion Algebras

Let M be a free quadratic module of rank 2 over R and continue the notation from the previous section. The canonical involution $^-$ of C(M) is an anti-automorphism of C(M). So it is an isomorphism $^-: C(M) \to C(M)^\circ$ onto the opposite algebra of C(M). We therefore identify $C(M)^\circ$ with C(M) and write $\bar{x}$ in place of x°. Recall the algebra $C(M)^e = C(M) \otimes_R C(M)$ and the map $\phi = \phi_{C(M)} : C(M)^e \to C(M)$, defined by $\phi(c \otimes \bar{d}) = cd$, from Chapter 2A.

(6.6). *Suppose M is nonsingular. Then C(M) is separable over R.*

Proof. In view of Proposition (2.1), it suffices to produce a separability idempotent e for C(M). Since M is nonsingular, $4s_1 s_2 + t^2 \in R^*$. Let $u = (4s_1 s_2 + t^2)^{-1}$ and set

$$e = u(s_1 s_2 (1 \otimes 1) + s_2 x_1 \otimes \bar{x}_1 - s_1 x_2 \otimes \bar{x}_2 + t x_2 \otimes \bar{x}_1 + x_1 x_2 \otimes \overline{x_1 x_2}).$$

Since

$$\phi(e) = u \left(s_1 s_2 + s_2 s_1 + s_1 s_2 + t x_2 x_1 + x_1 x_2 x_1 x_2 \right)$$

$$= u \, (3s_1s_2 + tx_2x_1 + x_1(t - x_1x_2)x_2)$$

$$= u \, (4s_1s_2 + t(x_2x_1 + x_1x_2)) = u(4s_1s_2 + t^2) = 1,$$

e satisfies the first requirement for a separability idempotent. Since $1, x_1, x_2,$ and $x_3 = x_1x_2$ span $C(M)$ as an R-module, it remains to show that

$$(x_i \otimes 1)e = (1 \otimes \bar{x}_i)e$$

for $i = 1, 2,$ and 3. This is a tedious process. We verify the case x_3, which is the lengthiest. To check the equality we may drop the scalar u. Expanding the left side, we get:

$$(s_1s_2x_1x_2) \otimes 1 + (s_2x_1x_2x_1) \otimes \bar{x}_1 - (s_1x_1x_2x_2) \otimes \bar{x}_2$$

$$+ (tx_1x_2x_2) \otimes \bar{x}_1 + (x_1x_2x_1x_2) \otimes \overline{x_1x_2}$$

$$= s_1s_2(x_1x_2 \otimes 1) + s_2x_1(t - x_1x_2) \otimes \bar{x}_1 + s_1s_2x_1 \otimes \bar{x}_2$$

$$- ts_2x_1 \otimes \bar{x}_1 + x_1(t - x_1x_2)x_2 \otimes \overline{x_1x_2}$$

$$= x_1x_2 \otimes s_1s_2 + ts_2x_1 \otimes \bar{x}_1 - s_1s_2x_2 \otimes \bar{x}_1 + s_1s_2x_1 \otimes \bar{x}_2$$

$$- ts_2x_1 \otimes \bar{x}_1 + tx_1x_2 \otimes \overline{x_1x_2} + s_1s_2 \otimes \overline{x_1x_2}$$

$$= x_1x_2 \otimes s_1s_2 - s_1s_2x_2 \otimes \bar{x}_1 + s_1s_2x_1 \otimes \bar{x}_2$$

$$+ tx_1x_2 \otimes \overline{x_1x_2} + s_1s_2 \otimes \overline{x_1x_2}.$$

On the right side,

$$s_1s_2 \otimes \overline{x_1x_2} + s_2x_1 \otimes (\overline{x_1x_2x_1}) - s_1x_2 \otimes \overline{x_2x_1x_2}$$

$$+ tx_2 \otimes s_1\bar{x}_2 + x_1x_2 \otimes \overline{x_1x_2x_1x_2}$$

$$= s_1s_2 \otimes \overline{x_1x_2} + s_2x_1 \otimes s_1\bar{x}_2 - s_1x_2 \otimes \overline{x_2(t - x_2x_1)}$$

$$+ ts_1x_2 \otimes \bar{x}_2 + x_1x_2 \otimes \overline{x_1(t - x_1x_2)x_2}$$

$$= s_1s_2 \otimes \overline{x_1x_2} + s_2x_1 \otimes s_1\bar{x}_2 - s_1s_2x_2 \otimes \bar{x}_1$$

$$+ tx_1x_2 \otimes \overline{x_1x_2} + x_1x_2 \otimes s_1s_2.$$

It is now clear that $(x_3 \otimes 1)e = (1 \otimes \bar{x}_3)e$. The other two cases are similar and are left to the reader.   QED.

D. Nonsingular Algebras

A glance at the various formulas of Section A shows that an algebra with standard involution is a structure in which the multiplication, the forms nr and f, and the involution are tightly interwoven. Under the additional assumption that A is a finitely generated projective module and that f is nonsingular, the situation becomes so rigid that it can occur only in ranks ≤ 4. The "freedom of movement" that higher ranks would provide cannot occur.

(6.7). *Theorem. Let* R *be a local ring and let* A *be an* R-*algebra with standard involution* σ. *Assume that* A *is free of finite rank and that* f *is nonsingular. Then* A *is isomorphic as an algebra with involution to either the trivial algebra* R, *a separable free quadratic algebra, or a separable free quaternion algebra.*

Proof. Let rank A = n. Note that $f(1, 1) = 2nr(1)$. So R1 is a nonsingular submodule of A if and only if $2 \in R^*$.

Assume first that n = 1. By (4.18), $2 \in R^*$. So R1 is nonsingular. By (4.9), $A = R1 \perp (R1)^{\perp}$ and it follows that $(R1)^{\perp} = \{0\}$. So A is trivial.

Assume that $n \geq 2$. We begin by showing that there is an element c in A such that $c - c^{\sigma} \in A^*$. Assume first that R1 is nonsingular. So $(R1)^{\perp}$ is nonsingular, and by Exercise 16 of Chapter 4E, there is a c in $(R1)^{\perp}$ such that $nr(c) \in R^*$. Observe that $c - c^{\sigma} = 2c$ is invertible. Assume that R1 is not nonsingular. In this case, $f(1, 1) = 2$ is in the maximal ideal $\mathfrak{m}$ of R. If $c + c^{\sigma} = tr(c) \in \mathfrak{m}$ for all c of A, then $f(a, b) \in \mathfrak{m}$ for all a and b in A. This contradicts the nonsingularity of f. So there is a c with $c + c^{\sigma} \in R^*$. Since $(c + c^{\sigma})^2 \in R^*$, $(c + c^{\sigma})^2 - 4cc^{\sigma} \in R^*$. Therefore, $(c - c^{\sigma})^2 \in R^*$. So $c - c^{\sigma}$ is invertible by (6.1). Now let $A = A_0 \oplus A_1$, where $A_0 = R \oplus Rc$ and $A_1 = (A_0)^{\perp}$ be the grading provided by Theorem (6.2). If n = 2, then $A_1 = \{0\}$. So $A = A_0$ is a free separable quadratic algebra.

From now on, $n \geq 3$. So $A_1 \neq \{0\}$. Since f is nonsingular, both A_0 and A_1 are nonsingular submodules of A. In particular, by Exercise 16 of Chapter 4E there is an element x_1 in A_1 such that $nr(x_1) = x_1 x_1^{\sigma} \in R^*$. Clearly, x_1 is invertible in A. By (6.2), $x_1^{\sigma} = -x_1$ and $A_0 x_1 \subseteq A_1$. Put $B = A_0 \perp A_0 x_1$

and check that $\{1, c, x_1, x_2\}$, where $x_2 = cx_1$, is a basis of B. Since A_0 is nonsingular, the matrix of f in the base $\{1, c\}$ has determinant in R^*. Use of equality (iv) of Section A shows that the same is true of the matrix of f in the basis $\{1, c, x_1, x_2\}$. So B is nonsingular. Since $x_1 A_0 \subseteq A_1$, $a^\sigma = -a$ for all a in $x_1 A_0$. Therefore, $A_0(x_1 A_0) \subseteq A_0(x_1 A_0)^\sigma \subseteq A_0 x_1$, and it follows that B is a subalgebra of A. With the restriction of σ, B is an algebra with standard involution. Since (6.2) applies to B, Cen B = R. Put $A = B \perp (B)^\perp$. We claim that $(B)^\perp = \{0\}$. Assume the contrary. Again by Exercise 16 of Chapter 4E, we can choose $d \in (B)^\perp$ such that $nr(d) \in R^*$. Since $d \in A_1$, $d^\sigma = -d$. For any b in B, $0 = f(d, b^\sigma) = db + b^\sigma d^\sigma = db - b^\sigma d$. So $db = b^\sigma d$. Hence for b and b' in B, $d(bb') = (bb')^\sigma d = b'^\sigma b^\sigma d = b'^\sigma db = d(b'b)$. Since d is invertible, $bb' = b'b$. But this contradicts the fact that Cen B = R. Therefore, $(B)^\perp = \{0\}$. So A = B and $n = 4$.

To complete the proof it must be shown that A is isomorphic (as an algebra with involution) to the Clifford algebra of a nonsingular quadratic module of rank 2. Let M be the quadratic module $Rx_1 \oplus Rx_2$ equipped with the restriction of the form $-nr$ to M. The associated bilinear form is the restriction of $-f$. Let $\alpha : M \to A$ be the inclusion map. We claim that the pair (A, α) is compatible with M. To see this, note first that $(\alpha x_1)^2 = -x_1 x_1^\sigma = -nr(x_1)$.

Since $0 = f(c, x_1) = cx_1^\sigma + x_1 c^\sigma = -cx_1 + x_1 c^\sigma$, $cx_1 = x_1 c^\sigma$. Therefore, $(\alpha x_2)^2 =$
$(cx_1)(cx_1) = cx_1 x_1 c^\sigma = -(cc^\sigma)(x_1 x_1^\sigma) = -nr(c)nr(x_1) = -nr(x_2)$. By equality (iv) of Section A, $nr(x_1)f(1, c) = f(x_1, x_2)$. Therefore,

$$\alpha(x_1)\alpha(x_2) + \alpha(x_2)\alpha(x_1) = x_1 cx_1 + cx_1 x_1$$
$$= x_1 x_1 c^\sigma - nr(x_1)c = -nr(x_1)(c^\sigma + c) = -f(x_1, x_2).$$

The equalities just established and linearity provide the required compatibility. Therefore there is an algebra homomorphism, $\varphi : C(M) \to A$ such that $\varphi(x_1) = x_1$ and $\varphi(x_2) = x_2$. Since $\varphi(x_2 \cdot x_1) = cx_1 x_1 = nr(x_1)c$ and $\{1, x_1, x_2, x_2 \cdot x_1\}$ is a basis of C(M), φ takes a basis to a basis. So φ is an algebra isomorphism. Since $\varphi(\bar{x}_1) = -\varphi x_1 = -x_1 = x_1^\sigma$, and similarly for $\bar{x}_2$, φ preserves the standard involutions. It remains to show that C(M) is separable, or in view of (6.4), that M is nonsingular. To see this, observe from the equalities that the matrix of $-f$ in the basis $\{x_1, x_2\}$ differs from that of f in the basis $\{1, c\}$ by the multiple $-nr(x_1)$. This second matrix is invertible since A_0 is nonsingular. So M is nonsingular.   QED.

Let M be a finitely generated projective module over R. By Theorem 7.7 in Jacobson [1985, II] there is an integer n such that rank $M_{\mathfrak{p}} \leq n$ for all prime ideals $\mathfrak{p}$ of R. We write this condition as rank $M \leq n$.

(6.8). *Theorem. Let* A *be an* R-*algebra with standard involution* σ. *Suppose that* A *is finitely generated projective and that* f *is nonsingular. Then* rank $A \leq 4$. *Moreover,* A *is commutative if and only if* rank $A \leq 2$.

Proof. Let $\mathfrak{p}$ be any prime ideal of R. The $R_{\mathfrak{p}}$-algebra $A_{\mathfrak{p}} = A \otimes_R R_{\mathfrak{p}}$ has standard involution $\sigma_{\mathfrak{p}} = \sigma \otimes 1$. Check that the norm and bilinear form corresponding to $\sigma_{\mathfrak{p}}$ are the respective localizations $nr_{\mathfrak{p}}$ and $f_{\mathfrak{p}}$ of nr and f. By (4.7), $A_{\mathfrak{p}}$ is nonsingular relative to $f_{\mathfrak{p}}$. By Theorem (6.7), rank $A_{\mathfrak{p}} \leq 4$. This proves the first part. The second statement is a consequence of (6.7) and the following observation: A is commutative if and only if $A_{\mathfrak{p}}$ is commutative for all $\mathfrak{p}$. To verify this, assume first that $A_{\mathfrak{p}}$ is commutative for all $\mathfrak{p}$. Fix $a \in A$, and consider the R-homomorphism $A \to A$ given by $x \to ax - xa$. The localization of this map at any $\mathfrak{p}$ is zero. So the map is the zero map. It follows that A is commutative. The other direction is trivial.   QED.

Remarks. It is a corollary of Theorem (6.7), the proof of Theorem (6.8), and Theorem (9.6) of Chapter 9 that an algebra with standard involution that satisfies the conditions of Theorem (6.8) is separable. Also, Theorem (9.6) can be used along with Proposition (7.3.6) in Chapter I of Knus [1991] to establish the folowing for any finitely generated projective quadratic module M of rank 2: M is nonsingular $\Leftrightarrow C(M)$ is separable $\Leftrightarrow C_0(M)$ is separable.

E. Exercises

1. Suppose that M is a hyperbolic plane and consider the isomorphism $C(M) \to \text{Mat}_2(R)$ of Exercise 1 of Chapter 5E. What matrix map does the canonical involution of $C(M)$ correspond to ?

2. Let A and A' be R-algebras with standard involutions σ and σ'. Let $\varphi : A \to A'$ be an algebra isomorphism which preserves standard involutions. Then φ is an isometry relative to the respective norms.

3. Let A be an R-algebra and assume that $\sigma = \mathrm{id}_A$ is a standard involution. If 2 is not a zero divisor in A, then A is commutative. If $2 \in R^*$, then $A = R$.

4. Let A be an R-algebra and assume that $A = R1 \oplus B$ (see Corollary 1.4 in Chapter 1 of Orzech-Small for the fact that $R1$ is a direct summand of A whenever A is faithful and finitely generated projective). Determine conditions that insure that $\sigma = \mathrm{id}_{R1} \oplus -\mathrm{id}_B$ is a standard involution on A.

5. Let $R\{X_1,..., X_n\}$ be the algebra of polynomials in the noncommuting variables $X_1,..., X_n$. Let $\mathfrak{a}$ be the two-sided ideal generated by the elements X_i^2, $X_i X_j + X_j X_i$ and let A be the algebra $A = R\{X_1,..., X_n\}/\mathfrak{a}$. Show that A is a free R-module with basis $\{1, X_1,..., X_n; X_1 X_2,..., X_1 X_n; X_2 X_3,..., X_2 X_n; X_3 X_4,..., X_3 X_n; ... X_{n-1} X_n; X_1 X_2 X_3,... ..., X_1 X_2 \cdots X_{n-1} X_n\}$. Let B be the submodule spanned by all these vectors except 1 and show that $\sigma = \mathrm{id}_R \oplus -\mathrm{id}_B$ is a standard involution on A. Does this contradict the conclusion of Theorem (6.8) ?

6. Let A_0 be a commutative R-algebra, let A_1 be an (A_0, A_0)-bimodule, and let $f : A_1 \times A_1 \to A_0$ be a bilinear form on A_1. Put $A = A_0 \oplus A_1$ and define a product $A \times A \to A$ on A as follows: $(a_0, a_1)(b_0, b_1) = (a_0 b_0 + f(a_1, b_1), a_0 b_1 + a_1 b_0)$. Under what conditions on f is A an R-algebra? Note that if A is an algebra, then $A = A_0 \oplus A_1$ is a grading of A. Suppose that the left and right A_0-actions on A_1 are the same and that A_1 is a free A_0-module. Let C be the matrix of f in a basis of A_1. What conditions on C insure that A is an algebra?

7. Let A_0, A_1, f, and A be as in Exercise 6. Assume that A_0 is a free quadratic algebra and that τ is its conjugation. Given that A is an algebra, find conditions that will insure that $\sigma = \tau \oplus -\mathrm{id}_{A_1}$ is a standard involution.

8. Consider the R-algebra $A = \mathrm{Mat}_2(R)$. Refer to Proposition (6.2) and determine the grading $A = A_0 \oplus A_1$ that $c = \begin{bmatrix} 1 & 0 \\ 0 & 0 \end{bmatrix}$ induces on A.

9. Let A be an R-algebra with standard involution σ and associated nr and f. Assume that $(c - c^\sigma)^2 = 0$ for all $c \in A$. Deduce that if 2 is not a zero divisor in A, then both nr and f are identically equal to zero on $(R1)^\perp$.

There is considerable variation in the definition of a quaternion algebra over a commutative ring. See Knus [1991], Knus-Ojanguren [1974], and Nelis-Oystaeyen, for example. Bourbaki [Algebra I, 1975] has another variant. This is discussed in the next four exercises.

10. The quaternion algebra of Bourbaki is given as follows: Let A be a free R-module with basis $\{1, i, j, k\}$. Make A into an algebra by letting 1 be the identity element and defining

$$\begin{cases} i^2 = \alpha 1 + \beta i, & ij = k, & ik = \alpha j + \beta k, \\ ji = \beta j - k, & j^2 = \gamma 1, & jk = \beta \gamma 1 - \gamma i, \\ ki = -\alpha j, & kj = \gamma i, & k^2 = -\alpha \gamma 1. \end{cases}$$

For $u = \rho 1 + \xi i + \eta j + \zeta k \in A$, define $u^\sigma = (\rho + \beta \xi)1 - \xi i - \eta j - \zeta k$. Check that $uu^\sigma = \rho^2 + \beta \rho \xi - \alpha \xi^2 - \gamma(\eta^2 + \beta \eta \zeta - \alpha \xi^2)$ and in particular that σ is a standard involution on A. The matrix of the corresponding symmetric bilinear form in the basis $\{1, i, j, k\}$ is

$$\begin{bmatrix} 2 & \beta & 0 & 0 \\ \beta & -2\alpha & 0 & 0 \\ 0 & 0 & -2\gamma & -\beta \gamma \\ 0 & 0 & -\beta \gamma & 2\alpha \gamma \end{bmatrix}.$$

An algebra of this type will be called a B-quaternion algebra.

11. Let A be an R-algebra with standard involution σ. Prove that A is a B-quaternion algebra if and only if

i) $A = A_0 \oplus A_1$ is graded, with A_0 a free quadratic algebra, $\sigma_{|A_0}$ the conjugation of A_0, and A_1 free of rank 1 as a left A_0-module, and

ii) $A = A_0 \perp A_1$ relative to the symmetric bilinear form defined by σ.

12. Let M be a free quadratic module over R of rank 2. Then $C(M)$, with its standard involution, is a B-quaternion algebra if and only if M represents a unit of R, i.e., if and only if $q(z) \in R^*$ for some z in M.

13. Suppose that R is a local ring. Let M be a nonsingular quadratic module which is free of rank 2 over R. Then $C(M)$ is a B-quaternion algebra.

14. Let M be a finitely generated projective quadratic module which is non-singular and has rank 2. Show that $C_0(M)$ is commutative.

Hints:

11. If A is B-quaternion, let $A_0 = R1 \oplus Ri$ and $A_0 = Rj \oplus Rk$. This also shows what to do in the other direction.

12. Let $z \in M$. Show that $q(z) \in R^*$ if and only if $M = C_0(M)z$.

13. Apply Exercise 16 of Chapter 4E.

14. Use the fact that this is true in the free case and apply the strategy of the proof of Theorem (6.8).

7
Arf Algebras and Special Elements

Overview

Let R be any commutative ring. Let M be a quadratic module over R and let $C(M)$ be its Clifford algebra. The centralizer of the even subalgebra $C_0(M)$ in $C(M)$ is a graded algebra which carries important information. We will see in Chapter 8 that it controls the relationship between the structures of $C(M)$ and $C_0(M)$, provides the connection between the tensor product and graded tensor product of Clifford algebras, and that it has consequences for the representations of $C(M)$. The (graded) isomorphism class of this centralizer leads to an invariant for quadratic forms over R, which has significant impact on their structure. This is pursued in Chapters 13 and 14. Since this invariant reduces to the classical Arf invariant when R is a field of characteristic 2, we will denote the centralizer by $A(M)$ and call it the Arf algebra of M. (It is also known as the discriminant algebra of M in the literature.) The present chapter develops some of the very basic properties of $A(M)$. It is assumed throughout that M is finitely generated projective and nonsingular. The quadratic and bilinear forms of M are denoted by q and h, respectively.

A. The Arf Algebra

As already indicated, the *Arf algebra* of M is defined by

$$A(M) = \mathrm{Cen}_{C(M)} C_0(M) = \{c \in C(M) \mid cd = dc \text{ for all } d \text{ in } C_0(M)\}.$$

This subalgebra of $C(M)$ clearly contains both $\mathrm{Cen}\, C(M)$ and $\mathrm{Cen}\, C_0(M)$. In fact,

$$\mathrm{Cen}\, C_0(M) = C_0(M) \cap A(M).$$

The grading $C(M) = C_0(M) \oplus C_1(M)$ induces one on $A(M)$. This is easy to see. Let $c = c_0 + c_1$, with c_i in $C_i(M)$, be in $A(M)$ and let d in $C_0(M)$ be arbitrary. Since

$$c_0 d + c_1 d = cd = dc = dc_0 + dc_1,$$

we find that $c_0 d = dc_0$ and $c_1 d = dc_1$. In particular, both c_0 and c_1 are in $A(M)$. So $A(M)$ has the grading

$$A(M) = A_0(M) \oplus A_1(M)$$

where $A_i(M) = A(M) \cap C_i(M)$.

The construction of an important involution of $A(M)$ follows next. It requires that M is faithful and is based on the following lemma:

(7.1). *Suppose that* M *is faithful. There are elements* $x_1,..., x_k$ *in* M *and* $r_1,..., r_k$ *in* R *such that* $r_1 q(x_1) + ... + r_k q(x_k) = 1.$

Proof. Let $\mathfrak{a}$ be the ideal generated by the set $\{q(x) \mid x \in M\}$. Since every element in $\mathfrak{a}$ has the form $r_1 q(x_1) + ... + r_k q(x_k)$, it must be shown that $\mathfrak{a} = R$. Suppose that this is false. So $\mathfrak{a}$ is contained in some maximal ideal $\mathfrak{p}$ of R and the localized quadratic module $M_{\mathfrak{p}} = (M \otimes_R R_{\mathfrak{p}}, q_{\mathfrak{p}})$ over $R_{\mathfrak{p}}$ satisfies $q_{\mathfrak{p}}(M_{\mathfrak{p}}) \subseteq \mathfrak{p} R_{\mathfrak{p}}$, where $\mathfrak{p} R_{\mathfrak{p}}$ is the maximal ideal of $R_{\mathfrak{p}}$. Let $h_{\mathfrak{p}}$ be the bilinear form associated to $q_{\mathfrak{p}}$. By (4.3), $M_{\mathfrak{p}}$ is nonzero and free of finite rank. Let $\mathcal{X}$ be a basis of $M_{\mathfrak{p}}$. Since $h_{\mathfrak{p}}(x, y) = q_{\mathfrak{p}}(x + y) - q_{\mathfrak{p}}(x) - q_{\mathfrak{p}}(y)$ for all x and y in $M_{\mathfrak{p}}$, it follows that $\det_{\mathcal{X}} h_{\mathfrak{p}}$ is in $\mathfrak{p} R_{\mathfrak{p}}$. By (4.7) on the other hand, $h_{\mathfrak{p}}$ is nonsingular so that $\det_{\mathcal{X}} h_{\mathfrak{p}}$ is in $(R_{\mathfrak{p}})^*$. This contradiction completes the proof.   QED.

(7.2). *Suppose that* M *is faithful. There is a unique algebra automorphism* μ *of* $A(M)$ *such that* $\mu^2 = \mathrm{id}_{A(M)}$ *and* $xc = c^{\mu}x$ *for all* x *in* M *and* c *in* $A(M)$. *In addition,*

$$\mathrm{Cen}\ C(M) = \{c \in A(M) \mid c^{\mu} = c\}.$$

Proof. Use the previous proposition to fix elements $r_1,..., r_k$ in R and $x_1,..., x_k$ in M with the property that $r_1 q(x_1) + ... + r_k q(x_k) = 1$. That μ is unique if it exists is easy: Let c in $A(M)$ be arbitrary. Since $x_i c x_i = c^{\mu} x_i x_i = c^{\mu} q(x_i)$, we find that

$$c^{\mu} = c^{\mu} \sum_i r_i q(x_i) = \sum_i r_i x_i c x_i ,$$

so that c^{μ} is uniquely determined. We now define μ by the preceding equation and show that it has the required properties. Let x in M be arbitrary and observe that

$$c^{\mu}x = \sum_i r_i x_i c(x_i x) = \sum_i r_i q(x_i) x c = xc .$$

So $c^{\mu}x = xc$ and similarly $xc^{\mu} = cx$. It follows that $c^{\mu}xy = xyc^{\mu}$ for all x and y in M, and in particular, that $c^{\mu} \in A(M)$. That μ is additive is obvious. For c and d in $A(M)$,

$$c^{\mu}d^{\mu} = \left(\sum_i r_i x_i c x_i\right)\left(\sum_j r_j x_j d x_j\right) = \sum_{i,j} r_i r_j q(x_i) x_j c d x_j$$

$$= \left(\sum_i r_i q(x_i)\right)\left(\sum_j r_j x_j c d x_j\right) = (cd)^{\mu} .$$

Since $r^{\mu} = r$ for r in R, it only remains to verify that $\mu^2 = $ id:

$$c^{\mu^2} = \sum_j r_j x_j \left(\sum_i r_i x_i c x_i\right) x_j = \sum_{i,j} r_j r_i x_j x_i c x_i x_j c = \left(\sum_j r_j q(x_j)\right)\left(\sum_i r_i q(x_i)\right)c = c .$$

This completes the proof.   QED.

Let $c \in A_0(M)$. Then $c^{\mu}x = xc \in C_1(M)$ for all x in M. Therefore, $c^{\mu}x^2 \in C_0(M)$ for all x in M. So $c^{\mu} \in C_0(M)$ by (7.1), and hence $c^{\mu} \in A_0(M)$. If $c \in A_1(M)$, then the same argument shows that $c^{\mu} \in A_1(M)$. It follows that μ is a graded automorphism of $A(M)$.

The involution $\beta = 1_{C_0(M)} \oplus -1_{C_1(M)}$ of $C(M)$ restricts to an algebra automorphism of $A(M)$. This involution of $A(M)$ will also be denoted by β. Observe that it preserves the grading of $A(M)$. The uniqueness statement of (7.2) implies that $\beta\mu\beta^{-1} = \mu$. So $\beta\mu = \mu\beta$. Define the automorphism α of $A(M)$ by

$$\alpha = \beta\mu = \mu\beta.$$

Note that α is also an involution. Since β and μ are graded, α is also. We call α the *conjugation* of $A(M)$. In the trivial case $M = \{0\}$, $A(M) = C(M) = R$ and $\beta = \text{id}_R$, and we set $\mu = \alpha = \text{id}_R$.

The rest of this section analyzes $A(M)$ and α in the cases where is M free of rank 1 or 2.

Suppose first that M is free of rank 1 and let $\{x\}$ be a basis of M. Since M is nonsingular, $h(x, x) = 2q(x)$ is in R^*. So 2 and $q(x) = b$ are in R^*. Several easy consequences of Example 1 of Chapter 5A and Example 2 of Chapter 5B follow. Since $C(M)$ is commutative, $A(M) = C(M)$ and μ is trivial. The conjugation of $A(M)$ is $\alpha = \beta$ and the grading of $A(M)$ is $A(M) = R \oplus M$. Finally, there is an isomorphism

$$\psi_1 : A(M) \to R[X]/(X^2 - b)$$

of R-algebras given by $\psi_1(xr) = rv$ for all r in R, where $v = X + (X^2 - b)$. By (2.2), $R[X]/(X^2 - b)$ is separable. Refer to Chapter 1C and provide $R[X]/(X^2 - b)$ with the graded structure $(0, b)^{odd}$. The definition of this grading implies that ψ_1 is graded. Recall from Chapter 1B that the conjugation σ of $R[X]/(X^2 - b)$ is defined by $v^\sigma = -v$. Since $\psi_1(x^\alpha) = \psi_1(-x) = -v = v^\sigma = \psi_1(x)^\sigma$, ψ_1 is an isomorphism of algebras which preserves conjugations. In summary:

(7.3). *The map* $\psi_1 : A(M) \to (0, b)^{odd}$ *is an isomorphism of graded algebras which preserves conjugations. In particular,* $A(M) = C(M)$ *is a separable free quadratic algebra.*

Now to the case where M is free of rank 2 with basis $\{x_1, x_2\}$. Refer to Chapter 6B for the facts used in the discussion that follows. By (6.5), $A(M) = C_0(M) = \text{Cen } C_0(M)$. In particular, the grading of $A(M)$ is trivial. The involution $\bar{}$ of $C(M)$ stabilizes $C_0(M)$ and hence $A(M)$. Since $A(M)$ is commutative, $\bar{}$ is an automorphism of $A(M)$. Observe that

$$x_1(x_1 x_2) = \overline{(x_1 x_2)}x_1 \text{ and } x_2(x_1 x_2) = \overline{(x_1 x_2)}x_2.$$

By the uniqueness assertion of (7.2), $\mu = \bar{}$. Since $A(M) = C_0(M)$, β is trivial on $A(M)$. Therefore, the conjugation of $A(M)$ is $\alpha = \bar{}$.

Set $a = h(x_1, x_2)$ and $q(x_1)q(x_2) = -b$. Since M is nonsingular, $a^2 + 4b$ is in R^*. By (2.2), the algebra $(a, b) = R[X]/(X^2 - aX - b)$ is separable. Consider the basis $\{1, v = X + (X^2 - aX - b)\}$ and define the map

$$\psi_2 : A(M) \to (a, b)$$

by $\psi_2(1) = 1$, $\psi_2(x_1 x_2) = v$ and by extending linearly. This R-linear map is bijective, since it takes a basis onto a basis. Since

$$\psi_2((x_1 x_2)^2) = \psi_2(x_1(a - x_1 x_2)x_2) = \psi_2(ax_1 x_2 + b)$$
$$= av + b = (\psi_2(x_1 x_2))^2,$$

ψ_2 is an R-algebra isomorphism. Recall from Chapter 1B that the conjugation σ of (a, b) is given by $v^\sigma = a - v$. The computation

$$\psi_2(\overline{x_1 x_2}) = \psi_2(t - x_1 x_2) = a - v = v^\sigma = \psi_2(x_1 x_2)^\sigma$$

shows that ψ_2 preserves conjugations. Supplying (a, b) with the grading $(a, b)^{even}$, i.e., the trivial grading, we now have

(7.4). *The map* $\psi_2 : A(M) \to (a, b)^{even}$ *is an isomorphism of graded algebras which preserves conjugations. In particular, $A(M) = C_0(M)$ is a separable free quadratic algebra.*

B. The Arf Algebra of an Orthogonal Sum

Continue to assume that M is faithful. Suppose that $M = M_1 \perp M_2$. Since M is finitely generated projective and nonsingular, so are M_1 and M_2. Since M_1 and M_2 determine M, it should not come as a surprise that the Arf algebras $A(M_1)$ and $A(M_2)$ together determine $A(M)$. This section will make this precise. In the process it provides a fact which will be important in unraveling the structure of $A(M)$ in general.

Let i be either 1 or 2. Let $A(M_i)$ be the Arf algebra of M_i and let $\alpha_i = \beta_i \mu_i = \mu_i \beta_i$ be the conjugation of $A(M_i)$. Since α_i is a graded automorphism of $A(M_i)$, it follows that $\alpha_1 \otimes \alpha_2$ is a graded automorphism of $A(M_1) \hat{\otimes}_R A(M_2)$. Define the R-algebra $A(M_1) * A(M_2)$ by

$$A(M_1) * A(M_2) = \{c \in A(M_1) \hat{\otimes}_R A(M_2) \mid c^{(\alpha_1 \otimes \alpha_2)} = c\}.$$

We will see later that the parallels between this definition and the discussion in Chapter 3D are no coincidence. Consider the sequence of R-module homomorphisms

$$A(M_1) \otimes_R A(M_2) \rightarrow A(M_1) \otimes_R C(M_2) \rightarrow C(M_1) \otimes_R C(M_2)$$

provided by the appropriate inclusions. It is easy to see that the composite is a homomorphism

$$A(M_1) \hat{\otimes}_R A(M_2) \rightarrow C(M_1) \hat{\otimes}_R C(M_2)$$

of graded R-algebras. Recall from Chapter 5C the graded algebra isomorphism

$$\varphi : C(M_1) \hat{\otimes}_R C(M_2) \rightarrow C(M_1 \perp M_2) = C(M)$$

given by $\varphi(c_1 \otimes c_2) = c_1 c_2$ for all c_1 in $C(M_1)$ and c_2 in $C(M_2)$. Notice that $\beta_1 \otimes \beta_2$ corresponds to β under this isomorphism. One more composition provides the graded algebra homomorphism

$$A(M_1) \hat{\otimes}_R A(M_2) \rightarrow C(M),$$

which we also denote by φ.

(7.5). *Suppose that* M *is the orthogonal direct sum* $M = M_1 \perp M_2$ *of two free nonsingular submodules* M_1 *and* M_2 *of finite ranks. Assume that the R-modules* $A(M_1)$ *and* $A(M_2)$ *are also both free of finite rank. Then the restriction of* φ *is an algebra isomorphism*

$$A(M_1) * A(M_2) \rightarrow A(M).$$

The involution $\alpha_1 \otimes id = id \otimes \alpha_2$ *of* $A(M_1) * A(M_2)$ *corresponds to the conjugation* α *of* $A(M)$.

Proof. The map $A(M_1) \hat{\otimes}_R A(M_2) \rightarrow C(M_1) \hat{\otimes}_R C(M_2)$ is injective, since $A(M_1)$ and $C(M_2)$ are both free. We will therefore consider $A(M_1) \hat{\otimes}_R A(M_2)$ as a subalgebra of $C(M_1) \hat{\otimes}_R C(M_2)$.

Put $D = C(M_1) \hat{\otimes}_R C(M_2)$ and let $D = D_0 \oplus D_1$ be the grading of D. Observe that $\varphi : C(M_1) \hat{\otimes}_R C(M_2) \rightarrow C(M)$ takes $Cen_D(D_0)$ to $A(M)$. To verify that the restriction of φ has the required property, it must therefore be shown that

$$Cen_D(D_0) = \{c \in A(M_1) \otimes_R A(M_2) \mid c^{(\mu_1 \otimes \beta_2)} = c^{(\beta_1 \otimes \mu_2)}\} .$$

Since $D_0 = ((C_0(M_1) \otimes C_0(M_2)) \oplus (C_1(M_1) \otimes C_1(M_2))$, the subsets $C_0(M_1) \otimes 1, 1 \otimes C_0(M_2)$ and $\{x \otimes y \mid x \in M_1 \text{ and } y \in M_2\}$ together generate D_0 as an algebra.

1) We will prove that $c \in D$ centralizes $C_0(M_1) \otimes 1$ and $1 \otimes C_0(M_2)$ if and only if $c \in A(M_1) \otimes A(M_2)$. Assume that $c \in D$ centralizes both $C_0(M_1) \otimes 1$ and $1 \otimes C_0(M_2)$. Let $\{v_1,..., v_k\}$ be a basis of $C(M_2)$ and put $c = \sum_i (u_i \otimes v_i)$ with $u_i \in C(M_1)$. Clearly,

$$\sum_i (u_i c_0 \otimes v_i) = c(c_0 \otimes 1) = (c_0 \otimes 1)c = \sum_i ((c_0 u_i \otimes v_i),$$

for all $c_0 \in C_0(M_1)$. By the independence of the v_j, $u_i c_0 = c_0 u_i$ for all $c_0 \in C_0(M_1)$ and all i. So all u_i are in $A(M_1)$ and $c \in A(M_1) \otimes C(M_2)$. A repetition of this argument shows that $c \in A(M_1) \otimes A(M_2)$. The other direction is trivial and 1) is established.

The next step completes the required characterization of $\text{Cen}_D(D_0)$.

2) Let $c \in A(M_1) \otimes A(M_2)$. We will prove that c commutes with all $x \otimes y$, where $x \in M_1$ and $y \in M_2$, if and only if $c^{(\mu_1 \otimes \beta_2)} = c^{(\beta_1 \otimes \mu_2)}$.

Put $c = \Sigma(u \otimes v)$, where the us are in $A(M_1)$ and the vs are in $A(M_2)$. Recall the grading of $A(M_1)$ and set each $u = u_0 + u_1$ with $u_0 \in A_0(M_1)$ and $u_1 \in A_1(M_1)$. Similarly, set each $v = v_0 + v_1$ with $v_0 \in A_0(M_2)$ and $v_1 \in A_1(M_2)$. So

$$c = \sum ((u_0 \otimes v_0) + (u_0 \otimes v_1) + (u_1 \otimes v_0) + (u_1 \otimes v_1)).$$

Let x in M_1 and y in M_2 be arbitrary. Making use of (7.2), we find that

$c(x \otimes y) =$
$$\Sigma((x(u_0)^{\mu_1} \otimes v_0 y) + (x(u_0)^{\mu_1} \otimes -v_1 y) + (x(u_1)^{\mu_1} \otimes v_0 y) + (x(u_1)^{\mu_1} \otimes -v_1 y))$$
$$= (x \otimes 1) c^{(\mu_1 \otimes \beta_2)} (1 \otimes y).$$

Completely analogously, $(x \otimes y)c = (x \otimes 1) c^{(\beta_1 \otimes \mu_2)} (1 \otimes y)$. It follows that if $c^{(\mu_1 \otimes \beta_2)} = c^{(\beta_1 \otimes \mu_2)}$, then c commutes with all $x \otimes y$. Assume, conversely, that c commutes with all $x \otimes y$. From above,

$$(x \otimes 1)\, c^{(\mu_1 \otimes \beta_2)}\, (1 \otimes y^2) = (x \otimes 1)\, c^{(\beta_1 \otimes \mu_2)}\, (1 \otimes y^2)$$

for any x in M_1 and y in M_2. Fixing x and applying (7.1) to M_2 shows that

$$(x \otimes 1)\, c^{(\mu_1 \otimes \beta_2)} = (x \otimes 1)\, c^{(\beta_1 \otimes \mu_2)}.$$

A repetition of this step proves that $c^{(\mu_1 \otimes \beta_2)} = c^{(\beta_1 \otimes \mu_2)}$.

3) It remains to prove the indicated correspondence of the involutions. That $\alpha_1 \otimes \mathrm{id} = \mathrm{id} \otimes \alpha_2$ on $A(M_1) * A(M_2)$ is clear. Let $d \in A(M)$ be arbitrary. So $d^{(\varphi^{-1})} \in A(M_1) * A(M_2)$. It follows from the equations in step 2) that

$$(x \otimes 1)\, d^{(\varphi^{-1})} = (x \otimes 1)\, d^{(\varphi^{-1})(\mu_1 \otimes \beta_2)^2} = d^{(\varphi^{-1})(\mu_1 \otimes \beta_2)}\, (x \otimes 1) \quad \text{and}$$

$$(1 \otimes y)\, d^{(\varphi^{-1})} = d^{(\varphi^{-1})(\beta_1 \otimes \mu_2)}\, (1 \otimes y),$$

for all x in M_1 and y in M_2. Since $d^{(\varphi^{-1})(\mu_1 \otimes \beta_2)} = d^{(\varphi^{-1})(\beta_1 \otimes \mu_2)}$, we get

$$xd = d^{(\varphi^{-1})(\mu_1 \otimes \beta_2)\varphi}\, x \quad \text{and} \quad yd = d^{(\varphi^{-1})(\mu_1 \otimes \beta_2)\varphi}\, y$$

by applying φ to these equations. By the uniqueness assertion of (7.2), $\mu = \varphi^{-1}(\mu_1 \otimes \beta_2)\varphi$. Since $\beta = \varphi^{-1}(\beta_1 \otimes \beta_2)\varphi$, $\alpha = \beta\mu = \varphi^{-1}(\alpha_1 \otimes \mathrm{id})\varphi$. QED.

Remark. The hypothesis that M_1, M_2, $A(M_1)$, and $A(M_2)$ are all free of finite rank is much stronger than what is needed for the conclusion of (7.5). The assumption that M is finitely generated projective, nonsingular, and faithful turns out to be sufficient. This more general fact follows from (7.5) by localization and Theorem (9.11, 1) of Chapter 9. However, the "free" case is all that will be needed in this volume.

C. Special Elements

To the conditions on M already in effect, namely, that it is a nonsingular finitely generated projective and faithful quadratic module, we now add the assumption that rank M is either odd or even. We will see later in Chapter 9C that the study of $C(M)$ can be reduced to these two cases.

The existence in $C(M)$ of certain "special" elements, not only has direct impact on the structure of the Arf algebra $A(M)$, but, as we will see later, e.g., in Chapter 8, on several structural features of $C(M)$.

An element $z \in C(M)$ is *special* if $\{1, z\}$ is a basis of the R-module $A(M)$ and

(i)   if rank M is odd, then $z \in C_1(M)$, $z^\alpha = -z$ and $z^2 = b$ with b in R^*,

(ii) if rank M is even, then $z \in C_0(M)$, $z^\alpha = a - z$ and $z^2 = az + b$ with a and b in R and $a^2 + 4b \in R^*$.

It is of course implicit in the definition of a special element that $A(M)$ is free of rank 2. We point out in passing that special elements are closely related to the volume elements used in the analytic theory of Clifford algebras. See the book of Lawson-Michelsohn and Exercise 6 here.

Suppose that z is special in $C(M)$. The polynomial $X^2 - aX - b \in R[X]$, where it is understood that $a = 0$ if rank M is odd, is the *polynomial* of z. If rank M is odd, then by (4.18), $2 \in R^*$. It follows that $a^2 + 4b \in R^*$ in the odd rank case also. Since $(a - 2z)^2 = a^2 - 4az + 4(az + b) = a^2 + 4b \in R^*$, $a - 2z$ is a unit in $A(M)$. If rank M is odd, then $2z$ and therefore z, are units in $A(M)$.

(7.6). *Let z be a special element of $C(M)$ with polynomial $X^2 - aX - b$. Let z' in $C(M)$ be arbitrary.*

(1) *If rank M is odd, then z' is special if and only if $z' = tz$ with t in R^*.*

(2) *If rank M is even, then z' is special if and only if $z' = r + tz$ with r in R and t in R^*.*

In either case, the element $(a^2 + 4b)(R^*)^2 \in R^*/(R^*)^2$ is independent of the choice of special element.

Proof. If z' is to be a special element, $\{1, z'\}$ must be a basis of $A(M)$ so that $z' = r + tz$ with r in R and t in R^*. Suppose rank M is odd. If z' is special, then $(z')^\alpha = -z'$. So $r = 0$ and $z' = tz$. It is clear that any such element is special with polynomial $X^2 - t^2 b$. Suppose rank M is even. For any $z' = r + tz$ with t in R^*, put $a' = 2r + ta$ and $b' = t^2 b - rta - r^2$. Since

$$(z')^\alpha = r + t(a - z) = 2r + ta - (r + tz) = a' - z' \quad \text{and}$$

$$(z')^2 = r^2 + 2rtz + t^2(az + b) = (2r + ta)(r + tz) + t^2 b - rta - r^2 = a'z' + b',$$

z' is a special element with polynomial $X^2 - a'X - b'$. In either case, $(a')^2 + 4b' = t^2(a^2 + 4b)$.   QED.

Special elements provide a precise connection between the Arf algebra and the separable free quadratic algebras of Chapter 2B. This is easy to see and follows next.

Let $\varepsilon \in \mathbb{Z}_2$ be equal to 1 if rank M is odd and to 0 if rank M is even. Let a and b be elements in R such that $a^2 + 4b \in R^*$, where $a = 0$ if $\varepsilon = 1$. Consider the free separable algebra $(a, b) = R[X]/(X^2 - aX - b)$ and provide it with the graded structure $(a, b)^\varepsilon$. Let $v = X + (X^2 - aX - b)$.

Assume that z is a special element of C(M) with polynomial $X^2 - aX - b$. Define $\phi : (a, b) \to A(M)$ by extending $1 \to 1$ and $v \to z$ linearly. The properties of z directly imply that $\phi : (a, b)^\varepsilon \to A(M)$ is an isomorphism of graded algebras which preserves the conjugations σ and α. Assume conversely that $\phi : (a, b)^\varepsilon \to A(M)$ is an isomorphism of graded algebras with conjugation. Then $z = \phi v$ is a special element of C(M) with poynomial $X^2 - aX - b$. That $\{1, z\}$ is a basis of A(M) is clear. If $\varepsilon = 1$, then z is in the second component of the grading of A(M), i.e., $z \in C_1(M)$. If $\varepsilon = 0$, $(a, b)^\varepsilon$ has the trivial grading and hence z is in $C_0(M)$. The other requirements for z follow directly. We have proved:

(7.7). *The following statements are equivalent:*

   (1)   *C(M) has a special element with polynomial* $X^2 - aX - b$.

   (2)   *There is an isomorphism* $(R[X]/(X^2 - aX - b)^\varepsilon \to A(M)$ *of graded algebras with conjugation.*

Combining this with (7.3) and (7.4) provides the following examples and proposition.

Example 1. Let rank $M = 1$ and let $\{z\}$ be a basis of M. Then z is a special element with polynomial $X^2 - q(z)$.

Example 2. Suppose rank $M = 2$ and let $\{x_1, x_2\}$ be a basis. Put $a = h(x_1, x_2)$ and $b = -q(x_1)q(x_2)$. Then $z = x_1x_2$ is a special element with polynomial $X^2 - aX - b$. Note that if M is a hyperbolic plane and $\{x_1, x_2\}$ a hyperbolic basis, then the polynomial of z is $X^2 - X$.

(7.8). *If M is free of rank 1 or 2, then C(M) has a special element.*

The following result is an important ingredient in this book's approach to the structure theory of the Clifford algebras as developed in subsequent chapters.

(7.9). *Suppose that* M *has an orthogonal splitting* $M = M_1 \perp M_2$, *where* M_1 *and* M_2 *are both nonsingular and free of finite ranks* n_1 *and* n_2, *respectively. Suppose that* $C(M_1)$ *has a special element* z_1 *with polynomial* $X^2 - a_1 X - b_1$, *and that* $C(M_2)$ *has a special element* z_2 *with polynomial* $X^2 - a_2 X - b_2$. *Then* $a_1 a_2 1 - a_2 z_1 - a_1 z_2 + 2 z_1 z_2$ *is a special element of* $C(M)$ *with polynomial* $X^2 - aX - b$, *where*

$$a = a_1 a_2 \quad and \quad b = a_1^2 b_2 + a_2^2 b_1 + (-1)^{(n_1 n_2)} (4 b_1 b_2).$$

Proof. Let ε_i be the residue class of n_i mod 2 and set $\varepsilon = \varepsilon_1 + \varepsilon_2$. Refer to the discussion that precedes (7.7). Let $v = X + (X^2 - a_1 X - b_1)$, $w = X + (X^2 - a_2 X - b_2)$, and let $\phi_i : (a_i, b_i)^{\varepsilon i} \to A(M_i)$ be the respective graded isomorphisms with conjugation given by $v \to z_1$ and $w \to z_2$. Refer to Chapter 3D, in particular to the algebra P, the element z, and the graded conjugation-preserving isomorphism $\theta : (a, b)^{\varepsilon} \to P$ of (3.9). Consider next the isomorphisms $P \to A(M_1) * A(M_2)$ induced by ϕ_1 and ϕ_2 and $A(M_1) * A(M_2) \to A(M)$ given by (7.5). Note that their composite $P \to A(M)$ is an isomorphism that takes the element z to $a_1 a_2 1 - a_2 z_1 - a_1 z_2 + 2 z_1 z_2$. Refer to the respective gradings and conjugations of P and $A(M)$ and check that the isomorphisms preserve both. Therefore, the composite $(a, b)^{\varepsilon} \to A(M)$ is an algebra isomorphism which preserves gradings and conjugations and takes $X + (X^2 - aX - b)$ to $a_1 a_2 1 - a_2 z_1 - a_1 z_2 + 2 z_1 z_2$. Another look at the discussion that precedes (7.7) completes the proof.   QED.

(7.10). *Suppose that* R *is a local ring and that* M *is nonzero. Then* $C(M)$ *has a special element.*

Proof. Do an induction using (7.8), (7.9), and (4.16). QED.

Continue to assume that R is a local ring. The preceding proposition provides an important invariant of the quadratic module M. Suppose $M \neq \{0\}$ and let $X^2 - aX - b$ be the polynomial of a special element in $C(M)$. Let $\varepsilon \in \mathbb{Z}_2$ be equal to 1 or 0 according to whether the rank of M is odd or even. The *Arf invariant* Arf M of M is the isomorphism class of

$(R[X]/(X^2 - aX - b)^\varepsilon$. By (7.7), this algebra is isomorphic to the Arf algebra $A(M)$. It follows that the element

$$\text{Arf } M \in QU_f(R)$$

is independent of the choice of the special element. If $M = \{0\}$, we set $\text{Arf } M = 1 \in QU_f(R)$.

Suppose that $M = M_1 \perp M_2$. Since M_1 and M_2 are finitely generated projective, they are free of finite rank. A combination of (7.9) and the definition of the product of $QU_f(R)$ shows that

$$\text{Arf } M = (\text{Arf } M_1)(\text{Arf } M_2).$$

The Arf invariant will be defined for any nonsingular finitely generated projective quadratic modules over a commutative ring R in Chapter 13C, and it will be seen to play a major role in the study of quadratic forms over arithmetic domains in Chapter 14C. In Chapter 10D, we will see that it is closely related to the classical discriminant of a free quadratic module.

Example 3. Let R be a local ring and let M be free of rank 2. Then M is a hyperbolic plane if and only if $\text{Arf } M = 1$. In view of Example 2, it remains to assume that $\text{Arf } M = 1$ and to show that M is hyperbolic. By Exercise 16 of Chapter 4E, there is an $x \in M$ with $q(x) \in R^*$. Assume first that $2 \in R^*$. So $h(x, x) = 2q(x) \in R^*$. It follows by (4.9), that M has an orthogonal basis $\{x, y\}$. Let $q(x) = c$ and $q(y) = d$. By Example 2, xy is a special element with polynomial $X^2 + cd$. In particular, $cd \in R^*$. Since $\text{Arf } M = 1$, $-cd \in (R^*)^2$ by (3.3). So $d\gamma^2 = -c$ for some $\gamma \in R^*$. Put $x_1 = x + y\gamma$ and check that $\{x_1, y\}$ is a basis of M with $q(x_1) = 0$. The nonsingularity of M implies that $h(x_1, y) \in R^*$. So $h(x_1, y\delta) = 1$ with $\delta \in R^*$. Let $\eta = -q(y\delta)$, set $x_2 = x_1\eta + y\delta$, and verify that $\{x_1, x_2\}$ is a hyperbolic basis of M. The argument in the case $2 \notin R^*$, i.e., where 2 is in the maximal ideal $\mathfrak{m}$, is a little different, but no more difficult. See Exercise 8.

We conclude this chapter with a consequence of the existence of special elements for the structures of $\text{Cen } C(M)$ and $\text{Cen } C_0(M)$. The underlying commutative ring R is again arbitrary.

Suppose that z is a special element in $C(M)$. Assume rank M is odd. Since $z \in C_1(M)$, $z^\beta = -z$. It follows that $\alpha = \beta$ and hence that μ is the identity. Therefore by (7.2), $\text{Cen } C(M) = A(M)$. Since z is in $C_1(M)$, $\text{Cen } C_0(M) = C_0(M) \cap A(M) = R$. Assume that rank M is even. Now

$A(M) \subseteq C_0(M)$. So β is the identity and $\alpha = \mu$. Observe also that Cen $C_0(M) = A(M)$. Let $r + r'z$, with r and r' in R, be any element in Cen $C(M)$. By (7.2), $r + r'z = (r + r'z)^\alpha = (r + r'a) - r'z$. Hence, $r'a = 0$ and $2r' = 0$. So $r'(a^2 + 4b) = 0$, and hence $r' = 0$. Therefore Cen $C(M) = R$.

(7.11). *Theorem. Suppose z is a special element in $C(M)$ with polynomial $X^2 - aX - b$.*

(1) *If* rank M *is odd, then* $a = 0$, Cen $C(M) = A(M) \cong R[X]/(X^2 - b)^{odd}$
and

$$\text{Cen } C_0(M) = R.$$

(2) *If* rank M *is even, then* Cen $C(M) = R$ *and*

$$\text{Cen } C_0(M) = A(M) \cong R[X]/(X^2 - aX - b)^{even}.$$

If $X^2 - aX - b$ *has a root in* R, *then* Cen $C(M) \cong R \oplus R$ *in case* (1), *and* Cen $C_0(M) \cong R \oplus R$ *in case* (2).

Proof. In both (1) and (2), the isomorphism for $A(M)$ is a direct consequence of (7.7). Everything else has already been verified above except the very last statement. But this follows from (3.3) since $(1, 0) \cong R \oplus R$.   QED.

D. Exercises

1. Let M have basis $\{x, y\}$. Refer to (7.1) and its proof and produce elements $x_1,..., x_k$ in M and $r_1,..., r_k$ in R that satisfy the requirements.

2. Let M be as in Exercise 1 of Chapter 5E and let $C(M) \cong Mat_2(R)$ be the isomorphism established there. Give a matricial analysis of $A(M)$, its conjugation, and find a special element.

3. Let $S = R[X]/(X^2 - aX - b)$ with $a^2 + 4b \in R^*$ be a separable free quadratic algebra and refer to the Example of Chapter 4D. Denote the non-singular quadratic module (S, q) by M. Show that $A(M) \cong (a, b) = S$. In other words, the Arf algebra of S considered as a quadratic module is isomorphic to S. How are the respective conjugations related?

4. Let M have basis $\{x_1, x_2\}$. Use (7.2) to check that Cen $C(M) = R$.

5. Let M have basis $\mathcal{X} = \{x_1, x_2\}$. Let z be a special element of $C(M)$ with polynomial $X^2 - aX - b$. Compare the scalars $a^2 + 4b$ and $\det_{\mathcal{X}} h$.

6. Suppose that $\{x_1, ..., x_n\}$ is an *orthogonal basis* of M. This means that $h(x_i, x_j) = 0$ for $i \neq j$. Show that $z = x_1 \cdots x_n$ is a special element of $C(M)$ with polynomial $X^2 - b$, where $b = (-1)^{n(n-1)/2} q(x_1) \cdots q(x_n)$.

7. Supply all the details in the proof of (7.9).

8. Establish Example 3 in the case $2 \notin R^*$.

Hints:

2. Make use of Chapter 6B.

4. Use the fact that $A(M) = C_0(M)$ and $\mu = ^-$.

6. Use Example 1 of Section C, induction, (7.9), and (7.6) and its proof. Since M is nonsingular, $h(x_i, x_i) = 2q(x_i)$ is in R^*. So $2 \in R^*$.

7. To see that $(a, b)^\varepsilon \to A(M)$ preserves gradings consider the four cases corresponding to the possible values of ε_1 and ε_2.

8. As in the other case, it suffices to produce a basis $\{x_1, y\}$ of M with $q(x_1) = 0$. Start with any basis $\{x, y\}$ of M. The nonsingularity of M implies that $h(x, y) \in R^*$, so we can ssume that $h(x, y) = 1$. Let $q(x) = c$ and $q(y) = d$ and show that $X^2 + X + cd$ has a root, say γ, in R. Note that $\gamma' = -1 - \gamma$ is also a root, and that either γ or γ' is in R^*. Assume that it is γ (otherwise use γ') and set $x_1 = x\gamma + yc$. Check that $q(x_1) = 0$ and that $\{x_1, y\}$ is a basis of M.

8
Consequences of the Existence of Special Elements

Overview

The purpose of this chapter is the investigation of the consequences that the existence of special elements has for the structure of $C(M)$, particularly as to the interplay between $C(M)$ and its subalgebra $C_0(M)$. If rank M is odd, we will see that the structure of $C_0(M)$ completely determines that of $C(M)$; if rank M is even, this situation is reversed, and it is $C(M)$ that determines $C_0(M)$.

Throughout the chapter, R is a commutative ring and M is a quadratic module over R with quadratic form q and associated symmetric bilinear form h. An isomorphism between algebras for which gradings are specified will be understood to preserve the gradings. For an R-algebra A, recall that $<A>$ denotes the algebra A supplied with the trivial grading.

A. Connections between $C(M)$ and $C_0(M)$

In this section, M is finitely generated projective and nonsingular and rank M is either odd or even. Also, z is a special element in $C(M)$ with polynomial $X^2 - aX - b$. Note that $a^2 + 4b \in R^*$. If rank M is odd, then $a = 0$, so that 2 and b are in R^*.

(8.1). *If* rank M *is odd, then* $C_0(M) = zC_1(M)$ *and* $C_1(M) = zC_0(M)$. *If* rank M *is even,* $C_0(M) = \mathrm{Cen}_{C(M)}(z)$ *and* $C_1(M) = \{c \in C(M) \mid cz + zc = ac\}$.

Proof. Assume rank M is odd. Since z is in $C_1(M)$,

$$C_0(M) = bC_0(M) = z^2C_0(M) \subseteq zC_1(M) \subseteq C_0(M).$$

So $C_0(M) = zC_1(M)$ and in the same way, $C_1(M) = zC_0(M)$. Suppose rank M is even. Since $z \in$ Cen $C_0(M)$, $C_0(M) \subseteq \text{Cen}_{C(M)}(z)$. We verify next that $C_1(M) \subseteq \{c \in C(M) \mid cz + zc = ac\}$. For x in M, $(a - z)x = z^\alpha x = z^\mu x = xz$, by (7.2). So $xz + zx = ax$. The inclusion follows, since any element of $C_1(M)$ is a sum of products of odd numbers of elements of M. Let c be any element in $\text{Cen}_{C(M)}(z) \cap \{c \in C(M) \mid cz + zc = ac\}$. Since $(2z - a)c = 0$ and $2z - a$ is invertible in $C(M)$, $c = 0$. Therefore,

$$\text{Cen}_{C(M)}(z) \cap \{c \in C(M) \mid cz + zc = ac\} = 0.$$

Let c in $\text{Cen}_{C(M)}(z)$ be arbitrary. Put $c = c_0 + c_1$ with c_i in $C_i(M)$. Notice that $c_0 \in C_0(M) \cap \text{Cen}_{C(M)}(z)$, so that $c_1 \in \text{Cen}_{C(M)}(z) \cap C_1(M)$. Therefore, $c_1 = 0$ and $c \in C_0(M)$. So $C_0(M) = \text{Cen}_{C(M)}(z)$. That $C_1(M) = \{c \in C(M) \mid cz + zc = ac\}$ follows in the same way.   QED.

(8.2). *Theorem. Assume that* rank M *is odd. Then*

(1) $C(M) \cong C_0(M) \otimes_R R[X]/(X^2 - b)$ *and*

$$C(M) \cong <C_0(M)> \hat{\otimes}_R R[X]/(X^2 - b)^{\text{odd}}.$$

(2) *If* $X^2 - b$ *has a root in* R, *i.e., if* $b \in (R^*)^2$, *then*

$$C(M) \cong C_0(M) \oplus C_0(M).$$

Proof. The assignment $C_0(M) \times \text{Cen } C(M) \to C(M)$ given by $(c,d) \to cd$ induces an R-homomorphism $C_0(M) \otimes_R \text{Cen } C(M) \to C(M)$ which satisfies $c \otimes d \to cd$. An inverse of this map is constructed as follows. Recall that z is a unit in A(M). By (7.11), z is a unit in Cen C(M) and by (8.1), $C_1(M) = C_0(M)z$. So every element in C(M) is uniquely of the form $c + dz$ with c and d in $C_0(M)$. Define

$$C(M) \to C_0(M) \otimes_R \text{Cen } C(M)$$

by $c + dz \to (c \otimes 1) + (d \otimes z)$. Check that this is the inverse of the earlier map. Since z is in Cen C(M), it is an algebra isomorphism. That the gradings are preserved follows from the definitions. We have proved (1). The (ungraded) isomorphism of (2) is an easy consequence of (1), (3.3), and the fact that $(1, 0) \cong R \oplus R$ (refer to the beginning of Chapter 1A). QED.

In the situation of a free M with rank M odd, this theorem shows that under the assumption of the existence of special elements, the structure of

$C_0(M)$ completely determines that of $C(M)$. In the even rank case, this situation is reversed: it is $C(M)$ that determines $C_0(M)$. We will see this in Section D via an analysis of $C_0(M)$ under the action of a representation of $C(M)$. The next two sections consist of preparations for Section D.

B. Gradings Defined by Roots of $X^2 - aX - b$

The conclusion of the "even" part of (8.1) is a particular case of a more general phenomenon which is explored next. Let $X^2 - aX - b$ in $R[X]$ be a polynomial with $a^2 + 4b \in R^*$. Let S be any R-algebra.

(8.3). *If γ is a root of $X^2 - aX - b$ in S, then S has the grading $S = S_0 \oplus S_1$, where*

$$S_0 = \mathrm{Cen}_S(\gamma) = \{s \in S \mid \gamma s = s\gamma\} \ and \ S_1 = \{s \in S \mid s\gamma + \gamma s = as\}.$$

Proof. That S_0 and S_1 are R-submodules of S is clear. Since $(2\gamma - a)^2 = a^2 + 4b$, $2\gamma - a$ is a unit in S. As direct consequence, $S_0 \cap S_1 = 0$. We show next that $S = S_0 + S_1$. Let $s \in S$ be arbitrary. It suffices to show that $S_0 + S_1$ contains $(2\gamma - a)s$, for any element of S has this form. Observe that $(2\gamma - a)s = (s\gamma + \gamma s - as) + (\gamma s - s\gamma)$. Since

$$(s\gamma + \gamma s - as)\gamma = \gamma s\gamma + s(a\gamma + b) - as\gamma = \gamma s\gamma + sb$$

$$= (a\gamma + b)s + \gamma s\gamma - a\gamma s = \gamma(s\gamma + \gamma s - as),$$

$s\gamma + \gamma s - as \in S_0$. A similar computation shows that $(\gamma s - s\gamma)\gamma + \gamma(\gamma s - s\gamma)$ $= a(\gamma s - s\gamma)$, and hence that $\gamma s - s\gamma$ is in S_1. We have now verified that $S = S_0 \oplus S_1$. The multiplicative properties remain. That $S_0 S_0 \subseteq S_0$ is clear. If s is in S_0 and t in S_1, then $\gamma(st) + (st)\gamma = s\gamma t + st\gamma = s(\gamma t + t\gamma) = a(st)$. So $st \in S_1$. Similarly, $ts \in S_1$. If s and t are both in S_1, then $s\gamma t + st\gamma = ast$ $= \gamma st + s\gamma t$. So $st\gamma = \gamma st$, and st is in S_0.   QED.

The grading $S = S_0 \oplus S_1$ is the *grading defined by* γ. This grading is trivial precisely if γ is in the center of S.

Example 1. Let $S = C(M)$ and let z be a special element in $C(M)$ with polynomial $X^2 - aX - b$. So z is a root of $X^2 - aX - b$. If rank M is odd, then by (7.11), z is central, so the grading defined by z is trivial. If rank M is even, then by (8.1), the grading defined by z is $C_0(M) \oplus C_1(M)$.

Example 2. Let A be an algebra with standard involution σ and assume that there is an element c in A such that $c - c^\sigma \in A^*$. Refer to Theorem (6.2) and its proof. Observe that there exist a and b in R with $a^2 + 4b \in R^*$ such that c is a root of $X^2 - aX - b$ and the grading $A = A_0 \oplus A_1$ is the grading defined by c.

Example 3. Let $S = \text{Mat}_2(R)$ and observe that $\gamma = \begin{bmatrix} 0 & b \\ 1 & a \end{bmatrix}$ is a root of $X^2 - aX - b$. Let $S = S_0 \oplus S_1$ be the grading defined by γ. By easy computations $\{I, \gamma\}$, where I is the identity matrix, is a basis of the R-module $S_0 = \text{Cen}_S(\gamma)$. The assignment $r \to rI$ and $X \to \gamma$ defines an R-algebra isomorphism from $R[X]/(X^2 - aX - b)$ onto S_0. In view of (2.2), S_0 is a separable free quadratic algebra. Let $w_1 = \begin{bmatrix} 1 & a \\ 0 & -1 \end{bmatrix}$ and $w_2 = \begin{bmatrix} 0 & -b \\ 1 & 0 \end{bmatrix}$ and check that $\{w_1, w_2\}$ is a basis for S_1. Later, other bases of S_0 and S_1 will be useful. Set $u = a^2 + 4b \in R^*$. Since

$$2(a\gamma + 2bI) - a(2\gamma - aI) = uI \quad \text{and} \quad a(a\gamma + 2bI) + 2b(2\gamma - aI) = u\gamma,$$

$\{a\gamma + 2bI, 2\gamma - aI\}$ is a basis for S_0. Put $w_3 = w_1\gamma - \gamma w_1$ and $w_4 = w_2\gamma - \gamma w_2$. Check that $w_3 = \begin{bmatrix} a & a^2 + 2b \\ -2 & -a \end{bmatrix}$ and $w_4 = \begin{bmatrix} -2b & -ab \\ -a & 2b \end{bmatrix}$, note that both w_3 and w_4 are in S_1, and verify that $aw_3 - 2w_4 = uw_1$ and $-2bw_3 - aw_4 = uw_2$. So $\{w_3, w_4\}$ is a basis of S_1.

C. Linear Maps with Polynomial $X^2 - aX - b$

Let S be an R-algebra and P a right S-module. Let $\text{End}_S P$ be the R-algebra of S-homomorphisms of P. The elements of $\text{End}_S P$ act on P on the left.

Let $S = S_0 \oplus S_1$ be any grading of S and suppose that $P = P_0 \oplus P_1$ is a *grading* of P. This means that $P = P_0 \oplus P_1$ as R-modules and that $P_i S_j \subseteq P_{i+j}$ for i, j in $\mathbb{Z}_2$. In particular, both P_0 and P_1 are S_0-modules. For f in $\text{End}_S P$, define f_0 and f_1 by

$$f_0(p + q) = pr_0\, f(p) + pr_1\, f(q) \quad \text{and} \quad f_1(p + q) = pr_1\, f(p) + pr_0\, f(q),$$

where $p \in P_0$, $q \in P_1$, and $pr_i : P \to P_i$ is the projection map. It is easy to check that f_0 and f_1 are both in $\text{End}_S\, P$ and that $f = f_0 + f_1$. As a consequence,

$$\{f \in \text{End}_S\, P \mid fP_0 \subseteq P_0,\, fP_1 \subseteq P_1\} \ \oplus\ \{f \in \text{End}_S\, P \mid fP_0 \subseteq P_1,\, fP_1 \subseteq P_0\}$$

is a grading of the R-algebra $\text{End}_S\, P$. When supplied with this grading, we denote $\text{End}_S\, P$ by

$$\text{END}_S\, (P_0 \oplus P_1) \quad \text{or} \quad \text{END}_S\, P.$$

The two components are denoted $(\text{END}_S\, P)_0$ and $(\text{END}_S\, P)_1$ respectively. Note that if the grading on S is trivial, i.e., if $S = S_0$, then $(\text{END}_S\, P)_0$ is isomorphic to $(\text{End}_S\, P_0) \oplus (\text{End}_S\, P_1)$.

Example 4. We consider the preceding construction in a special case. Let Q be a right S_0-module and let P be the right S-module $P = Q \otimes_{S_0} S$. Put $P_0 = Q \otimes_{S_0} S_0$ and $P_1 = Q \otimes_{S_0} S_1$. Since $Q \otimes_{S_0} S \cong Q \otimes_{S_0} (S_0 \oplus S_1) \cong P_0 \oplus P_1$, $P_0 \oplus P_1$ is a grading of P. Observe that the image of the homomorphism of R-algebras

$$\theta : \text{End}_{S_0}\, Q \ \to\ \text{End}_S\, P \,,$$

given by $\theta g = g \otimes \text{id}_S$, is contained in the 0-component of $\text{END}_S\, (P_0 \oplus P_1)$. We will show that θ maps $\text{End}_{S_0}\, Q$ isomorphically onto this 0-component by constructing an inverse for θ. Let $f \in \text{End}_S\, (P)$ satisfy

$$f(Q \otimes_{S_0} S_0) \subseteq Q \otimes_{S_0} S_0 \quad \text{and} \quad f(Q \otimes_{S_0} S_1) \subseteq Q \otimes_{S_0} S_1.$$

Let g be the composite $Q \to Q \otimes_{S_0} S_0 \to Q \otimes_{S_0} S_0 \to Q$, where the first and last maps are given by the natural isomorphism $Q \to Q \otimes_{S_0} S_0$ and its inverse, and the map in the middle is the restriction of f. Let q in Q be arbitrary and put $f(q \otimes 1) = \sum_i (q_i \otimes s_i)$ with $q_i \in Q$ and $s_i \in S_0$. Since g acts on q as $q \to q \otimes 1 \to \sum_i (q_i \otimes s_i) \to \sum_i q_i s_i$, $(g \otimes \text{id}_S)(q \otimes 1) = g(q) \otimes 1$

$= \sum_i (q_i \otimes s_i)$. Therefore, $\theta g = f$ and it follows that $f \to g$ defines an inverse of θ. So θ is an isomorphism from $\text{End}_{S_0} Q$ onto the 0-component of $\text{END}_S (P_0 \oplus P_1)$ as asserted.

Now let $X^2 - aX - b$ in $R[X]$ be a polynomial with $a^2 + 4b \in R^*$. Return to an arbitrary S-module P and let $T \in \text{End}_S P$ satisfy $X^2 - aX - b$, i.e., assume that $T^2 - aT - b(\text{id}_P) = 0$.

(8.4). *Assume that* γ *is a root of* $X^2 - aX - b$ *in* S *and let* $S = S_0 \oplus S_1$ *be the grading defined by* γ. *Then there is a grading* $P = P_0 \oplus P_1$ *of* P *such that*

$$T = \gamma(\text{id}_{P_0}) \oplus (a - \gamma)(\text{id}_{P_1}) \in \text{End}_{S_0} P.$$

In addition, the grading on $\text{End}_S P$ *defined by* T *is* $\text{END}_S (P_0 \oplus P_1)$.

Proof. In $S_0[X]$, $X^2 - aX - b = (X - \gamma)(X - (a - \gamma))$. Clearly, $\text{End}_S P \subseteq \text{End}_{S_0} P$. In $\text{End}_{S_0} P$, $(T - \gamma)(T - (a - \gamma)) = 0$, where γ acts on P by multiplication on the right. Put $P_0 = (T - (a - \gamma))P$ and $P_1 = (T - \gamma)P$. These are S_0-submodules of P. Notice that $P_0 \subseteq \ker (T - \gamma)$ and $P_1 \subseteq \ker (T - (a - \gamma))$. Let $p \in P$. Clearly, $(T - \gamma)p + (T - (a - \gamma))p = (2T - a)p$. Since $(2T - a)^2 = (a^2 + 4b)\text{id}_p$, $2T - a$ is invertible in $\text{End}_S P$, and

$$p = (2T - a)^{-1}(T - \gamma)p + (2T - a)^{-1}(T - (a - \gamma))p.$$

Since $2T - a$ commutes with any polynomial in T with coefficients in S_0, $p \in P_0 + P_1$. If $p \in \ker (T - \gamma) \cap \ker (T - (a - \gamma))$, then $(2T - a)p = 0$ and $p = 0$. We have proved that $P = P_0 \oplus P_1$ as S_0-modules, $P_0 = \ker (T - \gamma)$, and $P_1 = \ker (T - (a - \gamma))$. Since $P_i S_j \subseteq P_{i+j}$ for all i and j in $\mathbb{Z}_2$ is easily checked, $P = P_0 \oplus P_1$ is a grading of P. Only the last statement remains to be verified. Since $T = \gamma\text{id}_{P_0} + (a - \gamma)\text{id}_{P_1}$ in $\text{End}_{S_0} P$,

$$(\text{End}_S P)_0 \subseteq \{f \in \text{End}_S P \mid fP_0 \subseteq P_0, fP_1 \subseteq P_1\} = (\text{END}_S P)_0 .$$

Now take U in $(\text{End}_S P)_1$. By (8.3), $UT + TU = aU$. Therefore in $\text{End}_{S_0} P$, $U(T - \gamma) = -(TU - aU + \gamma U) = -(T - (a - \gamma))U$. So $UP_1 \subseteq P_0$. Similarly, $UP_0 \subseteq P_1$. Therefore,

$$(\text{End}_S P)_1 \subseteq \{f \in \text{End}_S P \mid fP_0 \subseteq P_1, fP_0 \subseteq P_1\} = (\text{END}_S P)_0 .$$

It follows that the grading defined by T on $\text{End}_S P$ and the grading $\text{END}_S (P_0 \oplus P_1)$ are the same.   QED.

We continue to assume that T is an element in $\text{End}_S P$ which satisfies $T^2 - aT - b(\text{id}_P) = 0$. The rest of this section investigates T in the situation where $X^2 - aX - b$ does not necessarily have a root in S. We make the following assumptions: $b1_S$ is not a zero divisor in S, P is finitely generated projective, and there is an S-submodule Q of P which satisfies $P = Q \oplus TQ$. In the discussion that follows, $b1_S$ will also be denoted by b. Note that if $X^2 - aX - b$ does not have a root in S then the condition on b is satisfied if S is a domain; and the required Q exists if S is a division ring — see the proof of Corollary (8.9).

Let q be any element in Q. If $Tq = 0$, then $bq = T^2q - aTq = 0$. Since Q is finitely generated projective, embed Q in a free S-module of finite rank, and express q as a linear combination of a basis. Since $bq = 0$, and b is not a zero divisor in S, $q = 0$. Therefore, $T : Q \to TQ$ is an isomorphism.

Note that P and hence Q have natural $(R\text{-}S)$-bimodule structures. Consider the composite of $(R\text{-}S)$-isomorphisms

$$P = Q \oplus TQ \xrightarrow{\text{id} \oplus T^{-1}} Q \oplus Q \xrightarrow{\;\sim\;} (R \oplus R) \otimes_R Q .$$

Next consider the R-algebra isomorphisms

$$\text{End}_S P \xrightarrow{\;\sim\;} \text{End}_S((R \oplus R) \otimes_R Q) \xrightarrow{\;\sim\;}$$

$$\text{End}_R((R \oplus R) \otimes_R \text{End}_S Q \xrightarrow{\;\sim\;} \text{Mat}_2(R) \otimes_R \text{End}_S Q,$$

where the first is induced by the composite, the last is obvious, and the one in the middle is the inverse of the isomorphism

$$\text{End}_R(R \oplus R) \otimes_R \text{End}_S Q \xrightarrow{\;\sim\;} \text{End}_{(R \otimes_R S)}((R \oplus R) \otimes_R Q)$$

established in Exercise 5 of Section F. Denote the composite

$$\text{End}_S P \rightarrow \text{Mat}_2(R) \otimes_R \text{End}_S Q$$

by θ_T. This is the algebra isomorphism that we will analyze.

We show first that $\theta_T(T) = \begin{bmatrix} 0 & b \\ 1 & a \end{bmatrix} \otimes \text{id}_Q$. Let $(r, t) \in R \oplus R$ and $q \in Q$ and verify that the action of $\theta_T(T)$ on $(r, t) \otimes q$ is given by:

$$(r, t) \otimes q \rightarrow (rq, tq) \xrightarrow{\text{id} \oplus T} rq + tTq \xrightarrow{T}$$

$$tbq + (r + ta)Tq \xrightarrow{\text{id} \oplus T^{-1}} (tbq, (r+ta)q) \rightarrow (tb, r+ta) \otimes q .$$

The fact that $\theta_T(T) = \begin{bmatrix} 0 & b \\ 1 & a \end{bmatrix} \otimes \text{id}_Q$ is now clear.

Put $\gamma = \begin{bmatrix} 0 & b \\ 1 & a \end{bmatrix}$ and note that $\theta_T(T) = \gamma \otimes \text{id}_Q$ in $\text{Mat}_2(R) \otimes_R \text{End}_S Q$ is a root of $X^2 - aX - b$. We will study the grading on $\text{Mat}_2(R) \otimes_R \text{End}_S Q$ defined by $\theta_T(T)$. Our discussion will make use of Example 3 of Section B. Let U in $\text{Mat}_2(R) \otimes_R \text{End}_S Q$ be arbitrary. Since $\{I, \gamma, w_1, w_2\}$ is a basis of $\text{Mat}_2(R)$, U is uniquely of the form

$$U = I \otimes B_1 + \gamma \otimes B_2 + w_1 \otimes B_3 + w_2 \otimes B_4,$$

with B_i in $\text{End}_S Q$. Consider the elements

$$U_0 = I \otimes B_1 + \gamma \otimes B_2 \quad \text{and} \quad U_1 = w_1 \otimes B_3 + w_2 \otimes B_4.$$

Assume that U commutes with $\gamma \otimes \text{id}_Q$. Since U_0 commutes with $\gamma \otimes \text{id}_Q$, so does U_1. Expanding $U_1(\gamma \otimes \text{id}_Q) = (\gamma \otimes \text{id}_Q)U_1$, we find that $w_3 \otimes B_3 + w_4 \otimes B_4 = 0$. Since $\{w_3, w_4\}$ is part of a basis of $\text{Mat}_2(R)$, it follows that $B_3 = B_4 = 0$. We have proved that

$$U \in (R[X]/(X^2 - aX - b)) \otimes_R \text{End}_S Q.$$

Conversely, if U is in this set, then clearly U commutes with $\gamma \otimes \text{id}_Q$. Next assume that $U(\gamma \otimes \text{id}_Q) + (\gamma \otimes \text{id}_Q)U = aU$. Since w_1 and w_2 are in S_1, U_1 satisfies this equality and hence $U_0 = U - U_1$ does also. Since U_0 commutes with $\gamma \otimes \text{id}_Q$, $2U_0(\gamma \otimes \text{id}_Q) = aU_0$. An expansion leads to the equality $(2\gamma - aI) \otimes B_1 + (a\gamma + 2bI) \otimes B_2 = 0$. Since $\{2\gamma - aI, a\gamma + 2bI\}$ is part of a basis of $Mat_2(R)$, $B_1 = B_2 = 0$. So U is in $S_1 \otimes_R End_S Q$.

A moment's deliberation shows that we have proved the following.

(8.5). *Suppose* $b1_S$ *is not a zero divisor in* S, P *is finitely generated projective, and there is an* S-submodule Q *of* P *which satisfies* $P = Q \oplus TQ$. *Supply* $End_S P$ *and* $Mat_2(R)$ *with the gradings defined by the respective roots*

T *and* $\begin{bmatrix} 0 & b \\ 1 & a \end{bmatrix}$ *of* $X^2 - aX - b$. *Then there exists an isomorphism*

$$\theta_T : End_S P \rightarrow Mat_2(R) \hat{\otimes}_R <End_S Q>$$

of graded R-*algebras that satisfies* $\theta_T(T) = \begin{bmatrix} 0 & b \\ 1 & a \end{bmatrix} \otimes id_Q$.

D. Graded Properties of Representations

Let M be a quadratic module over R. Let S be an R-algebra and P a right S-module. A homomorphism of R-algebras

$$C(M) \rightarrow End_S P$$

is called a *representation of* $C(M)$, or an *S-representation*, when S is emphasized. Such a representation provides a left action of $C(M)$ on P. In this way, P has the structure of a $(C(M)-S)$-bimodule and is called a *Clifford module*. One can check conversely that a Clifford module defines a representation of $C(M)$.

Clifford modules – in the special cases where $R = \mathbb{R}$ and S is $\mathbb{R}, \mathbb{C}$, or Hamilton's quaternions – are of fundamental importance in both differential geometry and topology. Refer to Chapter 15 for a glimpse of these connections. In this section we will use Clifford modules to analyze the role that $C_0(M)$ plays inside $C(M)$ when rank M is even.

Assume that M is free and fix a basis $\mathcal{X} = \{x_1,..., x_n\}$. In this case, all Clifford modules arise as follows. Let $\{\varphi_1,..., \varphi_n\}$ be a subset of $End_S P$ with

$$\varphi_i^2 = q(x_i)id_P \text{ and } \varphi_i\varphi_j + \varphi_j\varphi_i = h(x_i, x_j)id_P$$

for all i and j. Such a set $\{\varphi_1,..., \varphi_n\}$ is called a *Clifford system for* $C(M)$ on P. Define $\alpha : M \rightarrow End_S P$ by $\alpha x_i = \varphi_i$ and linear extension to M. Recall the equality

$$q(x_1r_1 + \ldots + x_kr_k) = \sum_{1\le i\le n} r_i^2 q_i(x_i) + \sum_{1\le i<j\le n} r_ir_jh_i(x_i, x_j),$$

with $r_1,\ldots, r_n$ in R, from Chapter 4D. It implies that the pair $(End_S P, \alpha)$ is compatible with M. Therefore, there exists a representation

$$\Phi : C(M) \rightarrow End_S P$$

which satisfies $\Phi(x_i) = \varphi_i$, for all i. Conversely, it is clear that if $\Phi : C(M) \rightarrow End_S P$ is a representation, then $\{\Phi(x_i) \mid 1 \le i \le n\}$ is a Clifford system. It follows that there is a one-to-one correspondence between representations of $C(M)$ on P and Clifford systems for $C(M)$ on P.

Example 5. Let M be a hyperbolic plane with hyperbolic basis $\{x_1, x_2\}$. The homomorphism $C(M) \rightarrow Mat_2(R)$ of Exercise 1 of Chapter 5E is the matrix version of a representation. The corresponding Clifford system is $\varphi_1 = \begin{bmatrix} 0 & 1 \\ 0 & 0 \end{bmatrix}$ and $\varphi_2 = \begin{bmatrix} 0 & 0 \\ 1 & 0 \end{bmatrix}$.

For the remainder of this section we assume that M is nonsingular and finitely generated projective of even rank. We assume also that $C(M)$ has a special element z with polynomial $X^2 - aX - b$. So $a^2 + 4b \in R^*$ and z is a root of $X^2 - aX - b$.

The following theorem is a direct consequence of a combination of Propositions (8.1) and (8.4).

(8.6). *Theorem. Let S be an R-algebra, P a right S-module, and*

$$\Phi : C(M) \rightarrow End_S P$$

a representation of $C(M)$. Set $\Phi z = T$. Assume that $X^2 - aX - b$ has a root γ in S, and let $S = S_0 \oplus S_1$ be the grading defined by γ. Then

$$P_0 = \{p \in P \mid Tp = p\gamma\} \text{ and } P_1 = \{p \in P \mid Tp = p(a - \gamma)\}$$
determine a grading $P = P_0 \oplus P_1$ of P and

$$\Phi : C(M) \rightarrow END_S (P_0 \oplus P_1)$$

is a graded representation.

Assume that M contains an x such that $q(x) \in R^*$. By Exercise 16 of Chapter 4E, this is the case if R is a local ring. Note that $x^2 = q(x)1$ and $x \in C_1(M)$. So $\Phi(x)$ is an invertible element in

$$\{f \in \mathrm{End}_S P \mid fP_0 \subseteq P_1,\ fP_1 \subseteq P_0\}.$$

Hence $\Phi(x)P_0 = P_1$, and P_0 and P_1 are isomorphic S_0-modules. It follows that if $\gamma \in \mathrm{Cen}\ S$ so that $S = S_0$, then the restriction of Φ provides an algebra homomorphism $C_0(M) \to \mathrm{End}_S P_0 \oplus \mathrm{End}_S P_0$. Part (1) of the next corollary follows as a special case of this observation.

(8.7). *Corollary. Let* R *be a field and* S *a division algebra over* R. *Let* P *be a finite-dimensional (right) vector space over* S *and*

$$\Phi : C(M) \to \mathrm{End}_S P$$

an isomorphism of R-*algebras.*

(1) *Assume that* $X^2 - aX - b$ *has a root* γ *in* S *and that* $\mathrm{Cen}_S(\gamma) = S_0 = S$. *Then* $P \cong P_0 \oplus P_0$ *and the restriction of* Φ *induces an isomorphism of* R-*algebras*

$$C_0(M) \to \mathrm{End}_S P_0 \oplus \mathrm{End}_S P_0.$$

(2) *Assume that* $X^2 - aX - b$ *has a root* γ *in* S *and that* $\mathrm{Cen}_S(\gamma) = S_0 \neq S$. *Then* S_0 *is a division algebra over* R, $\dim_{S_0} S = 2$, $\dim_{S_0} P = \dim_S P$, *and the restriction of* Φ *induces an isomorphism of* R-*algebras*

$$C_0(M) \to \mathrm{End}_{S_0} P_0.$$

Proof. Only (2) remains to be proved. Suppose that Proposition (8.6) has been applied to Φ. That $S_0 = \mathrm{Cen}_S(\gamma)$ is a division ring is easy. The right S-space $P_0 \otimes_{S_0} S$ has the grading $(P_0 \otimes_{S_0} S_0) \oplus (P_0 \otimes_{S_0} S_1)$. Consider the R-homomorphism $P_0 \otimes_{S_0} S \to P$ given by $p_0 \otimes s \to p_0 s$, for all p_0 in P_0 and s in S. Compare the gradings of $P_0 \otimes_{S_0} S$ and $P = P_0 \oplus P_1$ and note that it is a graded homomorphism of right S-modules. We claim that it is surjective. That the image contains P_0 is clear. Now let $p_1 \in P_1$ be arbitrary. Since $S_0 \neq S$, S_1 contains a nonzero, and hence invertible, element u of S. By (8.3), u^{-1} is in S_1 and hence $p_1 u^{-1}$ is in P_0. Since $p_1 u^{-1} \otimes u \to p_1$,

P_1 is in the image and the surjectivity follows. Let $s_1 \in S_1$ be arbitrary. Since $s_1 u^{-1} \in S_0$, $s_1 \in S_0 u$, and it follows that $S_1 = S_0 u$. So $\dim_{S_0} S_1 = 1$ and therefore, $\dim_{S_0} S = 2$. By the Remark after Theorem (8.6), $\dim_{S_0} P_0 = \dim_{S_0} P_1$. Since $P = P_0 \oplus P_1$, it follows that $\dim_{S_0} (P_0 \otimes_{S_0} S) = 2\dim_{S_0} P_0 = \dim_{S_0} P$. So $\dim_{S_0} P_0$ is finite and $\dim_S (P_0 \otimes_{S_0} S) = \dim_S P$. Since $P_0 \otimes_{S_0} S \to P$ is surjective, the equality of the dimensions implies that it is an isomorphism. The restriction of Φ provides an isomorphism between $C_0(M)$ and the 0-component of $END_S (P_0 \oplus P_1)$. By the graded isomorphism developed earlier and Example 4 of Section C, $C_0(M) \cong End_{S_0} P_0$ as required.

$$\text{QED.}$$

The next theorem analyzes a representation $\Phi : C(M) \to End_S P$ in the case where $X^2 - aX - b$ does not necessarily have a root in S. It is a straightforward consequence of Propositions (8.1) and (8.5) and the proof of the latter.

(8.8). *Theorem. Let S be an R-algebra and let P be a finitely generated projective (right) S-module. Assume that $b1_S$ is not a zero divisor in S. Let*

$$\Phi : C(M) \to End_S P$$

be a representation of $C(M)$. Set $\Phi z = T$. Assume that there is an S-submodule Q of P which satisfies $P = Q \oplus TQ$. Then there exists an R-algebra isomorphism $\theta_T : End_S P \xrightarrow{\sim} Mat_2(R) \otimes_R End_S Q$, such that the composite

$$C(M) \to Mat_2(R) \overset{\wedge}{\otimes}_R <End_S Q>$$

of Φ and θ_T is a graded algebra isomorphism. In particular, by restriction

$$C_0(M) \to R[X]/(X^2 - aX - b) \otimes_R End_S Q.$$

Assume that R is a field, S a division algebra, and P a nonzero finite-dimensional vector space. Assume also that $X^2 - aX - b$ does not have a root in S. Then b is not a zero divisor of S and the required Q exists. This is easy to verify. If b is a zero divisor in S, then b is zero, and clearly $X^2 - aX - b$ has a root in S. Let $x \in P$ be nonzero. If $Tx = xs$ for some s in S, then $0 = (T^2 - aT - b)x = x(s^2 - as - b)$, but this is impossible. It follows that T moves all lines of the S-space P. Observe that $xS \oplus (Tx)S$ is a plane which is stabilized by T. If y lies outside this plane, then the planes $xS \oplus (Tx)S$ and $yS \oplus (Ty)S$ cannot intersect in a line, since such a line would be stabilized by T. These planes must therefore intersect trivially. Continuing in this way, we find that

$$P = (x_1 S \oplus (Tx_1)S) \oplus (x_2 S \oplus (Tx_2)S) \oplus \ldots \oplus (x_k S \oplus (Tx_k)S),$$

for some nonzero $x_1, x_2, \ldots, x_k$ in P. The existence of Q is now clear.

(8.9). *Corollary. Let* R *be a field,* S *a division algebra over* R, *and* P *a nonzero finite-dimensional (right) vector space over* S. *Assume that* $X^2 - aX - b$ *does not have a root in* S. *Suppose that*

$$\Phi : C(M) \to End_S P$$

is an isomorphism of R-*algebras. Set* $\Phi z = T$. *Then there exist an* S-*subspace* Q *of* P *and a graded isomorphism*

$$C(M) \to Mat_2(R) \, \hat{\otimes}_R \, <End_S Q>.$$

In particular,

$$C_0(M) \cong R[X]/(X^2 - aX - b) \otimes_R End_S Q.$$

E. Comparing the Tensor and Graded Tensor Products

Special elements also shed light on the connection between the ordinary tensor product and the graded tensor product of Clifford algebras. It turns out that they differ by a "twist" in the underlying quadratic form.

Let $u \in R^*$. The equation $^u q(x) = uq(x)$ defines a new quadratic form $^u q$ on M and M equipped with this *scaled* quadratic form is denoted $^u M$. If M is nonsingular, so is $^u M$.

(8.10). *If* $u \in (R^*)^2$, *then* $C(^u M) \cong C(M)$ *as graded algebras.*

Proof. Choose $r \in R^*$, such that $u = r^2$. Define $\delta : {}^u M \to M$ by $\delta x = xr$ for all x in M and check that it is an isometry. Now consider the algebra isomorphism $C(\delta) : C(^u M) \to C(M)$ given in Chapter 5B.   QED.

(8.11). *Suppose that* rank M *is even. Assume that* C(M) *has a special element* z *with polynomial* $X^2 - aX - b$, *and let* $u = a^2 + 4b$. *Then* $C(^{-u}M) \cong C(M)$ *as graded algebras.*

Proof. By (7.11), $A(M) \subseteq C_0(M)$. So $a - z = z^\alpha = z^{\beta\mu} = z^\mu$. Let $w = a - 2z$. So $w \in A(M)$ and $w^\mu = (a - 2z)^\mu = -w$. Note that $w^2 = a^2 - 4az + 4(az + b)$ $= u$. Consider the R-module map $\delta : {}^{-u}M \to C(M)$ given by $\delta(x) = wx$. Using (7.2), we find that $\delta(x)^2 = -w^2 x^2 = -uq(x) = {}^{-u}q(x)$ for all x in ${}^{-u}M$. So the pair $(C(M), \delta)$ is compatible with ${}^{-u}M$. Therefore, δ lifts to an algebra homomorphism $\varphi : C(^{-u}M) \to C(M)$. Since $\varphi(xy) = wxwy = -w^2 xy$ for all x and y in ${}^{-u}M$ and $w^2 \in R^*$, it follows that $C_0(M)$ is contained in the image of φ. Since $w \in C_0(M)$, we can choose $w' \in C(^{-u}M)$ such that $\varphi w' = w$. Now let x in M be arbitrary. Since $\varphi(w'x) = w^2 x = ux$, it follows that φ is surjective. By a combination of Propositions (5.6) and (4.5), φ is an isomorphism. It is clear that it is graded.   QED.

There are situations where the Clifford algebras of ${}^u M$ and M are not isomorphic. Chapter 11D provides examples.

(8.12). *Theorem. Suppose* $M = N \perp N'$ *and that* rank N *is either odd or even. Assume that* C(N) *has a special element* z *with polynomial* $X^2 - aX - b$. *Let* $u = a^2 + 4b$.

(1) *If* rank N *is odd, then*

$$C_0(M) \cong (C(N) \hat{\otimes}_R C(N'))_0 \cong C_0(N) \otimes_R C(^{-u}N').$$

(2) *If* rank N *is even, then*

$$C(M) \cong C(N) \hat{\otimes}_R C(N') \cong C(N) \otimes_R C(^u N').$$

Proof. In view of Theorem (5.5), it remains to establish the second isomorphism in each case. Consider the element $w = a - 2z \in A(N)$. As observed earlier, $w^2 = u = a^2 + 4b$.

1) Assume first that rank N is odd. In this case, $a = 0$ and $2 \in R^*$. Also, w is in $C_1(N)$. Let D be the subalgebra $(R1 \otimes C_0(N')) \oplus (Rw \otimes C_1(N'))$ of $(C(N) \hat{\otimes}_R C(N'))_0$. By (7.11), $w \in \text{Cen } C(N)$. Observe that the elements in D commute with those in $C_0(N) \otimes R1$. By (8.1), $C_1(N) = wC_0(N)$. Therefore D and $C_0(N) \otimes R1$ together generate the algebra

$$(C(N) \overset{\wedge}{\otimes}_R C(N'))_0 = (C_0(N) \otimes C_0(N')) \oplus (C_1(N) \otimes C_1(N')).$$

It follows that multiplication provides a surjective algebra homomorphism

$$C_0(N) \otimes_R D \to (C(N) \overset{\wedge}{\otimes}_R C(N'))_0 .$$

Propositions (5.6) and (4.5) and a comparison of ranks show that this is an isomorphism. Consider the map $\delta : {}^{-u}N' \to D$ defined by $\delta(y) = w \otimes y$ for all $y \in N'$. Since $\delta(y)^2 = (w \otimes y)^2 = -(w^2 \otimes y^2)) = -uq(y)(1 \otimes 1)$, (D, δ) is compatible with ${}^{-u}N'$. Therefore, δ lifts to an algebra homomorphism $\varphi : C({}^{-u}N') \to D$. Since $w^2 \in R^*$, the image of φ contains $R1 \otimes C_0(N')$ and $Rw \otimes C_1(N')$. So φ is onto. Another application of (5.6) and (4.5) shows that φ is an isomorphism.

2) Assume that rank N is even. By (7.11), z and w are both in $C_0(N)$. By (8.1),

$$wc = ac - 2zc = ac - 2(ac - cz) = -ac + 2cz = -cw,$$

for all c in $C_1(N)$. Let $A = C(N)$, $B = C(N')$, and let D be the subalgebra $R1 \otimes B_0 + Rw \otimes B_1$ of $A \overset{\wedge}{\otimes}_R B$. The elements in D commute with those in $A \otimes R1$. This follows from the fact that

$$(w \otimes b_1)(a_1 \otimes 1) = -wa_1 \otimes b_1 = a_1w \otimes b_1 = (a_1 \otimes 1)(w \otimes b_1),$$

for all $a_1 \in A_1$ and $b_1 \in B_1$. Also, D and $A \otimes R1$ generate $A \overset{\wedge}{\otimes}_R B$ as an R-algebra. To see this, it suffices to check that the algebra generated by D and $A \otimes R1$ contains $a_i \otimes b_j$ for any $a_i \in A_i$, any $b_j \in B_j$, and any $i, j \in \mathbb{Z}_2$. If $j = 0$, this is clear. In the two remaining cases, simply note that

$$a_0 \otimes b_1 = ua_0 \otimes u^{-1}b_1 = (a_0w \otimes 1)(w \otimes u^{-1}b_1) \quad \text{and}$$

$$a_1 \otimes b_1 = ua_1 \otimes u^{-1}b_1 = (a_1w \otimes 1)(w \otimes u^{-1}b_1) .$$

We now establish an algebra isomorphism $\eta : A \otimes_R D \to A \overset{\wedge}{\otimes}_R B$. Consider the map $A \times D \to A \otimes_R B$ given by $(a, d) \to (a \otimes 1)d$. This induces a homomorphism $\eta : A \otimes_R D \to A \overset{\wedge}{\otimes}_R B$ of R-modules. Since the elements of D and $A \otimes R1$ commute, η is an algebra homomorphism. Since D and

$A \otimes R1$ generate $A \overset{\wedge}{\otimes}_R B$ as an R-algebra, η is surjective. Another application of (5.6) and (4.5) shows that η is an isomorphism.

We show next that the algebras D and $C(^uN')$ are isomorphic. Define $\delta : {}^uN' \to D$ by $\delta(y) = w \otimes y$ for all y in $^uN'$. Since

$$\delta(y)^2 = (w \otimes y)^2 = w^2 \otimes y^2 = uy^2 = uq(y),$$

the pair (D, δ) is compatible with the quadratic module $^uN'$. So δ lifts to an algebra homomorphism $\varphi : C(^uN') \to D$. Since $(w \otimes y)(w \otimes y') = (u \otimes yy')$, it follows that $R1 \otimes B_0$ is in the image of φ, and since $(w \otimes y)(1 \otimes b_0) = w \otimes yb_0$ for all b_0 in B_0, $Rw \otimes B_1$ is in the image of φ. So φ is onto. By yet another application of (5.6) and (4.5), φ is an isomorphism.

Together, φ and η provide an algebra isomorphism

$$C(N) \otimes_R C(^uN') \to C(N) \overset{\wedge}{\otimes}_R C(N'),$$

which satisfies $a \otimes y \to (a \otimes 1)(w \otimes y)$ for all $a \in C(N)$ and $y \in {}^uN'$. A comparison of the gradings shows that it is graded.   QED.

F. Exercises

1. Let S be an R-algebra and let $S = S_0 \oplus S_1$ be a grading. Suppose that $u \in S^* \cap S_i$. Show that $u^{-1} \in S_i$. If $u \in S^* \cap S_1$, then $S_1 = uS_0 = S_0u$.

2. Let S be an R-algebra and let $S = S_0 \oplus S_1$ be a grading. Let n be a positive integer and consider the free S module S^n. Check that R-submodules $P_0 = (S_0)^n$ and $P_1 = (S_1)^n$ provide a grading $S^n = (S_0)^n \oplus (S_1)^n$ of S^n. Consider the isomorphism $End_S S^n \to Mat_n(S)$ given by the standard basis. What grading does $END_S (P_0 \oplus P_1)$ determine on $Mat_n(S)$? If $S = S_0 \oplus S_1$ is the grading defined by a root γ of $X^2 - aX - b \in R[X]$ with $a^2 + 4b \in R^*$, is the grading on $Mat_n(S)$ also defined by a root of $X^2 - aX - b$?

3. Let A be an R-algebra and let M, N, P, and Q be (right) A-modules. Refer to the definition of $END_S (P_0 \oplus P_1)$ and show that

$\operatorname{Hom}_A(M \oplus N, P \oplus Q) \cong$
$$\operatorname{Hom}_A(M, P) \oplus \operatorname{Hom}_A(M, Q) \oplus \operatorname{Hom}_A(N, P) \oplus \operatorname{Hom}_A(N, Q).$$

4. Let A and B be R-algebras, let M and M' be A-modules, and let N and N' be B-modules. Define a homomorphism of R-modules

$$\operatorname{Hom}_A(M, M') \otimes_R \operatorname{Hom}_B(N, N') \to \operatorname{Hom}_{A \otimes_R B}(M \otimes_R N, M' \otimes_R N')$$

via the map

$$\operatorname{Hom}_A(M, M') \times \operatorname{Hom}_B(N, N') \to \operatorname{Hom}_{A \otimes_R B}(M \otimes_R N, M' \otimes_R N')$$

defined by $(f, g) \to f \otimes g$. Using Exercise 3 show that this homomorphism is an isomorphism, first if M, M', N, and N' are all free of finite rank, and then if they are finitely generated projective.

5. Let A and B be R-algebras, let M be an A-module and N a B-module. Show that the map

$$\operatorname{End}_A(M) \otimes_R \operatorname{End}_B(N) \to \operatorname{End}_{A \otimes_R B}(M \otimes_R N)$$

defined in Exercise 4 is a homomorphism of R-algebras, and that it is an isomorphism if M and N are finitely generated projective.

6. Let M be a hyperbolic plane with hyperbolic basis $\{x_1, x_2\}$. Refer to Example 2 in Chapter 7C for the fact that $x_1 x_2$ is a special element of $C(M)$ with polynomial $X^2 - X$. What information does Theorem (8.6) provide about the isomorphism $C(M) \to \operatorname{Mat}_2(R)$ of Exercise 1 in Chapter 5E? See Section B and compare Example 1 with Example 3.

7. Refer to Exercises 1 and 2 of Chapter 5E. Show that the isomorphism $C(M) \to \operatorname{Mat}_2(C(N))$ in Exercise 2 is a consequence of Theorem (8.12) and Exercise 1.

Hints:

2. Consider the matrix $\begin{bmatrix} \gamma & 0 \\ 0 & a - \gamma \end{bmatrix}$.

4. Try $M = M' = A$ and $N = N' = B$ first.

9
Structure of Clifford and Arf Algebras

Overview

Let M be a finitely generated projective nonsingular quadratic module over a commutative ring. This chapter establishes the basic structure theory of the Clifford algebra $C(M)$ and its subalgebras $C_0(M)$, $A(M)$, Cen $C(M)$ and Cen $C_0(M)$. It will be proved that both $C(M)$ and $C_0(M)$ are separable. For a faithful M it will be shown that the Arf algebra $A(M)$ is separable quadratic, and that the following equivalences hold:

$$\text{Cen } C(M) = R \iff \text{rank } M \text{ is even} \iff \text{Cen } C_0(M) = A(M), \text{ and}$$

$$\text{Cen } C_0(M) = R \iff \text{rank } M \text{ is odd} \iff \text{Cen } C(M) = A(M).$$

The proofs rely heavily on the results of Chapters 7 and 8 and use localization arguments and facts from the theory of separable algebras. Throughout, R is an arbitrary commutative ring.

A. More on Separable Algebras

This section recalls some basic facts from the theory of separable algebras. This theory is well established in the literature, for example in DeMeyer-Ingraham, the lecture notes Knus-Ojanguren [1974], Orzech-Small, and Knus [1991]. We will sketch some of the easier proofs and leave the lengthier ones to the Exercises (and the references). Recall from the Notation and Terminology section that an R-algebra is said to be finitely generated, projective, or faithful, if it is so as R-module.

(9.1). *Let* A *and* B *be separable algebras over* R. *Then* $A \otimes_R B$ *is separable over* R *and* $\text{Cen} (A \otimes_R B) \cong (\text{Cen } A) \otimes_R (\text{Cen } B)$. *If* S *is a separable* R-*algebra, then* $\text{Mat}_n(S)$ *is a separable* R-*algebra and* $\text{Cen Mat}_n(S) \cong \text{Cen } S$.

Proof. Note first that $(A \otimes_R B)^\circ \cong A^\circ \otimes_R B^\circ$. Next, let e and f be separability idempotents for A and B, respectively, and show that $e \otimes f$ is one for $A \otimes_R B$. To show that $\text{Cen}(A \otimes_R B) \cong (\text{Cen } A) \otimes_R (\text{Cen } B)$, make use of the fact (see Chapter 2B) that $\text{Cen}(A \otimes_R B) = (e \otimes f)(A \otimes_R B)$. The last statement follows from Exercises 2 and 3 of Chapter 2E.   QED.

(9.2). *Let* A *be a separable algebra over* R *and let* S *be a commutative algebra over* R. *Then* $A \otimes_R S$ *is a separable algebra over* S *and the natural map* $(\text{Cen } A) \otimes_R S \to \text{Cen}(A \otimes_R S)$ *is an isomorphism.*

Proof. Let $e \in A \otimes_R A^\circ$ be a separability idempotent for A. Consider the algebra isomorphism $(A \otimes_R A^\circ) \otimes_R (S \otimes_R S) \to (A \otimes_R S) \otimes_R (A \otimes_R S)^\circ$ and show that the image f of $e \otimes (1 \otimes 1)$ is a separability idempotent for $A \otimes_R S$. To verify the isomorphism, show that $(\text{Cen } A) \otimes_R S = (eA \otimes_R S) \cong f(A \otimes_R S) = \text{Cen}(A \otimes_R S)$.   QED.

(9.3). *Let* S *be a commutative* R-*algebra and let* A *be an* S-*algebra. Then* A *is an* R-*algebra in a natural way. If* A *is a separable* R-*algebra, then* A *is a separable* S-*algebra. If* A *is a separable* S-*algebra and* S *is a separable* R-*algebra, then* A *is a separable* R-*algebra.*

Proof. The composite $R \to S \to \text{Cen } A$ provides A with the structure of an R-algebra. The map $A \times A^\circ \to A \otimes_S A^\circ$ given by $(a, b^\circ) \to a \otimes b^\circ$ induces an R-algebra map $A \otimes_R A^\circ \to A \otimes_S A^\circ$. Check that if $e = \sum_i c_i \otimes d_i^\circ$ in $A \otimes_R A^\circ$ is a separabilty idempotent for A as R-algebra, then the image of e in $A \otimes_S A^\circ$ is a separability idempotent for A as S-algebra. So A is a separable S-algebra. For the last statement refer to Lemma 4.4 on page 39 of Orzech-Small or to Proposition 1.12 on page 46 of DeMeyer-Ingraham.   QED.

(9.4). *If* A *and* B *are separable algebras, then* $A \oplus B$ *is a separable algebra.*

Proof. Let e and f be separability idempotents for A and B, respectively, and check that (e, f) is a separability idempotent for the $R \times R$ algebra $A \times B$. So $A \times B$ is a separable $(R \times R)$-algebra. Since the R-algebra $R \times R$ is isomorphic to the quadratic algebra $R[X]/(X^2 - X)$, it is separable by Proposition (2.2). Therefore by the last statement of (9.3), $A \times B$ is a separable R-algebra, i.e., the algebra $A \oplus B$ is separable.   QED.

An R-algebra A is called an *Azumaya algebra over* R if it is central and separable over R. By (9.3), any separable R-algebra A is an Azumaya algebra over Cen A. Refer to the literature for the many equivalent characterizations of Azumaya algebras.

(9.5). *Let* A *be an algebra over* R *which is faithful and finitely generated projective. Then* R *is a direct summand of* A. *If* A *is Azumaya over* R, *then* A *is faithful and finitely generated projective.*

Proof. See Corollary 1.4 of Chapter 1 and Proposition 2.10, page 16, both in Orzech-Small.

(9.6). *Theorem. Let* A *be an* R-*algebra which is finitely generated. Then the following are equivalent:*

(1) A *is separable over* R.

(2) $A \otimes_R R_{\mathfrak{m}}$ *is separable over the local ring* $R_{\mathfrak{m}}$ *for all maximal ideals* $\mathfrak{m}$ *of* R.

(3) $A \otimes_R R/\mathfrak{m}$ *is separable over the field* $R/\mathfrak{m}$ *for all maximal ideals* $\mathfrak{m}$ *of* R.

Proof. See Theorem 7.1 of DeMeyer-Ingraham, or Proposition 2.5 in Chapter III of the lecture notes Knus-Ojanguren [1974], or Lemma 5.1.10 in Knus [1991].

The preceding theorem focuses attention on finite-dimensional separable algebras over fields, and in particular on the famous theorem of Wedderburn.

(9.7). *Theorem. Let* F *be a field and* A *a finite-dimensional* F-*algebra. Then the following are equivalent:*

(1) A *is an Azumaya algebra over* F,

(2) A *is central simple over* F,

(3) A *is isomorphic to* $\text{Mat}_k(D)$ *for some* k *and some finite-dimensional central divison algebra* D *over* F.

Proof. Refer to the exercises for an outline of a proof.

B. The Separability of $C(M)$ and $C_0(M)$

We will make important use of the isomorphism

$$\psi : C(M \otimes_R A) \to C(M) \otimes_R A$$

and its restriction $\psi : C_0(M \otimes_R A) \to C_0(M) \otimes_R A$ of Chapter 5C, in two special situations: $A = R_{\mathfrak{p}}$ for a prime ideal $\mathfrak{p}$ of R and $A = R/\mathfrak{m}$ for $\mathfrak{m}$ a maximal ideal of R.

(9.8). *Theorem. Suppose that* R *is a commutative ring and that* M *is a nonsingular finitely generated projective quadratic module over* R.

(1) *If* rank M *is even, then* $C(M)$ *is an Azumaya algebra over* R.

(2) *If* rank M *is odd, then* $C_0(M)$ *is an Azumaya algebra over* R.

Proof. 1) Consider the case where rank M is even. It must be proved that $C(M)$ is separable and central. Assume first that R is a local ring. So M is free of rank $M = 2m$. If $m = 0$, the result follows by Example 0 of Chapter 5A, and if $m = 1$, by Propositions (6.5) and (6.6). So let $m \geq 2$. By an application of (4.16), $M = N_1 \perp ... \perp N_m$ where all the N_i free of rank 2 and nonsingular. Set $N = N_1$ and $N' = N_2 \perp ... \perp N_m$. By (7.10) and Theorem (8.12), $C(M) = C(N) \otimes_R C(\overset{u}{N'})$ for some u in R^*. By induction, both $C(N)$ and $C(\overset{u}{N'})$ are separable and central over R. So by (9.1), $C(M)$ is separable and central over R. Now let R be arbitrary and let $\mathfrak{p}$ be a prime ideal of R. The change-of-scalars isomorphism $\psi : C(M \otimes_R R_{\mathfrak{p}}) \to C(M) \otimes_R R_{\mathfrak{p}}$ and the local case already dealt with show that $C(M) \otimes_R R_{\mathfrak{p}}$ is separable and central over $R_{\mathfrak{p}}$. By Theorem (9.6), $C(M)$ is separable. Only the centrality of $C(M)$ remains. Consider the homomorphism $R \to \text{Cen } C(M)$. The localization of this map, the isomorphism $(\text{Cen } C(M)) \otimes_R R_{\mathfrak{p}} \to \text{Cen } (C(M) \otimes_R R_{\mathfrak{p}})$ given by (9.2), and the restriction $\text{Cen } (C(M) \otimes_R R_{\mathfrak{p}}) \cong \text{Cen } C(M \otimes_R R_{\mathfrak{p}})$ of ψ, together provide the composite

$$R_{\mathfrak{p}} \to (\text{Cen } C(M)) \otimes_R R_{\mathfrak{p}} \to \text{Cen}(C(M) \otimes_R R_{\mathfrak{p}}) \to \text{Cen } C(M \otimes_R R_{\mathfrak{p}}).$$

Check that this composite is the natural inclusion $R_{\mathfrak{p}} \to \text{Cen } C(M \otimes_R R_{\mathfrak{p}})$. Since $C(M \otimes_R R_{\mathfrak{p}})$ is central over $R_{\mathfrak{p}}$, it is an isomorphism. It follows that the localization of $R \to \text{Cen } C(M)$ is an isomorphism for any $\mathfrak{p}$. Therefore, $R \to \text{Cen } C(M)$ is an isomorphism and $C(M)$ is central over R.

2) We show that $C_0(M)$ is central and separable if rank M is odd. Using the isomorphism $\psi : C_0(M \otimes_R R_{\mathfrak{p}}) \to C(M)_0 \otimes_R R_{\mathfrak{p}}$ and the preceding argument, it suffices to consider the case where R is a local ring. So M is free of rank M = 2m + 1. By (4.16), $M = N_0 \perp N_1 \perp ... \perp N_m$ where N_0 is free of rank 1, N_i is free of rank 2 for $i \geq 1$, and all N_i are nonsingular. Put $N = N_0$ and $N' = N_1 \perp ... \perp N_m$. By (7.10) and Theorem (8.12), $C_0(M) \cong C_0(N) \otimes_R C(^uN')$ for some u in R^*. By Example 2 of Chapter 5B, $C_0(N) \cong R$, and therefore $C_0(M) \cong C(^uN')$. Since rank N' is even, $C_0(M)$ is central and separable by case 1).   QED.

Neither C(M) nor $C_0(M)$ is an Azumaya algebra over R in general. However, in view of the following theorem only the centrality requirement fails.

(9.9). *Theorem. Suppose that R is a commutative ring and M a nonsingular finitely generated projective quadratic module over R. Then both C(M) and $C_0(M)$ are separable R-algebras.*
Proof. Let $\mathfrak{m}$ be a maximal ideal of R. Let F be the field $R/\mathfrak{m}$ and let V be the finite-dimensional nonsingular quadratic module $M \otimes_R F$ over F. In view of Theorem (9.6) and the change-of-scalar isomorphism, it suffices to prove that C(V) and $C_0(V)$ are separable over F. By Proposition (7.10), C(V) has a special element. Let $X^2 - aX - b \in F[X]$ be its polynomial.
    1) Suppose dim V is even. By Theorem (9.8), C(V) is separable and central over F. The separability of $C_0(V)$ remains. By Theorem (9.7), there exists a central division algebra S over F, a finite-dimensional vector space P over S, and an F-algebra isomorphism $\Phi : C(V) \to End_S P$. The separabilty of $C_0(V)$ is now a consequence of Theorem (9.7) in combination with the results in Chapter 8D. If $X^2 - aX - b$ has a root γ in S and $Cen_S(\gamma) = S$, apply Corollary (8.7, 1) and (9.4). If $X^2 - aX - b$ has a root γ in S but $Cen_S(\gamma) \neq S$, apply Corollary (8.7, 2). Finally, if $X^2 - aX - b$ does not have a root in S, apply Corollary (8.9), (2.2), and (9.2).
    2) In the case dim V odd, $C_0(V)$ is central separable by Theorem (9.8) and the separability of C(V) follows from Theorem (8.2), (2.2), and (9.2).   QED.

C. The Even–Odd Splitting of C(M)

We will see in this section that the structure of the Clifford algebra C(M) of a finitely generated projective nonsingular quadratic module M over R is completely determined by its structure in two special cases, namely, the case where rank M is odd and the case where rank M is even. The reason is that in the general case, $C(M)$ "splits" as a product of an "odd" and an "even" part. This splitting, when applied to Cen $C(M)$ implies that $C(M)$ can be central only if its odd part is trivial, i.e., only if rank M is even.

Suppose that $R = R_0 \times R_1$ is a decomposition of R as Cartesian product of two rings R_0 and R_1. The typical element $r \in R$ has the form $r = (r_0, r_1)$ with $r_0 \in R_0$ and $r_1 \in R_1$. The projections $R \to R_0$ and $R \to R_1$ given by $r \to r_0$ and $r \to r_1$ are surjective ring homomorphisms. For a subset S of R, denote the images of S under the respective projections by S_0 and S_1. Notice that $S \subseteq S_0 \times S_1$. Let $\mathfrak{a}$ be an ideal of R. Then $\mathfrak{a}_0$ and $\mathfrak{a}_1$ are ideals of R_0 and R_1. Let $r = (r_0, r_1)$ in $\mathfrak{a}_0 \times \mathfrak{a}_1$ be arbitrary. So for some t, $(r_0, t) \in \mathfrak{a}$, and hence $(1, 0)(r_0, t) = (r_0, 0) \in \mathfrak{a}$. Similarly, $(0, r_1) \in \mathfrak{a}$, and therefore, $r \in \mathfrak{a}$. So if $\mathfrak{a}$ is an ideal of R, then $\mathfrak{a} = \mathfrak{a}_0 \times \mathfrak{a}_1$. Check that $\mathfrak{a} = \mathfrak{a}_0 \times \mathfrak{a}_1$ is prime if and only if either $\mathfrak{a}_0$ is a prime ideal of R_0 and $\mathfrak{a}_1 = R_1$, or $\mathfrak{a}_0 = R_0$ and $\mathfrak{a}_1$ is prime in R_1.

Let M be a finitely generated projective module over R. Consider the submodules $M_0 = (1, 0)M$ and $M_1 = (0, 1)M$ of M. Observe that $M_0 \cap M_1 = 0$. For if $(1, 0)y = x = (0, 1)z$, then $(0, 1)x = 0$ and $(1, 0)x = 0$, so that $x = (1, 1)x = 0$. For $x \in M$, put $x_0 = (1, 0)x$ and $x_1 = (0, 1)x$. Since $x = x_0 + x_1$, $M = M_0 + M_1$ and therefore, $M = M_0 \oplus M_1$. Note that M_0 and M_1 are also R_0 and R_1 modules, respectively, by appropriate restriction of the scalars, and that the Cartesian product $M_0 \times M_1$ is an R-module component-wise. Check that the assignment $x_0 + x_1 \to (x_0, x_1)$ is an R-module isomorphism from M onto $M_0 \times M_1$. Let $X = \{x^1, x^2, ..., x^n\}$ be a subset of M. Set $X_0 = \{x_0^1, x_0^2, ..., x_0^n\}$ and $X_1 = \{x_1^1, x_1^2, ..., x_1^n\}$. It is easy to see that X spans M over R if and only if X_0 and X_1 span M_0 and M_1 over R_0 and R_1, respectively, and that X is a basis of M if and only if X_0 and X_1 are bases of M_0 and M_1. Check that if $M = P \oplus Q$, then $M_0 = P_0 \oplus Q_0$ and $M_1 = P_1 \oplus Q_1$. Since M is finitely generated projective over R, it follows that M_0 and M_1 are finitely generated projective over R_0 and R_1, respectively.

Suppose that $\mathfrak{p}$ is a prime ideal of the form $\mathfrak{p} = \mathfrak{p}_0 \times R_1$, with $\mathfrak{p}_0$ a prime ideal of R_0. Observe that $R - \mathfrak{p} = (R_0 - \mathfrak{p}_0) \times R_1$. So the image of $R - \mathfrak{p}$ under $R \to R_0$ is $R_0 - \mathfrak{p}_0$. Verify that $R_\mathfrak{p} \to (R_0)_{\mathfrak{p}_0}$ given by $r/s \to r_0/s_0$ is a ring homomorphism. Denote it by pr_0. Since $r_0/s_0 \to (r_0, 0)/(s_0, 0)$ is an inverse, pr_0 is an isomorphism. We use it to compare $M \otimes_R R_\mathfrak{p}$ and

$M_0 \otimes_{R_0} (R_0)_{\mathfrak{p}_0}$. Consider the map $M \times R_{\mathfrak{p}} \to M_0 \otimes_{R_0} (R_0)_{\mathfrak{p}_0}$ given by $(x, t) \to x_0 \otimes \mathrm{pr}_0(t)$. It gives an additive map $M \otimes_R R_{\mathfrak{p}} \to M_0 \otimes_{R_0} (R_0)_{\mathfrak{p}_0}$. In the other direction, the assignment $(x_0, t_0) \to (x_0, 0) \otimes \mathrm{pr}_0^{-1}(t)$ induces an additive map $M_0 \otimes_{R_0} (R_0)_{\mathfrak{p}_0} \to M \otimes_R R_{\mathfrak{p}}$. Use pr_0 to identify $R_{\mathfrak{p}}$ and $(R_0)_{\mathfrak{p}_0}$, and observe that $M_0 \otimes_{R_0} (R_0)_{\mathfrak{p}_0}$ and $M \otimes_R R_{\mathfrak{p}}$ are isomorphic. If $\mathfrak{p}$ has the form $\mathfrak{p} = R_0 \times \mathfrak{p}_1$ with $\mathfrak{p}_1$ a prime ideal of R_1, then in the same way, $\mathrm{pr}_1 : R_{\mathfrak{p}} \to (R_1)_{\mathfrak{p}_1}$ given by $\mathrm{pr}_1(r/s) = r_1/s_1$ is a ring isomorphism, and after identifying $R_{\mathfrak{p}}$ and $(R_1)_{\mathfrak{p}_1}$ via pr_1, $M_1 \otimes_{R_1} (R_1)_{\mathfrak{p}_1}$ and $M \otimes_R R_{\mathfrak{p}}$ become isomorphic.

Suppose that q is a nonsingular quadratic form on M with associated bilinear form h. Let $x_0 \in M_0$ and $x_1 \in M_1$. Notice that $q(x_0) = q((1, 0)x_0)$ $= (1, 0)q(x_0) \in R_0$. Similarly, $q(x_1) \in R_1$, $h(x_0, x_0) \in R_0$, $h(x_1, x_1) \in R_1$ and $h(x_0, x_1) = 0$. It follows that for $x = (x_0, x_1)$, $q(x) = (q(x_0), q(x_1))$. Restricting q to M_0 and M_1 respectively provides quadratic forms $q_0 : M_0 \to R_0$ and $q_1 : M_1 \to R_1$. Since $M = M_0 \perp M_1$ as R-modules, it follows that both q_0 and q_1 are nonsingular. Therefore, (M_0, q_0) and (M_1, q_1) are finitely generated projective nonsingular quadratic modules over R_0 and R_1 respectively.

Let $C(M_0)$ and $C(M_1)$ be the Clifford algebras of (M_0, q_0) and (M_1, q_1) respectively. The Cartesian product $C(M_0) \times C(M_1)$ is an R-algebra component-wise. Define $M \to C(M_0) \times C(M_1)$ by $x \to (x_0, x_1)$. Since $(x_0, x_1)^2 = (x_0^2, x_1^2) = (q(x_0), q(x_1)) = q(x)(1, 1)$, this map provides a compatible pair for (M, q) and therefore a homomorphism $C(M) \to C(M_0) \times C(M_1)$ of R-algebras. Consider the R-module map $M \to C(M)$. Restricted to M_0 it is a map $M_0 \to C(M)$ of R_0-modules which satisfies $x_0 \to (x_0, 0)$. Since this defines a compatible pair for (M_0, q_0), there is a corresponding homomorphism $C(M_0) \to C(M)$ of R_0-algebras. Combining this with the same construction for (M_1, q_1) produces an inverse $C(M_0) \times C(M_1) \to C(M)$ of the R-algebra map above. We have therefore proved that

$$C(M) \cong C(M_0) \times C(M_1).$$

Check that this isomorphism sends $C_0(M)$ onto $C_0(M_0) \times C_0(M_1)$.

We now construct a specific decomposition $R = R_0 \times R_1$ of R. Recall from Chapter 4A that the map rank M : Spec R $\to \mathbb{Z}$ is continuous. Therefore the respective inverse images of the set of even integers (including 0) and the set of odd integers partition Spec(R) into a disjoint union of two open sets. By basic facts about Spec(R), in particular by Theorem 7.3, page 406, of Jacobson [1985, II], there exists a unique idempotent e in R such that

$$\text{rank } M_{\mathfrak{p}} \text{ is even} \Leftrightarrow \mathfrak{p} \supseteq Re \quad \text{and} \quad \text{rank } M_{\mathfrak{p}} \text{ is odd} \Leftrightarrow \mathfrak{p} \supseteq R(1-e) .$$

Put $d = 1 - e$, and observe that d is an idempotent that satisfies de = 0. The ideals Rd and Re of R are commutative rings with respective identities d and e. Let $r \in Rd \cap Re$. So re = 0, rd = 0, and hence r = 0. Since $R = Rd + Re$, $R = Rd \oplus Re$. Now let $r = r_0 d + r_1 e$ and $s = s_0 d + s_1 e$ be two elements of R. By easy computations $r + s = (r_0 + s_0)d + (r_1 + s_1)e$ and $rs = r_0 s_0 d + r_1 s_1 e$. Denote the rings Rd and Re by R_0 and R_1 respectively. The preceding equalities show that R is isomorphic to the Cartesian product $R_0 \times R_1$. We therefore identify R with $R_0 \times R_1$. In particular, d = (1, 0) and e = (0, 1).

Consider the decomposition $M_0 \times M_1$ of M determined by this splitting. Let $\mathfrak{p}_0$ be any prime of R_0 and put $\mathfrak{p} = \mathfrak{p}_0 \times R_1$. Since $\mathfrak{p} \supseteq R_1 = Re$, rank $M_{\mathfrak{p}} = \text{rank } (M \otimes_R R_{\mathfrak{p}})$ is even. So by an identification already established, rank $(M_0 \otimes_{R_0} (R_0)_{\mathfrak{p}_0})$ is even. Therefore, the R_0-module M_0 has even rank. Similarly, the R_1-module M_1 has odd rank. In reference to the R-module M, observe that rank M is even $\Leftrightarrow$ e is in all prime ideals $\mathfrak{p}$ of R $\Leftrightarrow$ e = 0 $\Leftrightarrow$ d = 1 $\Leftrightarrow$ $R_0 = R$ and $R_1 = \{0\}$. Similarly, rank M is odd $\Leftrightarrow$ e = 1 $\Leftrightarrow$ $R_1 = R$ and $R_0 = \{0\}$. A summary of the relevant results follows.

(9.10). *Therorem. Let* M *be a nonsingular finitely generated projective module over* R. *There is a splitting* $R = R_0 \times R_1$ *of* R *which induces*

(1) *a splitting* $M = M_0 \times M_1$ *of* M, *where* M_0 *and* M_1 *are non-singular finitely generated projective modules over* R_0 *and* R_1 *with the property that* rank M_0 *is even and* rank M_1 *is odd,*

(2) *an induced isomorphism of* R-*algebras* $C(M) \cong C(M_0) \times C(M_1)$ *which restricts to an isomorphism* $C_0(M) \cong C_0(M_0) \times C_0(M_1)$.

D. The Structures of Cen C(M), Cen $C_0(M)$, and A(M)

Continue to let M be a finitely generated projective nonsingular quadratic module over a commutative ring R. The main point of this section is the analysis − see Theorem (9.11) − of the structures of the Arf algebra A(M) = $\text{Cen}_{C(M)} C_0(M)$ and its subalgebras Cen C(M) and Cen $C_0(M)$.

The investigation of A(M), Cen C(M), and Cen $C_0(M)$ proceeds locally. The change-of-scalars isomorphism $\psi : C(M \otimes_R R_{\mathfrak{p}}) \to C(M) \otimes_R R_{\mathfrak{p}}$ will be used repeatedly.

(A) Assume first that rank $M_{\mathfrak{p}}$ = 0. By use of Example 0 of Chapter 5A,

$$R_{\mathfrak{p}} \; = \; C_0(M \otimes_R R_{\mathfrak{p}}) \; = \; C(M \otimes_R R_{\mathfrak{p}}) \; = \; A(M \otimes_R R_{\mathfrak{p}}).$$

Applying the graded isomorphism $\psi : C(M \otimes_R R_{\mathfrak{p}}) \to C(M) \otimes_R R_{\mathfrak{p}}$, we get

$$R_{\mathfrak{p}} \; = \; (\text{Cen } C_0(M)) \; \otimes_R R_{\mathfrak{p}} \; = \; (\text{Cen } C(M)) \; \otimes_R R_{\mathfrak{p}} \; = \; A(M) \; \otimes_R R_{\mathfrak{p}}.$$

(B) Suppose next that rank $M_{\mathfrak{p}}$ > 0 and even. By Theorems (9.8), (9.9), and (9.2), $R_{\mathfrak{p}}$ = (Cen C(M)) $\otimes_R R_{\mathfrak{p}}$. Consider the commutative diagram

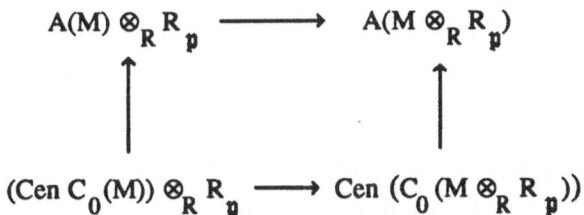

where the maps on the top and bottom are the injections obtained by restricting the isomorphism $\psi^{-1} : C(M) \otimes_R R_{\mathfrak{p}} \to C(M \otimes_R R_{\mathfrak{p}})$. The other two maps are the inclusion on the right and the localization of the inclusion on the left. By (7.10) and (7.11), the inclusion on the right is an equality. The map on the bottom is an isomorphism by (9.9) and (9.2). It follows that all maps of the diagram, and in particular, both of the maps

$$(\text{Cen } C_0(M)) \; \otimes_R R_{\mathfrak{p}} \; \to \; A(M) \otimes_R R_{\mathfrak{p}} \; \to \; A(M \otimes_R R_{\mathfrak{p}})$$

are isomorphisms. By (7.11), A(M $\otimes_R R_{\mathfrak{p}}$) is a separable free quadratic algebra over $R_{\mathfrak{p}}$. Therefore, (Cen $C_0(M)$) $\otimes_R R_{\mathfrak{p}}$ and A(M) $\otimes_R R_{\mathfrak{p}}$ are also.

(C) Suppose that rank $M_{\mathfrak{p}}$ is odd. In this case, by Theorems (9.8), (9.9), and (9.2), $R_{\mathfrak{p}} = (\text{Cen } C_0(M)) \otimes_R R_{\mathfrak{p}}$. Replacing $C_0(M)$ by $C(M)$ in the argument of (B), provides the isomorphisms

$$(\text{Cen } C(M)) \otimes_R R_{\mathfrak{p}} \to A(M) \otimes_R R_{\mathfrak{p}} \to A(M \otimes_R R_{\mathfrak{p}}).$$

By (7.10) and (7.11), $A(M \otimes_R R_{\mathfrak{p}})$ is a separable free quadratic algebra over $R_{\mathfrak{p}}$. So $(\text{Cen } C(M)) \otimes_R R_{\mathfrak{p}}$ and $A(M) \otimes_R R_{\mathfrak{p}}$ are also separable free quadratic over $R_{\mathfrak{p}}$.

What can we conclude ? Since $A(M) \otimes_R R_{\mathfrak{p}}$, $\text{Cen } C(M) \otimes_R R_{\mathfrak{p}}$ and $(\text{Cen } C_0(M)) \otimes_R R_{\mathfrak{p}}$ are all either $R_{\mathfrak{p}}$ or a separable free quadratic algebra over $R_{\mathfrak{p}}$ for any prime ideal $\mathfrak{p}$, it follows by Theorem (9.6) that

$$A(M), \text{ Cen } C(M), \text{ and } \text{Cen } C_0(M)$$

are all separable R-algebras.

Suppose that rank M is even. We already know from Theorem (9.8) that $\text{Cen } C(M) = R$. By (A) and (B), the localization

$$(\text{Cen } C_0(M)) \otimes_R R_{\mathfrak{p}} \to A(M) \otimes_R R_{\mathfrak{p}}$$

of the inclusion $\text{Cen } C_0(M) \to A(M)$ is an isomorphism for all $\mathfrak{p}$. Therefore,

$$\text{Cen } C_0(M) = A(M).$$

By Theorem (9.9) and (9.3), $C_0(M)$ is an Azumaya algebra over $\text{Cen } C_0(M)$. So by (9.5), $\text{Cen } C_0(M)$ is a direct summand of $C_0(M)$. By (5.6), $C_0(M)$ is finitely generated projective over R and it follows that $\text{Cen } C_0(M)$ is also finitely generated projective over R.

Suppose next that rank M is odd. By Theorem (9.8), $\text{Cen } C_0(M) = R$. By (C), the map

$$(\text{Cen } C(M)) \otimes_R R_{\mathfrak{p}} \to A(M) \otimes_R R_{\mathfrak{p}}$$

is an isomorphism for all $\mathfrak{p}$. So

$$\text{Cen } C(M) = A(M).$$

Arguing in the same way as in the even case, we get that $\text{Cen } C(M)$ is finitely generated projective over R.

Return to a general finitely generated projective nonsingular M. Combining the facts just collected with Theorem (9.10), we can conclude that

$$A(M),\ \ Cen\ C(M),\ \ and\ \ Cen\ C_0(M)$$

are all finitely generated projective R-modules. By (4.3), M is faithful if and only if case (A) does not arise. Therefore, M is faithful if and only if A(M) has rank 2.

The various properties satisfied by the algebra A(M) for a faithful M lead to a concept that will be important in several of the upcoming chapters: We call an R-algebra *separable quadratic* if it is separable and finitely generated projective of rank 2. If a separable quadratic algebra over R is free, then it is isomorphic to some $R[X]/(X^2 - aX - b)$ with $a^2 + 4b \in R^*$. This follows by Exercise 3 of Chapter 1E and Proposition (2.2).

We conclude this section by summarizing facts and consequences from the preceding discussion.

(9.11). *Theorem. Let* M *be a faithful finitely generated projective nonsingular quadratic module over* R. *Then*

(1) A(M) *is a separable quadratic algebra over* R.

(2) C(M) *is Azumaya over* R $\Leftrightarrow$ rank M *is even.*

(3) C_0(M) *is Azumaya over* R $\Leftrightarrow$ rank M *is odd.*

(4) Cen C(M) *is a separable quadratic* R-*algebra* $\Leftrightarrow$ Cen C(M) = A(M)
$\Leftrightarrow$ rank M *is odd.*

(5) Cen C_0(M) *is a separable quadratic* R-*algebra* $\Leftrightarrow$ Cen C_0(M) = A(M)
$\Leftrightarrow$ rank M *is even.*

E. Exercises

1. Let A and B be R-algebras. Suppose that A is separable over R, and that $\varphi : A \to B$ is a surjective algebra homomorphism. Then B is separable over R and φ(Cen A) = Cen B.

2. Suppose D is a division ring. Show that Mat_n(D) is a simple ring. Let I be the $n \times n$ identity matrix. Show that the center of Mat_n(D) is the set of matrices dI with $d \in$ Cen D.

3. Let M be a *simple* module (i.e., one which has no submodules other than $\{0\}$ and itself) over a ring S. Show that $\text{End}_S M$ is a division ring.

4. Let S be a simple ring and M a nonzero minimal right ideal of S. Then M is a simple S-module, $D = \text{End}_S M$ is a division ring, and M is a left D-module. For $s \in S$, define $\varphi(s) : M \to M$ by $\varphi(s)(x) = xs$ for all x in M.

   i)   Check that $\varphi(s) \in \text{End}_D M$ and show that $\varphi : S \to \text{End}_D M$ defines an injective ring homomorphism.

   ii) Let $y \in M$ and show that the map d_y defined by $d_y x = yx$ is in $\text{End}_S M = D$. So if $f \in \text{End}_D M$, $f(yx) = f(d_y x) = d_y f(x) = yf(x)$ for all x and y in M.

   iii) Show that $\varphi(S)$ is a right ideal of $\text{End}_D M$ and deduce that φ is an isomorphism.

5. Let F be a field and let A be a simple finite-dimensional F-algebra. Then $A \cong \text{Mat}_n(D)$ for some integer n and some finite-dimensional division algebra D over F.

The exercises that follow are more difficult. Refer to the Hints for references to the literature.

6. Let F be a field and let A be a separable F-algebra. Then $\dim_F A$ is finite and $A \cong \text{Mat}_{n_1}(D_1) \times \dots \times \text{Mat}_{n_k}(D_k)$, where D_i is a finite-dimensional division algebra over F for all $i, 1 \leq i \leq k$,

7. Let A be a finite-dimensional algebra over a field F. Then A is central simple if and only if A is Azumaya.

8. Let E/F be a field extension of finite degree. Show that if E/F is a separable field extension then E is a separable algebra over F. (The converse is also true. See the next exercise.)

9. Let F be a field and let A be an F-algebra. Then A is separable if and only if $A \cong \text{Mat}_{n_1}(D_1) \times \dots \times \text{Mat}_{n_k}(D_k)$, where $D_i, 1 \leq i \leq k$, is a finite-dimensional division algebra over F and $\text{Cen } D_i$ is a separable field extension of F.

10. Let A be an R-algebra with standard involution. Assume that A is finitely generated projective and that the form f defined by the involution is nonsingular. Show that if A is commutative, then there are splittings $R = R_1 \times R_2$ and $A = A_1 \times A_2$ such that the R_1-algebra A_1 is the trivial algebra and the R_2-algebra A_2 is separable quadratic. Is there an analogue of this if A is no longer assumed to be commutative?

11. Let A be an Azumaya algebra over R. Let C be a two-sided ideal of A and $\mathfrak{c}$ an ideal of R. Then $\mathfrak{c}A \cap R = \mathfrak{c}$ and $C = (C \cap R)A$. Further, $\mathfrak{c} \to \mathfrak{c}A$ is an isomorphism from the lattice of ideals of R onto the lattice of two-sided ideals of A, with inverse given by $C \to C \cap R$.

12. Let M be a finitely generated projective nonsingular quadratic module over a commutative R and consider the idempotent e constructed in Section C. Using the ideas of that section, show that the idempotents that arise in this way are precisely those that satisfy $2re = e$ for some r in R.

Hints:

1. Clearly, φ is also a homomorphism from A° onto B°. Denote it by φ°. Define $\varphi \times \varphi^\circ : A \times A^\circ \to B \otimes_R B^\circ$ by $(a, b^\circ) \to \varphi a \otimes \varphi^\circ b^\circ$. Check that the map $A^e = A \otimes_R A^\circ \to B \otimes_R B^\circ = B^e$ that it induces is a surjective algebra homomorphism. Let $e = \sum_i c_i \otimes d_i^\circ \in A^e$ be a separability idempotent for A, let e' be its image in B^e, and verify that e' is a separability idempotent for B. Since for any a in A, $ea = \sum_i c_i a d_i$, we find that $\varphi(\text{Cen } A) = \varphi(eA) = e'\varphi A = e'B = \text{Cen } B$.

2. Use the elements ε_{ij}. See Chapter 2E.

5. Use Exercise 4.

6. See Theorem 3.1 in Chapter III of the lectures Knus-Ojanguren [1974], or Theorem 2.5 in Chapter II of DeMeyer-Ingraham, in combination with Theorem 3.3 in Chapter IX of Hungerford. Refer also to Pierce.

8. Refer to Corollary 4.7 of Orzech-Small or to Proposition 3.4 in Chapter III of the lectures Knus-Ojanguren [1974].

9. Refer to the references given in the hint for Exercise 6.

10. Apply Theorem 7.7 on page 416 of Jacobson [1985, II] together with Theorem (6.8). There is an analogue involving quaternion algebras and three factors.

11. See Orzech-Small.

12. By applying (4.18) to the quadratic module M_1 over $R_1 = Re$, we see that e must satisfy this condition. To see that any e which satisfies this requirement arises, let $R_0 = R(1 - e)$ and $R_1 = Re$, take a free nonsingular quadratic module M_0 of rank 2 over R_0 and a free nonsingular quadratic module M_1 of rank 1 over R_1. Make $M = M_0 \times M_1$ into a nonsingular quadratic module over R.

10
The Existence of Special Elements

Overview

Let M be a nonsingular quadratic module over a commutative ring R. The primary goal of this chapter is the proof of the following fact: If M is free of finite rank, then $C(M)$ has a special element, or, equivalently, the Arf algebra $A(M)$ of M is isomorphic to $R[X](X^2 - aX - b)$ for some a and b in R with $a^2 + 4b \in R^*$. This has the important consequence that the results developed in Chapters 7 and 8 for $C(M)$ hold for any free M. In addition, we will establish that the discriminant $a^2 + 4b$ of $X^2 - aX - b$ is the classical discriminant of the quadratic module M (except for a factor 2 if rank M is odd). It was established in Chapter 9D that the Arf algebra $A(M)$ is a separable quadratic algebra. The structural features of this algebra will play a decisive role in the proofs. Throughout, R is an arbitrary commutative ring.

A. Separable Quadratic Algebras

Let S be a separable quadratic algebra over R. Recall that this means that S is separable and finitely generated projective of rank 2. If S is free, then, as already pointed out in Chapter 9D,

$$S \cong R[X](X^2 - aX - b)$$

for some a and b in R with $a^2 + 4b \in R^*$. It is the aim of this section and the next to develop for a general S the various properties observed in Chapters 2C and D for the free case.

Return to an arbitrary S. Since S is faithful by (4.3), the defining homomorphism $R \to S$ is injective. As on previous occasions, we therefore identify R with $R1$, and consider $R \subseteq S$. By (9.5), R is an R-direct summand in S. Let $\mathfrak{p}$ be any prime ideal of R, and consider the $R_{\mathfrak{p}}$-algebra $S_{\mathfrak{p}} = S \otimes_R R_{\mathfrak{p}}$. By Theorem (9.6), $S_{\mathfrak{p}}$ is a separable quadratic algebra over $R_{\mathfrak{p}}$.

Since $S_{\mathfrak{p}}$ is free over $R_{\mathfrak{p}}$, $S_{\mathfrak{p}}$ is commutative by the isomorphism above. Now fix t in S and consider the map $\gamma : S \to S$ defined by $s \to st - ts$. The commutativity of $S_{\mathfrak{p}}$ implies that $\gamma_{\mathfrak{p}}$ is the zero map for every $\mathfrak{p}$. Therefore, γ is the zero map, and it follows that S is commutative. We summarize the facts observed so far.

(10.1). *Let S be a separable quadratic algebra over R. Then S is commutative and R is a direct summand of S. If S is free, then* $S \cong R[X](X^2 - aX - b)$ *with* $a^2 + 4b$ *in* R^*.

Any involution of S fixes all the elements of R by definition. It is the goal of the remainder of this section to construct an involution σ on S with the property that

$$R = \{s \in S \mid s^{\sigma} = s\}.$$

The construction of σ proceeds by localization to the free case, where (see Chapter 2D) it already exists.

Since S is commutative, the enveloping algebra $S \otimes_R S^o$ of S is $S \otimes_R S$ and the map $\phi = \phi_S$, given by $\phi(s \otimes s') = ss'$, is a homomorphism of R-algebras. Consider the exact sequence

$$0 \to \ker \phi \to S \otimes_R S \xrightarrow{\phi} S \to 0.$$

Fix a prime ideal $\mathfrak{p}$ of R. On the one hand, there is the localization

$$0 \to (\ker \phi)_{\mathfrak{p}} \to (S \otimes_R S)_{\mathfrak{p}} \xrightarrow{\phi_{\mathfrak{p}}} S_{\mathfrak{p}} \to 0$$

of the sequence, and on the other, there is its analogue

$$0 \to \ker \phi' \to S_{\mathfrak{p}} \otimes_{R_{\mathfrak{p}}} S_{\mathfrak{p}} \xrightarrow{\phi'} S_{\mathfrak{p}} \to 0$$

for the $R_{\mathfrak{p}}$-algebra $S_{\mathfrak{p}}$. We will check that these two sequences are in essence the same.

For s and s' in S, denote the elements $s \otimes 1$ in $S_{\mathfrak{p}}$ and $(s \otimes s') \otimes 1$ in $(S \otimes_R S)_{\mathfrak{p}}$ respectively by $s_{\mathfrak{p}}$ and $(s \otimes s')_{\mathfrak{p}}$. Consider

$$S \times R_{\mathfrak{p}} \to (S \otimes_R S) \otimes_R R_{\mathfrak{p}}$$

in two ways: first by $(s, r') \to (s \otimes 1) \otimes r'$ and then by $(s, r') \to (1 \otimes s) \otimes r'$. This gives rise to two maps $S_{\mathfrak{p}} \to (S \otimes_R S) \otimes_R R_{\mathfrak{p}}$. "Multiplying" them together provides an $R_{\mathfrak{p}}$-algebra map $S_{\mathfrak{p}} \otimes_{R_{\mathfrak{p}}} S_{\mathfrak{p}} \to (S \otimes_R S)_{\mathfrak{p}}$, which sends $s_{\mathfrak{p}} \otimes s'_{\mathfrak{p}}$ to $(s \otimes 1)_{\mathfrak{p}} (1 \otimes s')_{\mathfrak{p}} = (s \otimes s')_{\mathfrak{p}}$ for any s and s' in S. We show that it is an isomorphism. Check that the map $S \times S \to S_{\mathfrak{p}} \otimes_{R_{\mathfrak{p}}} S_{\mathfrak{p}}$ provided by $S \to S_{\mathfrak{p}}$ induces an additive map

$$\gamma : S \otimes_R S \to S_{\mathfrak{p}} \otimes_{R_{\mathfrak{p}}} S_{\mathfrak{p}}.$$

Defining $(S \otimes_R S) \times R_{\mathfrak{p}} \to S_{\mathfrak{p}} \otimes_{R_{\mathfrak{p}}} S_{\mathfrak{p}}$ by $(z, r') \to (\gamma z) r'$ for all z in $S \otimes_R S$ and r' in $R_{\mathfrak{p}}$ gives an $R_{\mathfrak{p}}$-module map $(S \otimes_R S)_{\mathfrak{p}} \to S_{\mathfrak{p}} \otimes_{R_{\mathfrak{p}}} S_{\mathfrak{p}}$, which sends $(s \otimes s')_{\mathfrak{p}}$ into $s_{\mathfrak{p}} \otimes s'_{\mathfrak{p}}$ for any s and s' in S. Since this is the inverse of the earlier map,

$$S_{\mathfrak{p}} \otimes_{R_{\mathfrak{p}}} S_{\mathfrak{p}} \to (S \otimes_R S)_{\mathfrak{p}}$$

is an algebra isomorphism as asserted. It links the two exact sequences in the commutative diagram

$$
\begin{array}{ccccccccc}
0 & \to & \ker \phi' & \to & S_{\mathfrak{p}} \otimes_{R_{\mathfrak{p}}} S_{\mathfrak{p}} & \overset{\phi'}{\longrightarrow} & S_{\mathfrak{p}} & \to & 0 \\
 & & \downarrow & & \downarrow & & \downarrow {\scriptstyle \mathrm{id}} & & \\
0 & \to & (\ker \phi)_{\mathfrak{p}} & \to & (S \otimes_R S)_{\mathfrak{p}} & \overset{\phi_{\mathfrak{p}}}{\longrightarrow} & S_{\mathfrak{p}} & \to & 0
\end{array}
$$

The map on the left is, of course, the restriction of that in the center. Observe that it is an isomorphism.

(10.2). *Let S be a separable quadratic algebra. Then S has a unique separability idempotent e.*

Proof. Since S is separable, it has a separability idempotent. So the issue is the uniqueness. Suppose that e and e' are two such idempotents in $S \otimes_R S$. Let $\mathfrak{p}$ be any prime ideal of R. Since $S_{\mathfrak{p}}$ is generated as an $R_{\mathfrak{p}}$-module by the image of S under $S \to S_{\mathfrak{p}}$, it follows that the images γe and $\gamma e'$ under $\gamma : S \otimes_R S \to S_{\mathfrak{p}} \otimes_{R_{\mathfrak{p}}} S_{\mathfrak{p}}$ are separability idempotents for $S_{\mathfrak{p}}$. In view of (2.3), $\gamma e = \gamma e'$. By applying the isomorphism $S_{\mathfrak{p}} \otimes_{R_{\mathfrak{p}}} S_{\mathfrak{p}} \to (S \otimes_R S)_{\mathfrak{p}}$ to

$\gamma e = \gamma e'$, we find that the images of e and e' under $S \otimes_R S \rightarrow (S \otimes_R S)_{\mathfrak{p}}$ are the same. Since $\mathfrak{p}$ is arbitrary, $e = e'$ by (4.1).   QED.

Since S is separable, $0 \rightarrow \ker \phi \rightarrow S \otimes_R S \xrightarrow{\phi} S \rightarrow 0$ is split by the map

$$\theta : S \rightarrow S \otimes_R S$$

given by $\theta s = (1 \otimes s)e = (s \otimes 1)e$. Put $T = \ker \phi$ and note that $S \otimes_R S = \theta S \oplus T$. Define the map $\varphi : S \rightarrow S \otimes_R S$ by $s \rightarrow (s \otimes 1)(1 - e)$. Since $\phi((s \otimes 1)(1 - e)) = s(1 - 1) = 0$, φ maps into T. We will now prove that

$$\varphi : S \rightarrow T$$

is an isomorphism of R-modules. That φ is linear is clear. To prove that it is an isomorphism it suffices to prove that the localization $\varphi_{\mathfrak{p}} : S_{\mathfrak{p}} \rightarrow T_{\mathfrak{p}}$ is an isomorphism for any prime ideal $\mathfrak{p}$ of R.

As already observed in the proof of (10.2), the image γe of e under $\gamma : S \otimes_R S \rightarrow S_{\mathfrak{p}} \otimes_{R_{\mathfrak{p}}} S_{\mathfrak{p}}$ is the unique separability idempotent for $S_{\mathfrak{p}}$. By (2.4), the map $S_{\mathfrak{p}} \rightarrow \ker \phi'$ given by $s' \rightarrow (s' \otimes 1)(1 - \gamma e)$ for any s' in $S_{\mathfrak{p}}$ is an isomorphism of $R_{\mathfrak{p}}$-modules. Observe that γe goes to $e_{\mathfrak{p}} = e \otimes 1$ under $S_{\mathfrak{p}} \otimes_{R_{\mathfrak{p}}} S_{\mathfrak{p}} \rightarrow (S \otimes_R S)_{\mathfrak{p}}$. Consider the composite $S_{\mathfrak{p}} \rightarrow \ker \phi' \rightarrow T_{\mathfrak{p}}$ of the isomorphism $S_{\mathfrak{p}} \rightarrow \ker \phi'$ and the isomorphism $\ker \phi' \rightarrow (\ker \phi)_{\mathfrak{p}} = T_{\mathfrak{p}}$ given by the commutative diagram. Let $s \in S$. Under this composite,

$$s_{\mathfrak{p}} \rightarrow (s_{\mathfrak{p}} \otimes 1_{\mathfrak{p}})(1 - \gamma e) \rightarrow (s \otimes 1)_{\mathfrak{p}}((1 \otimes 1)_{\mathfrak{p}} - e_{\mathfrak{p}}),$$

and under $\varphi_{\mathfrak{p}} : S_{\mathfrak{p}} \rightarrow T_{\mathfrak{p}}$,

$$s_{\mathfrak{p}} \rightarrow (\varphi s)_{\mathfrak{p}} = ((s \otimes 1)(1 - e))_{\mathfrak{p}} = (s \otimes 1)_{\mathfrak{p}}((1 \otimes 1)_{\mathfrak{p}} - e_{\mathfrak{p}}).$$

It follows that the composite is equal to $\varphi_{\mathfrak{p}}$ and therefore that $\varphi_{\mathfrak{p}}$ is an isomorphism. So $\varphi : S \rightarrow T$ is an isomorphism as asserted.

Note that $1 - e$ is in T and that $(1 - e)(1 - e) = 1 - e$. Since any element of T has the form $(s \otimes 1)(1 - e)$ for s in S, it follows that T is closed under the product of $S \otimes_R S$ and that $1 - e$ is an identity element for T. So T has the structure of an R-algebra. Since it is easy to check that $\varphi : S \rightarrow T$ is an algebra map, we now find that

(10.3). *The map* $\varphi : S \to T$ *is an isomorphism of* R-*algebras*.

We are now in a position to construct the involution σ. As in Chapter 2D, the "switch" $sw : S \otimes_R S \to S \otimes_R S$, given by $sw(s \otimes s') = (s' \otimes s)$ for s and s' in S, is an involution on $S \otimes_R S$ which restricts to an involution on the algebra T. Since the separability idempotent e of S is unique, note that $sw(e) = e$ and hence that $sw(1 - e) = 1 - e$. Now define σ to be the unique involution on S which makes the diagram

(*)

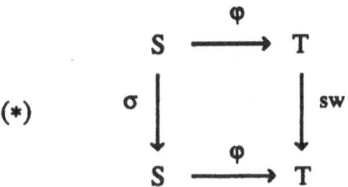

commute. The σ is involution is called the *conjugation* of the separable quadratic algebra S.

Let $\mathfrak{p}$ be a prime ideal of R. By (10.1), we may assume that $S_{\mathfrak{p}} = R_{\mathfrak{p}}[X]/(X^2 - aX - b)$ with a and b in $R_{\mathfrak{p}}$ and $a^2 + 4b$ in $(R_{\mathfrak{p}})^*$. Comparing the localized commutative diagram

$(*)_{\mathfrak{p}}$

$$
\begin{array}{ccc}
S_{\mathfrak{p}} & \xrightarrow{\varphi_{\mathfrak{p}}} & T_{\mathfrak{p}} \\
\sigma_{\mathfrak{p}} \downarrow & & \downarrow sw_{\mathfrak{p}} \\
S_{\mathfrak{p}} & \xrightarrow{\varphi_{\mathfrak{p}}} & T_{\mathfrak{p}}
\end{array}
$$

with the commutative diagram of (2.4) shows that $\sigma_{\mathfrak{p}}$ is the unique involution of $S_{\mathfrak{p}}$ with fixed point set $R_{\mathfrak{p}}$, i.e., that $\sigma_{\mathfrak{p}}$ is the conjugation of $S_{\mathfrak{p}}$.

(10.4). *The conjugation* σ *of* S *is the unique involution on* S *which has fixed point set* R.

Proof. Let $S \to S$ be the R-module map defined by $s \to s - s^{\sigma}$ and consider the induced map of R-modules $\alpha : S/R \to S$. We must show that α is injective. Fix a prime ideal $\mathfrak{p}$ of R. Since the fixed point set of $\sigma_{\mathfrak{p}}$ is $R_{\mathfrak{p}}$, the $R_{\mathfrak{p}}$-module map $S_{\mathfrak{p}} \to S_{\mathfrak{p}}$ given by $s' \to s' - s'^{(\sigma_{\mathfrak{p}})}$ induces an injection

$S_\mathfrak{p}/R_\mathfrak{p} \to S_\mathfrak{p}$. Note that there is an isomorphism $(S/R)_\mathfrak{p} \to S_\mathfrak{p}/R_\mathfrak{p}$ of $R_\mathfrak{p}$-modules which has the property that $(s + R)_\mathfrak{p} \to s_\mathfrak{p} + R_\mathfrak{p}$ for s in S. Check that its inverse is induced by the localization of the quotient map $S \to S/R$. Since the composite $(S/R)_\mathfrak{p} \to S_\mathfrak{p}/R_\mathfrak{p} \to S_\mathfrak{p}$ is the localization $\alpha_\mathfrak{p} : (S/R)_\mathfrak{p} \to S_\mathfrak{p}$, it follows that $\alpha_\mathfrak{p}$ is injective. Since $\mathfrak{p}$ is arbitrary, α is injective. We have shown that σ has fixed point set R. That σ is unique with this property follows from the uniqueness of $\sigma_\mathfrak{p}$ in all $(*)_\mathfrak{p}$.   QED.

(10.5). *Let S be a commutative algebra which is finitely generated projective and has an involution σ. Then S is a separable quadratic algebra over R and σ its conjugation if and only if for all prime ideals $\mathfrak{p}$ of R, $S_\mathfrak{p}$ is a separable quadratic algebra over $R_\mathfrak{p}$ and $\sigma_\mathfrak{p}$ its conjugation.*

Proof. Assume that for all prime ideals $\mathfrak{p}$ of R, $S_\mathfrak{p}$ is separable quadratic over $R_\mathfrak{p}$ and $\sigma_\mathfrak{p}$ is its conjugation. Observe that diagram $(*)_\mathfrak{p}$ commutes for all $\mathfrak{p}$. By Theorem (9.6), S is separable quadratic over R. Let τ be its conjugation. Since diagram $(*)$ commutes with τ in place of σ, the same is true for $(*)_\mathfrak{p}$. This means that $\sigma_\mathfrak{p} = \tau_\mathfrak{p}$ for all $\mathfrak{p}$, and it follows that $\sigma = \tau$. Similar considerations prove the argument in the other direction.   QED.

B. The Discriminant Module of S

Continue to let S be a separable quadratic algebra over R and let σ be its conjugation. For $s \in S$, define the *trace* of s by $tr(s) = s + s^\sigma$. Since $(s + s^\sigma)^\sigma = (s + s^\sigma)$, $tr(s)$ is in R. Clearly, $tr : S \to R$ is a homomorphism of R-modules. Define the *discriminant* dis S of S by

$$\text{dis } S = \{s \in S \mid tr(s) = 0\}.$$

Example. Let $S = R[X](X^2 - aX - b)$ with $a^2 + 4b \in R^*$. Put $v = X + (X^2 - aX - b)$. Let $s = r + tv$ in S be arbitrary. Recall from Chapter 2D that $(r + tv)^\sigma = (r + ta) - tv$. So $tr(s) = (r + tv) + (r + ta) - tv = 2r + ta$. It follows that dis $S = \{r + tv \in S \mid 2r + ta = 0\}$. In particular, $a - 2v \in$ dis S. Let $d = r + tv \in$ dis S be arbitrary. Since $2r = -ta$, $a^2 d = a^2 r - 2arv = ar(a - 2v)$ and $2d = -ta + 2tv = -t(a - 2v)$ are both in $R(a - 2v)$. Therefore, $(a^2 + 4b)d = a^2 d + 2b(2d) \in R(a - 2v)$ and hence $d \in R(a - 2v)$. As a consequence, dis $S = R(a - 2v)$. If $r(a - 2v) = 0$, then $ra = 0$ and $2r = 0$. So $r(a^2 + 4b) = 0$ and hence $r = 0$. We have verified that

$$\text{dis } S = R(a - 2v)$$

is a free submodule of S with basis $a - 2v$. Note that $(a - 2v)^2 = a^2 - 4av + 4(av + b) = a^2 + 4b$. The product of S provides a form

$$p : \text{dis } S \times \text{dis } S \to R$$

on dis S by $p(c, d) = cd$. It is clear that this is a symmetric bilinear form on dis S. It is nonsingular, since $a^2 + 4b$ is in R^*. So (dis S, p) is a discriminant module. Put $u = a^2 + 4b$, and check that (dis S, p) is isometric to the discriminant module (R, f_u) via the map which sends $a - 2v$ to 1.

(10.6). *The trace* $tr : S \to R$ *is surjective.*

Proof. By a typical localization argument it suffices to assume that R is a local ring. So S is free, and by (10.1), we can take $S = R[X]/(X^2 - aX - b)$ with $a^2 + 4b \in R^*$. By the preceding example, $tr(r + tv) = 2r + ta$. With $r = 2b$ and $t = a$, we see that $a^2 + 4b$ is in the image of tr. Since tr is R-linear, it is onto.   QED.

(10.7). *Let* t *in* S *satisfy* $tr(t) = 1$. *Then* Rt *is free and* $S = Rt \oplus \text{dis } S$. *In particular,* dis S *is finitely generated projective of rank* 1.

Proof. Consider the exact sequence

$$0 \to \text{dis } S \to S \xrightarrow{\ tr\ } R \to 0$$

and note that the map $R \to S$ given by $r \to rt$ splits tr.   QED.

(10.8). S *is free (of rank 2) over* R *if and only if* dis S *is free (of rank 1) over* R.

Proof. One direction is clear. In the other, use (5.7). QED.

(10.9). *The assignment* $p(c, d) = cd$ *defines a nonsingular symmetric bilinear form* $p : \text{dis } S \times \text{dis } S \to R$. *So the pair* (dis S, p) *is a discriminant module over* R.

Proof. By (10.7), dis S is finitely generated projective of rank 1. Let c and d be in dis S. So $(cd)^\sigma = cd$, and therefore, $cd \in R$. So p maps into R. Since p is clearly symmetric and bilinear, only the nonsingularity remains. If S is free, then p is nonsingular by (10.1) and the example above. A typical localization argument using (4.7) shows that p is nonsingular in the general case as well.   QED.

C. Criteria for the Existence of Special Elements

Let M be a faithful finitely generated projective and nonsingular quadratic module over R. Let $A = A(M)$ be the Arf algebra of M. By Theorem (9.11), A is separable quadratic over R. We begin by showing that the conjugation of A as a separable quadratic algebra, i.e. as constructed in Section A, is equal to the involution $\alpha = \beta\mu = \mu\beta$ introduced in Chapter 7A (and already called conjugation there).

(10.10). *The conjugation of* A *is equal to* $\alpha = \beta\mu$. *If* rank M *is odd, then* μ *is the identity and* $\alpha = \beta$. *If* rank M *is even, then* β *is the identity and* $\alpha = \mu$.

Proof. Suppose first that rank M is odd. By Theorem (9.11), $A = $ Cen $C(M)$. So by (7.2), μ is the identity. By (9.11) again, $A \cap C_0(M) = $ Cen $C_0(M) = R$. By (4.18), $C_0(M)$ is the fixed point set of the involution β of $C(M)$. Therefore, R is the fixed point set of β restricted to A. So by (10.4), $\alpha = \beta$ is the conjugation of A. Suppose that rank M is even. By (9.11), $A \subseteq C_0(M)$ and Cen $C(M) = R$. So β is the identity, and by (7.2) R is the fixed point set of μ. Again by (10.4), $\alpha = \mu$ is the conjugation of A. The proof of the first statement remains. Let σ be the conjugation of A as a quadratic algebra. By the results of the first part of the proof and (10.4), the localizations of α and σ are equal at all primes $\mathfrak{p}$ of R. So σ and α are equal by (10.5).   QED.

Recall from Chapter 7A that $A_0(M) = A \cap C_0(M)$ and $A_1(M) = A \cap C_1(M)$ defines a grading on A. If rank M is even, then by Theorem (9.11), $A = $ Cen $C_0(M) \subseteq C_0(M)$, and this grading is trivial. Suppose rank M is odd. By (4.18), $2 \in R^*$. Taking $t = \frac{1}{2}$ in (10.7) shows that $A_0(M) = R$. Since $A_1(M) = A \cap C_1(M) = \{a \in A \mid a^\beta = -a = 0\}$, an application of (10.10) shows that the grading is

$$A = R \oplus \text{dis } A .$$

(10.11). *Suppose that* rank M *is either odd or even. The following statements are equivalent:*

(1) $C(M)$ *has a special element.*

(2) $A \cong R[X]/(X^2 - aX - b)$ *with* a *and* b *in* R *and* $a^2 + 4b$ *in* R^*.

(3) A *is free (of rank 2).*

(4) dis A *is free (of rank 1).*

Proof. That (1) implies (2) is given by (7.7) and that (2) implies (3) is obvious. By (10.8), (3) implies (4). It remains to prove that (4) implies (1). Let α be the conjugation of A. By (10.7) and (10.9), there is a basis $\{t, s\}$ of A, where t is any element satisfying $tr(t) = t + t^\alpha = 1$ and s satisfies $tr(s) = s + s^\alpha = 0$ and $s^2 = u \in R^*$. Note that $s \in A^*$.

Suppose rank M is odd. By (4.18), we can take $t = \frac{1}{2}$. So $\{1, s\}$ is a basis of A. By (10.10), $s^\alpha = s^\beta = -s$. Therefore, $s \in C_1(M)$ meets all the conditions required of a special element.

Supppose that rank M is even. Since $t + t^\alpha = 1, t - t^2 = tt^\alpha \in R$. Set $c = t - t^2 = tt^\alpha$ and $b = u^{-1}c$. Choose r and a in R such that $1 = rt + as$. Since $2 = tr(rt + as) = r, 1 = 2t + as$. Observe that

$$a^2 + 4b = s^{-2}(1 - 2t)^2 + 4u^{-1}c = s^{-2}((1 - 4t + 4t^2) + 4(t - t^2)) = s^{-2} = u^{-1}.$$

Since the matrix $\begin{bmatrix} 2 & a \\ a & -2b \end{bmatrix}$ is invertible, $\{1 = 2t + as, at - 2bs\}$ is a basis of A. We claim that $z = at - 2bs$ is a special element of C(M). By Theorem (9.11), $A = \text{Cen } C_0(M)$. So $z \in C_0(M)$. Since $s \in \text{dis } A, z + z^\alpha = tr(z) = a$. It remains to verify that $z^2 = az + b$. Note that

$$ts = st = u(a^2 + 4b)st = ua(as)t + 4cst = ua(1 - 2t)t + 2cs(1 - as)$$

$$= uat - 2ua(t^2 + c) + 2cs = -uat + 2cs.$$

To check that $z^2 = az + b$, make use of the various equations, express both sides in the form $rt + r's$ with r and r' in R, and compare coefficients.   QED.

Remarks: 1) If char R = 2, then C(M) has a special element. This follows from the equivalence (1) $\Leftrightarrow$ (4) and the fact that dis A = R in this case. By (4.18), rank M is necessarily even in this case.

2) There is another statement equivalent to (1) – (4). It involves the involution β of C(M). See Exercise 11.

D. Special Elements and the Discriminant

Throughout this section, R is an arbitrary commutative ring and M is a non-singular quadratic module over R which is free of finite rank n. The quadratic and associated symmetric bilinear form of M are denoted by q and h respectively and A = A(M) is the Arf algebra of M. This section has two interrelated aims: They are the proof of the fact that C(M) has a special element and that the isometry class [dis A, p] ∈ Dis(R) of the discriminant module (dis A, p) is closely related to the classical (signed) discriminant of M.

Let $\mathfrak{X}$ be a basis of M. Observe that $\det_{\mathfrak{X}} h$ is in R^* and define the *(signed) discriminant* of M to be the element

$$\text{dis } M = ((-1)^{n(n-1)/2} \det_{\mathfrak{X}} h)(R^*)^2 \in R^*/(R^*)^2.$$

Refer to Milnor-Husemoller, Lam, or Scharlau [1985], for example. Notice that the image of dis M under the injection $R^*/(R^*)^2 \to \text{Dis}(R)$ of (4.12) is the isometry class [R, u] of the discriminant module (R, f_u), where $u = (-1)^{n(n-1)/2} (\det_{\mathfrak{X}} h)$.

(10.12). *Theorem. Let M be a nonzero free nonsingular quadratic module of rank n. Then C(M) has a special element. Let z be any special element, let* $X^2 - aX - b$ *be the polynomial of z, and let* $\mathfrak{X}$ *be any basis of M. Then*

$$[\text{dis } A(M), p] = [R, a^2 + 4b] = [R, \xi(-1)^{n(n-1)/2} (\det_{\mathfrak{X}} h)],$$

where $\xi \in R$ *is equal to* 1 *or* 2 *according to whether* n *is even or odd.*

Proof. (1) Assume first that R is a local ring. That C(M) has a special element z is given by (7.10). Let $X^2 - aX - b$ be the polynomial of z. The equality [dis A(M), p] = [R, a² + 4b] follows by (7.7) and the Example of Chapter 10B. It remains to verify that

$$[R, a^2 + 4b] = [R, \xi(-1)^{n(n-1)/2} (\det_{\mathfrak{X}} h)] .$$

Replacing z by another special element has, by (7.6), the effect of multiplying $a^2 + 4b$ by the square of a unit in R, and changing the basis $\mathfrak{X}$ has, by remarks in Chapter 4B, the same effect on $\det_{\mathfrak{X}} h$. So if the equality is proved for some special element z and some basis $\mathfrak{X}$, then it is proved for all z and $\mathfrak{X}$. Consider a splitting of M given by (4.16). By reordering and combining components if necessary, we may assume that $M = M_1 \perp ... \perp M_m \perp L$, where all M_i are free nonsingular submodules of M of rank 2, and L is 0 or free of rank 1 according to whether rank M is even or odd. For each i, let $\mathcal{X}_i = \{x_{2i-1}, x_{2i}\}$ be a basis of M_i. Put $q(x_{2i-1}) = r_i$, $q(x_{2i}) = s_i$ and $h(x_{2i-1}, x_{2i}) = a_i$. Let $b_i = -r_i s_i$. Denote the restriction of h to $M_i \times M_i$ by h_i and observe that $\det_{\mathcal{X}_i} h_i = -4b_i - a_i^2$.

(1a) Assume first that rank $M = n$ is even. So $n = 2m$. Note that $\mathcal{X} = \{x_1,..., x_{2m}\}$ is a basis of M. By Example 2 of Chapter 7C, each $C(M_i)$ has a special element with polynomial $X^2 - a_i X - b_i$. Therefore, by (7.9) and induction, $C(M)$ has a special element with polynomial $X^2 - aX - b$, where a and b satisfy

$$a^2 + 4b = \prod_i (a_i^2 + 4b_i) = (-1)^m \prod_i (-4b_i - a_i^2) = (-1)^{n(n-1)/2} \det_{\mathcal{X}} h .$$

So $[R, a^2 + 4b] = [R, (-1)^{n(n-1)/2} (\det_{\mathcal{X}} h)]$.

(1b) Assume that rank $M = n$ is odd. So $n = 2m + 1$. With $N = M_1 \perp ... \perp M_m$, $M = N \perp L$. By (1a), N has a special element with polynomial $X^2 - a'X - b'$ where a' and b' satisfy $a'^2 + 4b' = (-1)^m \prod_i (-4b_i - a_i^2)$. Let $\{y\}$ be a basis of L. So $\mathcal{X} = \{x_1,..., x_{2m}, y\}$ is a basis of M. By Example 1 of Chapter 7C, $C(L)$ has a special element with polynomial $X^2 - q(y)$. Applying (7.9) to the splitting $M = N \perp L$ shows that $C(M)$ has a special element with polynomial $X^2 - aX - b$, where a and b satisfy $a = 0$ and

$$4b = 4b''(a'^2 + 4b') = 2h(y, y)(-1)^m \prod_i (-4b_i - a_i^2)$$
$$= 2(-1)^{(n-1)(n-2)/2} h(y, y) \prod_i (-4b_i - a_i^2) = 2(-1)^{n(n-1)/2} \det_{\mathcal{X}} h .$$

Therefore, $[R, a^2 + 4b] = [R, 4b] = [R, 2(-1)^{n(n-1)/2} (\det_{\mathcal{X}} h)]$. The proof in the case of a local ring is now complete.

(2) We continue with two considerations required in the proof of the general case. Here R is arbitrary, $\mathcal{X} = \{x_1,..., x_n\}$ is any basis of M, and $B = (h(x_i, x_j))$ is the matrix of h in $\mathcal{X}$.

(2a) The first concern is the change-of-scalars construction. Let S be a commutative algebra over R with defining homomorphism $f : R \to S$. Consider the quadratic module $M \otimes_R S$ over S. The matrix of its underlying bilinear form h' in $\mathcal{X}' = \{x_1 \otimes 1,..., x_n \otimes 1\}$ is $(h'(x_i \otimes 1, x_j \otimes 1)) = (fB_{ij})$.

Let

$$\phi : C(M) \otimes_R S \to C(M \otimes_R S)$$

be the inverse of the isomorphism $\phi : C(M \otimes_R S) \to C(M) \otimes_R S$ of graded R-algebras given by Theorem (5.5). By Theorem (9.9), $C(M)$ and $C_0(M)$ are separable over R and by Theorem (9.11), $A = \text{Cen } C_0(M)$ if n is even and $A = \text{Cen } C(M)$ if n is odd. Applying (9.2) to the algebra $C_0(M)$ in the even case and to $C(M)$ in the odd case, we get $\phi(A \otimes_R S) = A(M \otimes_R S)$. The analogue for $C(M \otimes_R S)$ of the involution β of $C(M)$ is $\phi(\beta \otimes \text{id}_S)\phi^{-1}$ and the conjugation of $A(M \otimes_R S)$ is $\phi(\alpha \otimes \text{id}_S)\phi^{-1}$. Suppose z is a special element of $C(M)$ with polynomial $X^2 - aX - b$. Then $\phi(z \otimes 1)$ is a special element of $C(M \otimes_R S)$ with polynomial $X^2 - f(a)X - f(b)$. By (7.7) and the example of Chapter 10B,

$$[R, a^2 + 4b] = [\text{dis } A, p] \quad \text{and} \quad [S, (fa)^2 + 4f(b)] = [\text{dis } A(M \otimes_R S), p].$$

It is therefore not difficult to see that if $[\text{dis } A, p] = [R, \xi(-1)^{n(n-1)/2} (\det_{\mathcal{X}} h)]$, then $[\text{dis } A(M \otimes_R S), p] = [S, \xi(-1)^{n(n-1)/2} (\det_{\mathcal{X}'} h')]$.

(2b) Suppose next that M' is another free nonsingular quadratic module over R and that there is an isometry $M \to M'$. Let h' be the symmetric bilinear form of M' and let $\mathcal{X}'$ be the image of the basis $\mathcal{X}$. Observe that $\det_{\mathcal{X}} h = \det_{\mathcal{X}'} h'$. Consider the isomorphism $\phi : C(M) \to C(M')$ of graded R-algebras which $M \to M'$ induces. Note that $\phi A = A(M')$. Suppose that z is a special element of $C(M)$ with polynomial $X^2 - aX - b$. Then by (7.7), ϕz is a special element of $C(M')$ also with polynomial $X^2 - aX - b$.

(3) Let R be arbitrary and let $\mathcal{X} = \{x_1, ..., x_n\}$ be any basis of M. To prove the theorem in the general case, it suffices to show that

(*) $[\text{dis } A, p] = [R, \xi(-1)^{n(n-1)/2} (\det_{\mathcal{X}} h)]$

holds. For if this is established, then dis A is free, so that by (10.11), $C(M)$ has a special element z. In addition, if $X^2 - aX - b$ is the polynomial of z, then by (7.7) and the Example of Chapter 10B, $[\text{dis } A, p] = [R, a^2 + 4b]$.

(3a) Assume that R is a domain. Let $u = \xi(-1)^{n(n-1)/2} (\det_{\mathcal{X}} h)$. Consider the discriminant modules $(\text{dis } A, p)$ and $(R, 1_u)$. Applying (2a) and (1), we see that their localizations are isomorphic for every prime ideal of R. So by (4.13), (*) is satisfied and we are done.

(3b) Let R be arbitrary. Let $S = \mathbb{Z}[X_{11}, X_{12}, \ldots, X_{n-1,n}, X_{n,n}]$ be the polynomial ring over $\mathbb{Z}$ in the variables X_{ij}, $1 \le i \le j \le n$. Let B' be the $n \times n$ matrix

$$\begin{bmatrix} 2X_{11} & X_{12} & \cdots & X_{1n} \\ X_{12} & 2X_{22} & \cdots & X_{2n} \\ & & \cdot & \\ & & \cdot & \\ & & \cdot & \\ X_{1n} & X_{2n} & \cdots & 2X_{nn} \end{bmatrix}.$$

Consider the domain $T = S[(\det B')^{-1}]$ inside the quotient field of S. Let N be a free module over T with basis $\mathcal{Y} = \{y_1, \ldots, y_n\}$ and define a quadratic form q' on N by $q'(\sum_i y_i t_i) = \sum_i t_i^2 X_{ii} + \sum_{i<j} t_i t_j X_{ij}$. The associated symmetric bilinear form h' is given by

$$h'(\sum_i y_i t_i, \sum_j y_j u_j) = \sum_i 2(t_i u_i) X_{ii} + \sum_{i<j} (t_i u_j + t_j u_i) X_{ij}.$$

Since B' is the matrix of h' in the basis $\mathcal{Y}$, we have made N into a non-singular quadratic module over T. Define a ring homomorphism $\rho : S \to R$ by $\rho(m) = m1$ for m in $\mathbb{Z}$, $\rho(X_{ii}) = q(x_i)$, and $\rho(X_{ij}) = h(x_i, x_j)$, for $i < j$. Note that ρ applied to the matrix B' gives the matrix $B = (h(x_i, x_j))$ of M in the basis $\mathcal{X}$. So $\rho(\det B') = \det B$. Therefore, ρ induces a homomorphism $\rho : T \to R$ such that $\rho((\det B')^{-1}) = (\det B)^{-1}$. Since

$$q(\sum_i x_i r_i) = \sum_i r_i^2 q(x_i) + \sum_{i<j} r_i r_j h(x_i, x_j),$$

the assignment $y_i \otimes 1 \to x_i$ defines an isometry from the quadratic module $N \otimes_T R$ onto M. Since T is a domain, the required equality (*) holds for the quadratic module N. By (2a), it therefore holds for the quadratic module $N \otimes_T R$ over R. Consequently by (2b), it holds for M.   QED.

Remarks. 1) Let M be a free and nonsingular quadratic module of rank n. In view of Theorem (10.12), the *Arf invariant* of M can now be defined in exactly the same way as was done in the local case in Chapter 7C. Namely, let z be a special element with polynomial $X^2 - aX - b$, let ε be 0 or 1 according to whether n is even or odd, and define Arf $M \in QU_f(R)$ to be the graded isomorphism class of $(R[X]/(X^2 - aX - b))^\varepsilon$. The multiplicative property of Arf holds as in the local case. If n is even, then there is a direct connection between Arf M and dis M. For in this case, Arf $M \in Qu_f(R)$ and by Theorem (10.12), $\delta(\text{Arf } M) = \text{dis } M$, where δ is the discriminant map of Chapter 3B. These interconnections are developed further in Chapter 13C.

2) For a different approach to the existence of special elements, see Haag or the description of the work of Haag in §4.8 of Knus [1991]. Given any basis $\mathcal{X}$ of the underlying quadratic module, Haag provides a precise expression of such elements in terms of $\mathcal{X}$. This approach is more general, since it does not require the nonsingularity of M. Of course, for a singular M, the resulting elements do not satisfy all the properties of the special elements developed here.

E. Exercises

1. Let S be a separable quadratic algebra with conjugation σ. Show that σ is a standard involution.

2. Let S be a separable quadratic R-algebra, let σ be its conjugation and $nr : S \to R$, where $nr(s) = ss^\sigma$, the norm map. Show that nr is a quadratic form on the module S with associated symmetric bilinear form the h given by $h(s, s') = s(s')^\sigma + s's^\sigma$ for all s and s' in S. Show that h is nonsingular.

3. Let A be an R-algebra with standard involution σ and refer to (6.1) and (6.2). Do the conclusions of Theorem (6.2) hold if the condition *there exists an element* $c \in A$ *with* $c - c^\sigma \in A^*$ is replaced by the weaker condition A *contains a separable quadratic R-algebra* A_0?

4. In the case where $S = R[X](X^2 - aX - b)$ with $a^2 + 4b \in R^*$, find an explicit t that satisfies the requirements of (10.7).

For the rest of these Exercises, M is a faithful finitely generated projective nonsingular quadratic module over R whose rank is either even or odd, and β is the involution of $C(M)$ defined Chapter 5B (as well as its restriction to $A(M)$).

5. Show that dis $A = \{c \in C(M) \mid c^\beta x = -xc \text{ for all } x \text{ in } M\}$. If rank M is even, show that dis $A = \{c \in C(M) \mid cd = d^\beta c \text{ for all } d \text{ in } C(M)\}$.

Recall that an element of $C(M)$ is homogeneous if it is either in $C_0(M)$ or $C_1(M)$ and that the degree ∂c of a homogeneous element c is equal to 0 in the first case and 1 in the second. Let $\gamma : C(M) \to C(M)$ be a homomorphism of R-modules. Then γ is a *graded inner automorphism of* $C(M)$ if there is a homogeneous invertible element u in $C(M)$ such that for all homogeneous c in $C(M)$, $\gamma c = (-1)^{\partial c \partial u} ucu^{-1}$. In this case, γ is denoted by γ_u.

6. Let γ_u be a graded inner automorphism of $C(M)$. Show that γ_u is an automorphism of $C(M)$ as graded algebra. Note that if $u \in C_0(M)$ or if char $R = 2$, then γ_u is an inner automorphism in the usual sense, i.e. it is a conjugation by a unit of $C(M)$. Is the converse true?

7. Use Exercises 5 and 6, Theorem (9.8), and Proposition 1.2 on page 107 of the lecture notes Knus-Ojanguren [1974], to show that if rank M is even and β is a graded inner automorphism of $C(M)$, then dis A is free of rank 1.

8. Suppose that rank M is even. Assume that z is a special element of $C(M)$ with polynomial $X^2 - aX - b$. Let $u = a - 2z$ and show that β is the graded inner automorphism γ_u.

9. Suppose that rank M is odd. Assume that β is a graded inner automorphism $\beta = \gamma_u$. Show that dis A is free of rank 1.

10. Suppose that rank M is odd. Assume that z is a special element and show that β is the graded inner automorphism γ_z.

11. Show that the statement β *is a graded inner automorphism of* $C(M)$ is equivalent to statements (1) – (4) of (10.11).

12. Fix a positive integer n. Let M be a free nonsingular quadratic module of rank n with basis $\mathcal{X}$ and consider the element $(\det_{\mathcal{X}} h)(R^*)^2$ in $R^*/(R^*)^2$. Determine the elements of $R^*/(R^*)^2$ which have this form. What happens in the special case $R = \mathbb{Z}$?

13. Compare Theorem (10.12) with Exercise 6 of Chapter 7D.

Hints:

2. Use localization in combination with the Example in Chapter 4D and (4.7).

3. Try a localization argument.

5. Let $T = \{c \in C(M) \mid c^\beta x = -xc$ for all x in $M\}$. Let $c = c_0 + c_1 \in T$ with c_i in $C_i(M)$ be arbitrary. For x in M, $c_0 x = -xc_0$ and $c_1 x = xc_1$. So $T \subseteq A$. If rank M is odd, then $A = \mathrm{Cen}\, C(M)$ by (9.11). Let $c \in A$. By (10.10) and (7.2), $c \in T \Leftrightarrow c^\beta x = -cx$ for all x in $M \Leftrightarrow c + c^\beta = 0$ $\Leftrightarrow c \in \mathrm{dis}\, A$. Let rank M be even. Then by (10.10) and (7.2), $c \in T \Leftrightarrow$ $cx = -xc = -c^\mu x$ for all $x \in M \Leftrightarrow c \in \mathrm{dis}\, A$.

6. Put $\gamma = \gamma_u$. Since $c = c^\gamma = ucu^{-1}$ for all c in $C_0(M)$, $u \in \mathrm{Cen}\, C_0(M) = A(M)$. So $\partial u = 0$, and hence $c^\gamma = ucu^{-1}$ for all c in $C(M)$.

8. Put $u = a - 2z$. By remarks in Chapter 7C, u is a unit of $A(M)$, and by (7.11), u is in $\mathrm{Cen}\, C_0(M)$. Let x in M be arbitrary. By (7.2), $xu = (a - 2z)^\mu x = -ux$. So $ucu^{-1} = -c$ for all c in $C_1(M)$.

9. Note that u is in $A(M) = \mathrm{Cen}\, C(M)$. So, $c^\beta = (-1)^{\partial c \partial u} c$ for all homogeneous c. If $\partial u = 0$, then $-x = x^\beta = x$ for x in M. Since M is faithful, char $R = 2$. So dis A is free by Remark 1 in Section C. Assume $\partial u = 1$. So $u^\beta = -u$ and $u \in \mathrm{dis}\, A$. Let $c \in \mathrm{dis}\, A$. Then $c \in C_1(M)$. Let $x \in M$. By (7.2) and (10.10), $xc = c^\beta x = -cx$. So $(cu^{-1})x = -cxu^{-1} =$ $x(cu^{-1})$ and $cu^{-1} \in \mathrm{Cen}\, C_0(M)$. By Theorem (9.8), $c = Ru$. Hence dis A is free.

10. Let $X^2 - b$ be the polynomial of z. Here $z \in C_1(M) \cap A$ and $z^2 = b \in R^*$. So z is a homogeneous invertible element with $\partial z = 1$. By (7.11), $z \in \mathrm{Cen}\, C(M)$. So $c^\beta = c = zcz^{-1}$ for all c in $C_0(M)$ and $c^\beta = -c = (-1)^{\partial c \partial z} zcz^{-1}$ for all c in $C_1(M)$.

12. If n is even, they are the elements $((-1)^{n(n-1)/2}(a^2 + 4b))(R^*)^2$. To see this use Theorem (10.12), the Example in Chapter 4D, and the orthogonal sum of quadratic modules of rank 2. If n is odd, then by (4.18), $2 \in R^*$. It follows by considering the orthogonal sum of quadratic modules of rank 1, that any element in $R^*/(R^*)^2$ has this form. If $R = \mathbb{Z}$, then 1 is the only element in $R^* = \{\pm 1\}$ of the form $a^2 + 4b$.

11
Matrix Theory of Clifford Algebras over Fields

Overview

Throughout this chapter F is a field. Nonsingular finite-dimensional quadratic modules over F will be called *quadratic spaces*. By Theorems (9.8) and (9.7), the Clifford algebra $C(M)$ of an even dimensional quadratic space M over F is isomorphic to a matrix algebra $\mathrm{Mat}_k(D)$ over a finite-dimensional central division algebra D over F. What subalgebra of $\mathrm{Mat}_k(D)$ does $C_0(M)$ correspond to under such an isomorphism? If $\dim M$ is odd, then it is $C_0(M)$ that is isomorphic to some $\mathrm{Mat}_k(D)$. Does the connection between $C_0(M)$ and $C(M)$ have a matricial description? What division algebras D can arise in an isomorphism $C(M) \cong \mathrm{Mat}_k(D)$? Can one determine the D that corresponds to a given M? This is the circle of questions with which this chapter will concern itself. In the first three sections F is an arbitrary field. Thereafter, it will be specialized to the real numbers $\mathbb{R}$, the complex numbers $\mathbb{C}$, and later to local and global fields.

A. Matrix Connections between $C(M)$ and $C_0(M)$

A comparison of the even and odd rank cases in the results of Chapter 9 reveals a certain "dual" relationship between $C(M)$ and $C_0(M)$. The matrix analysis of this section will make this strikingly explicit.

Let $X^2 - aX - b$ be any polynomial over F which satisfies $a^2 + 4b \neq 0$. Set $E = F[X]/(X^2 - aX - b)$. By (2.2), E is a separable F-algebra. If $X^2 - aX - b$ is irreducible, then by the Example of Chapter 2C, E is a separable extension field of F.

Lemma. Suppose that $X^2 - aX - b$ *is irreducible. Let* D *be a finite dimensional central division algebra over* F.

(1) *If* $X^2 - aX - b$ *does not have a root in* D, *then* $D \otimes_F E$ *is a central division algebra over* E.

(2) *If* $X^2 - aX - b$ *has a root, say* γ, *in* D, *then the centralizer* D_0 *of* γ *in* D *is a finite dimensional central divison algebra over* E *and* $D \otimes_F E \cong \mathrm{Mat}_2(D_0)$ *as* E-*algebras.*

Proof. By a combination of Theorem (9.7) and (9.2), $D \otimes_F E$ is separable and central over $F \otimes_F E = E$, and there exist a finite-dimensional central division algebra S over E and a finite-dimensional right vector space V over S such that $D \otimes_F E \cong \mathrm{End}_S V$ as algebras over E. Set $d = \dim_F D$, $s = \dim_E S$, and $v = \dim_S V$. Notice that $\dim_F V = 2 \dim_E V = 2vs$, and

$$d = \dim_E (D \otimes_F E) = \dim_E (\mathrm{End}_S V) = s \dim_S (\mathrm{End}_S V) = v^2 s.$$

So $\dim_F V = \frac{2d}{v}$. The action of $\mathrm{End}_S V$ makes V into a left $\mathrm{End}_S V$ module, hence into a left $D \otimes_F E$ module, and therefore into a left vector space over D. Since $\dim_F V = d \cdot \dim_D V$, $\dim_F V$ is a multiple of d and hence either $v = 1$ or $v = 2$. Observe that $D \otimes_F E$ is a division algebra if and only if $v = 1$.

1) Assume that $X^2 - aX - b$ does not have a root in D. We must show that $v = 1$. Accordingly, we assume that $v = 2$ and produce a contradiction. So $\dim_F V = d$ and therefore, $\dim_D V = 1$. Let $\gamma = X + (X^2 - aX - b)$ in E and note that $E = F(\gamma)$. Since $1 \otimes \gamma \in D \otimes_F E$ commutes with $D \otimes_F F$, $1 \otimes \gamma$ acts D-linearly on V. So $\mathrm{End}_D V \cong D$ contains a root of $X^2 - aX - b$, contrary to the assumptions.

2) Suppose $\gamma \in D$ is a root of $X^2 - aX - b$. So D contains an isomorphic copy $F(\gamma)$ of E. We denote both fields by $E = F(\gamma)$. Let $D_0 = \{c \in D \mid c\gamma = \gamma c\}$. Observe that D_0 is a division subalgebra of D and that E is contained in its center. We show first that $v = 2$. Assume instead that $v = 1$. So $D \otimes_F E$ is a division algebra. The subalgebra $D_0 \otimes_F E$ is also a division algebra. Its center contains $E \otimes_F E$. Since $\{1 \otimes 1, 1 \otimes \gamma, \gamma \otimes 1, \gamma \otimes \gamma\}$ is an F-basis of $E \otimes_F E$, the elements $1 \otimes \gamma$, $1 \otimes (a - \gamma)$, and $\gamma \otimes 1$ are distinct (use the fact that if $a = 0$, then char $F \neq 2$). So the center of $D_0 \otimes_F E$ is a field which contains three distinct roots of $X^2 - aX - b$. This is impossible. So we are in the case $D \otimes_F E \cong \mathrm{End}_S V$, where $\dim_S V = v = 2$. Since $\gamma \otimes 1$ is a

root of $X^2 - aX - b$ in $D \otimes_F E$, it is easy to see that there is an S-basis of V such that the image of $\gamma \otimes 1$ has matrix $\begin{bmatrix} 0 & b \\ 1 & a \end{bmatrix}$. Therefore there is an isomorphism $D \otimes_F E \to \mathrm{Mat}_2(S)$ of E-algebras which takes $\gamma \otimes 1$ to $\begin{bmatrix} 0 & b \\ 1 & a \end{bmatrix}$. By lengthy (but easy) computations there are F-algebra isomorphisms

$$D_0 \otimes_F E = \mathrm{Cen}\,(\gamma \otimes 1) \cong \left\{ \begin{bmatrix} s & bt \\ t & s+at \end{bmatrix} \,\middle|\, s \text{ and } t \text{ in } S \right\} \cong S \otimes_F E,$$

where the last map is given by $\begin{bmatrix} s & bt \\ t & s+at \end{bmatrix} \to s \otimes 1 + t \otimes \gamma$. By Theorem 5.11 of Hungerford, for example, every element of $S \otimes_F E$ is uniquely such a sum. Observe by a dimension comparison that $\dim_F D_0 = \dim_F S$. Consider the composite $D_0 \to D_0 \otimes_F E \to S \otimes_F E \to S$, where the first map is given by $d_0 \to d_0 \otimes 1$ and the last map is induced by the E vector space structure S. Denote this composite by θ and notice that it is a homomorphism of F-algebras. Check that $\theta(\gamma d_0) = \gamma\theta(d_0)$. So θ is in fact a homomorphism of E-algebras. Since $\theta(1) = 1$ and D_0 is a division algebra, the kernel of θ is trivial. A dimension comparison shows that θ is an isomorphism. It now follows that the E-algebras $D \otimes_F E$ and $\mathrm{Mat}_2(D_0)$ are isomorphic.   QED.

Now let M be a nonzero quadratic space over F. By Theorem (10.12), the Clifford algebra $C(M)$ has a special element. Let $X^2 - aX - b$ be its polynomial. Recall that $a^2 + 4b \neq 0$ and $a = 0$ if dim M is odd.

(11.1). *Theorem. Assume that* dim M *is odd. Then there is a finite-dimensional central division algebra* D *over* F *such that* $C_0(M) \cong \mathrm{Mat}_k(D)$ *as F-algebras, where both* $\dim_F D$ *and* k *are powers of* 2. *In addition,*

(1) *If* $X^2 - b$ *has a root in* F, *i.e., if* $b \in (F^*)^2$, *then* $E = F \oplus F$ *and*

$$C(M) \cong \mathrm{Mat}_k(D) \oplus \mathrm{Mat}_k(D).$$

(2) *If* $X^2 - b$ *has a root* γ *in* D *but not in* F, *then* E *is a subfield of* D, *the centralizer* D_0 *of* γ *in* D *is a central division algebra over* E, *and*

$$C(M) \cong \mathrm{Mat}_{2k}(D_0).$$

(3) *If* $X^2 - b$ *does not have a root in* D, *then* E *is a field,* $D \otimes_F E$ *is a central division algebra over* E, *and*

$$C(M) \cong \text{Mat}_k(D) \otimes_F E \cong \text{Mat}_k(D \otimes_F E).$$

Proof. By Theorem (9.8), $C_0(M)$ is central and separable over F, and by Theorem (9.7), there is an isomorphism $C_0(M) \cong \text{Mat}_k(D)$ of F-algebras, where D is a finite-dimensional central division algebra over F and k is a positive integer. By the comments that follow (5.4), $\dim_F C_0(M)$ is a power of 2. So both k and $\dim_F D$ are powers of 2. Case (1) follows by (3.3), the fact that $(1, 0) \cong F \oplus F$, and Theorem (8.2, 2). In the other cases, $X^2 - b$ is irreducible over F, so E is a separable extension of F. By Theorem (8.2, 1) and Exercise 2 in Chapter 2E,

$$C(M) \cong C_0(M) \otimes_F E \cong \text{Mat}_k(D) \otimes_F E \cong \text{Mat}_k(D \otimes_F E).$$

If we are in case (3), then $D \otimes_F E$ is a central division algebra over E by the Lemma, and we are done. In case (2), we see by the same Lemma, that $D \otimes_F E \cong \text{Mat}_2(D_0)$ where $D_0 = \{c \in D \mid c\gamma = \gamma c\}$. Since $\text{Mat}_k(D \otimes_F E) \cong \text{Mat}_k(\text{Mat}_2(D_0)) \cong \text{Mat}_{2k}(D_0)$, the proof is complete.   QED.

(11.2). *Theorem. Assume that* $\dim M$ *is even. Then there is a finite-dimensional central division algebra* D *over* F *such that* $C(M) \cong \text{Mat}_k(D)$ *as F-algebras, where both* $\dim_F D$ *and* k *are powers of 2. In particular,* $k = 2m$ *is even. In addition,*

(1) *If* $X^2 - aX - b$ *has a root in* F, *then* $E = F \oplus F$ *and*

$$C_0(M) \cong \text{Mat}_m(D) \oplus \text{Mat}_m(D).$$

(2) *If* $X^2 - aX - b$ *has a root, say* γ, *in* D *but not in* F, *then* E *is a subfield of* D, *the centralizer* D_0 *of* γ *in* D *is a central division algebra over* E, *and*

$$C_0(M) \cong \text{Mat}_{2m}(D_0).$$

(3) *If* $X^2 - aX - b$ *does not have a root in* D, *then* $D \otimes_F E$ *is a central division algebra over* E *and*

$$C_0(M) \cong \text{Mat}_m(D) \otimes_F E \cong \text{Mat}_m(D \otimes_F E).$$

Proof. By Theorem (9.8), C(M) is central and separable. Therefore, by Theorem (9.7), there is a finite-dimensional central division algebra D over F, a positive integer k, and an F-algebra isomorphism C(M) $\cong$ Mat$_k$(D). Since dim$_F$ C(M) is a power of 2, dim$_F$ D and k are powers of 2. Switching to vector space notation, there is an isomorphism of F-algebras $\Phi : C(M) \rightarrow$ End$_D$ P, where P is a k-dimensional vector space over D. If $X^2 - aX - b$ has a root in D, apply Corollary (8.7) in combination with the Lemma. In case (3), apply Corollary (8.9) and Exercise 2 in Chapter 2E, again in combination with the Lemma.   QED.

Let M be a quadratic space with dim M even, and consider an isomorphism C(M) $\cong$ Mat$_k$(D) for some k and some finite-dimensional central division algebra D over F. We observe first that by Exercise 2, k and D are uniquely determined by M (the latter up to isomorphism). This raises the question as to which division algebras D can arise in such an isomorphism. The answer will be provided by a reformulation of a deep theorem of Merkurjev and Albert. This is Theorem (11.8). This reformulation and subsequent results over specialized fields will make essential use of some basic facts about quaternion algebras over F. These are developed in Section C. We turn first to some fundamental properties of quadratic spaces which are needed along the way.

B. Basics about Quadratic Spaces

Let M be a quadratic module over F with quadratic form q and associated bilinear form h. We say that M, or q, *represents* the scalar t $\in$ F, if there is a nonzero x in M such that q(x) = t. A nonzero x in M is *isotropic* if q(x) = 0. We say that M or q is *isotropic*, if M has isotropic vectors, i.e., if M or q represents 0; and M or q is *anisotropic* otherwise. If M is two-dimensional and {x, y} is a basis of M such that q(x) = q(y) = 0 and h(x, y) = 1, then M is a *hyperbolic plane* and {x, y} a *hyperbolic basis*. Such a plane is isotropic and, since the determinant of the matrix h in {x, y} is −1, nonsingular.

Let M be a quadratic space. Suppose x $\in$ M is isotropic. Since M is nonsingular, there is a y in M such that h(x, y) $\neq$ 0. Changing y by a scalar multiple allows the choice h(x, y) = 1. Let H = Fx $\oplus$ Fy. Put a = −q(y) and consider the basis {x, ax + y} of H. Since h(x, ax + y) = 2aq(x) + 1 = 1 and q(ax + y) = q(ax) + q(y) + h(ax, y) = 0, {x, ax + y} is a hyperbolic basis and H is a hyperbolic plane. We have shown that any isotropic vector x of a quadratic space M is contained in a hyperbolic plane H of M. We find by repeated application of this fact and (4.9) that M has a splitting

$$M = H_1 \perp ... \perp H_i \perp N,$$

where $H_1, \ldots, H_i$ are hyperbolic planes and N is anisotropic (of course, $M = N$ is a possibility). While this splitting is not unique, it is true that if $M = H_1 \perp \ldots \perp H_j \perp N'$ is another such splitting, then $j = i$ and N' is isometric to N. This a direct consequence of the

Witt Cancellation Theorem. Let M and M' be isometric quadratic spaces over F with orthogonal splittings $M = L \perp N$ and $M' = L' \perp N'$. If $L \cong L'$, then $N \cong N'$.

This fundamental fact about quadratic spaces over fields is proved in all the basic texts, for example, in O'Meara [1971], Lam, or Scharlau [1985], if char $F \neq 2$, and in Hahn-O'Meara for any F.

The uniquely determined integer i in a splitting $M = H_1 \perp \ldots \perp H_i \perp N$ of this form is called the *Witt index* of M, and the anisotropic subspace N of M is the *anisotropic part* of M. If $N = \{0\}$, i.e., if dim M is even and the Witt index is maximal, then M is a *hyperbolic space*. Relevant in the present context is the following observation.

(11.3). *Let $M = H_1 \perp \ldots \perp H_j \perp L$, where $H_1, \ldots, H_j$ are hyperbolic planes. If $C(L) \cong \mathrm{Mat}_{2^k}(D)$, then $C(M) \cong \mathrm{Mat}_{2^{k+j}}(D)$. In particular, if M is hyperbolic, then $C(M) \cong \mathrm{Mat}_{2^j}(F)$.*

Proof. Use Exercise 1 of Chapter 5E and induction.   QED.

We turn next to a different way of splitting M. Let u be any nonzero element in F and define a quadratic form on the space F by $a \rightarrow ua^2$. Denote the resulting quadratic module by $\langle u \rangle$. The associated symmetric bilinear form $F \times F \rightarrow F$ is given by $(a, b) \rightarrow 2uab$. Observe that $\langle u \rangle$ is nonsingular if and only if char $F \neq 2$. If t in F is nonzero, $\langle t^2 u \rangle \cong \langle u \rangle$. For nonzero elements $u_1, \ldots, u_m$ in F,

$$\langle u_1 \rangle \perp \ldots \perp \langle u_m \rangle$$

is the external orthogonal direct sum of the $\langle u_i \rangle$. See Chapter 4D.

Suppose that char $F \neq 2$. Let M be a nonzero quadratic space over F. Take x_1 in M with $q(x_1) \neq 0$. The subspace Fx_1 is nonsingular and isometric to $\langle q(x_1) \rangle$. By (4.9) and induction, M has a basis $\{x_1, \ldots, x_m\}$ such that $h(x_i, x_j) = 0$ for all $i \neq j$. Such a basis is an *orthogonal basis*. Putting $q(x_i) = u_i$, we find that

$$M \cong \langle u_1 \rangle \perp \ldots \perp \langle u_m \rangle.$$

The $\langle u_i \rangle$ can be put in any order by changing the ordering of the basis.

(11.4). *Any nonzero quadratic space over $\mathbb{C}$ is isometric to one of the form*

$$\langle 1 \rangle \perp \ldots \perp \langle 1 \rangle,$$

and any nonzero quadratic space over $\mathbb{R}$ is isometric to one of the form

$$\langle 1 \rangle \perp \ldots \perp \langle 1 \rangle \perp \langle -1 \rangle \perp \ldots \perp \langle -1 \rangle.$$

Proof. Combine the observations already made with the fact that $\mathbb{C}^*/(\mathbb{C}^*)^2 = 1$ and $\mathbb{R}^*/(\mathbb{R}^*)^2 = \{\mathbb{R}^{*2}, (-1)\mathbb{R}^{*2}\}$.   QED.

C. Quaternion Algebras

The field F continues to be arbitrary. Let M be a two-dimensional quadratic space over F and consider the quaternion algebra $C(M)$. By Theorems (9.8) and (9.7), $C(M) \cong \text{Mat}_k(D)$, for some finite-dimensional central division algebra D over F. Since $\dim C(M) = 4$, there are only two possibilities:

$$C(M) \cong D \quad \text{or} \quad C(M) \cong \text{Mat}_2(F).$$

By (6.4), $C(M)$ with its norm form nr is a quadratic space. Since $nr(y) = y\bar{y}$ for all y in $C(M)$, $C(M) \cong D$ if nr is anisotropic, and $C(M) \cong \text{Mat}_2(F)$ if nr is isotropic.

We return to Chapter 6B and specialize the considerations there to the field case. Let $\{x_1, x_2\}$ be a basis for M. Set $x_3 = x_1 x_2 \in C(M)$. Recall that $C_0(M) = F1 + Fx_3$ and $C_1(M) = M = Fx_1 + Fx_2$.

Assume first that char $F \neq 2$. By Section B, we can take $\{x_1, x_2\}$ to be orthogonal. Put $q(x_1) = c$ and $q(x_2) = d$. So $M \cong \langle c \rangle \perp \langle d \rangle$. Refer to the matrix of the symmetric bilinear form f in the basis $\{1, x_3, x_1, x_2\}$ and observe that the quadratic space $C(M)$ with nr is isometric to

$$\langle 1 \rangle \perp \langle -c \rangle \perp \langle -d \rangle \perp \langle cd \rangle.$$

Denote the quaternion algebra $C(M)$ in this case by

$$\left(\frac{c, d}{F} \right).$$

Observe that if $<1> \perp <-c> \perp <-d> \perp <cd>$ is anisotropic, then $\left(\frac{c,\,d}{F}\right)$ is a central division algebra over F, and if it is isotropic, then $\left(\frac{c,\,d}{F}\right)$ is isomorphic to $\mathrm{Mat}_2(F)$.

(11.5). *If* $d \in (F^*)^2$, *then* $\left(\frac{c,\,d}{F}\right) \cong \mathrm{Mat}_2(F)$. *If* $d \notin (F^*)^2$, *then* $\left(\frac{c,\,d}{F}\right) \cong$ $\mathrm{Mat}_2(F)$ *if and only if* c *is a norm from the quadratic extension* $F(\sqrt{d})$ *of* F.

Proof. If $d \in (F^*)^2$, then $<1> \perp <-d>$ is isotropic. It follows by the observation just made that $\left(\frac{c,\,d}{F}\right) \cong \mathrm{Mat}_2(F)$. Assume $d \notin (F^*)^2$. If $c = \alpha^2 - \beta^2 d$, for α and β in F, then again $<1> \perp <-d>$ is isotropic. For the converse, we assume that $<1> \perp <-c> \perp <-d> \perp <cd>$ is isotropic and show that c is a norm from $F(\sqrt{d})$ over F. Note that $c = \alpha^2 - \beta^2 d + \gamma^2 cd$, for some α, β, and γ in F. So $c(1 - \gamma^2 d) = \alpha^2 - \beta^2 d$. Since $\gamma^2 d \neq 1$ and the inverse of a norm is a norm, c is a product of two norms. It follows that c is a norm.   QED.

Suppose next that char $F = 2$. Since M is nonsingular and $h(x, x) = 2q(x) = 0$ for any x, $h(x_1, x_2) = t \neq 0$. By replacing either x_1 or x_2 by a scalar multiple of itself, we can take $t = 1$. Put $q(x_1) = c$ and $q(x_2) = c'$. Assume that $c \neq 0$ and denote M by $<c, c'>$. Refer to the matrix of f in the basis $\{1, x_3, x_1, x_2\}$ and check that C(M) with nr is isometric to

$$<1, d> \perp <c, c'>,$$

where $d = nr(x_3) = cc'$. Denote this quaternion algebra C(M) by

$$\left(\frac{c,\,d}{F}\right].$$

As before, $\left(\frac{c,\,d}{F}\right]$ is a central division algebra over F if $<1, d> \perp <c, c'>$ is anisotropic, and $\left(\frac{c,\,d}{F}\right] \cong \mathrm{Mat}_2(F)$ otherwise. Since

$$x_1 x_3 + x_3 x_1 = x_1 x_1 x_2 + x_1 x_2 x_1 = cx_2 + x_1(1 + x_1 x_2) = x_1 \text{ and}$$
$$x_3^2 + x_3 = x_1 x_2 x_1 x_2 + x_3 = x_1(1 + x_1 x_2)x_2 + x_3 = d1,$$

$\left(\frac{c, d}{F}\right]$ is the algebra that is designated $[d, c)$ on page 313 of Scharlau [1985]. Recall the subgroup $\wp(F) = \{t + t^2 \mid t \in F\}$ of the additive goup of F and refer to page 314 of Scharlau for the following analogue of (11.5).

(11.6). *If* $d \in \wp(F)$, *then* $\left(\frac{c, d}{F}\right] \cong \mathrm{Mat}_2(F)$. *If* $d \notin \wp(F)$, *then* $\left(\frac{c, d}{F}\right] \cong$ $\mathrm{Mat}_2(F)$ *if and only if* c *is a norm from the quadratic extension* $F(\gamma)$ *of* F, *where* γ *is a root of* $X^2 + X + d$.

The algebras $\left(\frac{c, d}{F}\right)$ and $\left(\frac{c, d}{F}\right]$ characterize the four-dimensional central simple algebras over F.

(11.7). *Theorem. Let* A *be a four-dimensional central simple algebra over* F.

(1) *If* char $F \neq 2$, *then* $A \cong \left(\frac{c, d}{F}\right)$ *for some* c *and* d *in* F^*.

(2) *If* char $F = 2$, *then* $A \cong \left(\frac{c, d}{F}\right]$ *for some* c *and* d *with* c *in* F^*.

In particular, any four-imensional central simple algebra over F *is a quaternion algebra.*

Proof. If char $F \neq 2$, this is the Theorem on page 236 of Pierce, for example. If char $F = 2$, this is done on pages 312 – 313 of Scharlau [1985].   QED.

The following reformulation of a theorem of Merkurjev and Albert settles the question raised earlier as to what the requirements on a division algebra D must be if it is to satisfy $C(M) \cong \mathrm{Mat}_k(D)$ for some M.

(11.8). *Theorem. Let* D *be a finite-dimensional central division algebra over* F. *Then there exists an integer* k *and a nonzero quadratic space* M *over* F *such that* $C(M) \cong \mathrm{Mat}_k(D)$ *if and only if* D *admits an anti-involution.*

Proof. Suppose $C(M) \cong \mathrm{Mat}_k(D)$ for some D and k. Since $C(M)$ has an anti-involution (e.g., the canonical involution $^-$), D has an anti-involution by Exercise 1. Conversely, if D has an anti-involution, then $D \cong D^o$. Therefore, by Merkurjev if char $F \neq 2$, and by Albert (pages 104–108) if char $F = 2$, there exist quaternion division algebras $A_1, \ldots, A_j$ over F such that $\mathrm{Mat}_n(D) \cong$ $\mathrm{Mat}_m(A_1 \otimes_F \cdots \otimes_F A_j)$ for some n and m. Put $A_i = C(N_i)$ for a two-dimensional quadratic space N_i over F. By (8.12, 2) and induction, there exists an even dimensional quadratic space M over F such that $C(M) \cong A_1 \otimes_F \cdots \otimes_F A_j$. Therefore, $\text{Mat}_n(D) \cong \text{Mat}_m(C(M))$. By Theorems (9.8) and (9.7), $C(M) \cong \text{Mat}_k(D')$ for some central division algebra D' over F. So $\text{Mat}_n(D) \cong \text{Mat}_m(\text{Mat}_k(D')) \cong \text{Mat}_{mk}(D')$. This implies by Exercise 2, that $D' \cong D$. So $C(M) \cong \text{Mat}_k(D)$.   QED.

Remarks: 1) Suppose $C(M) \cong \text{Mat}_k(D)$. By Theorems (9.11) and (9.7), $\dim_F M$ must be even. By (5.3), $k(\dim_F D) = 2^{\dim_F M}$. As a consequence, both $\dim_F D$ and k must be powers of 2.

2) If $\dim M$ is odd, there is a similar result with $C_0(M)$ in place of $C(M)$. This follows by an application of (8.12, 1).

D. Periodicity Phenomena

Let F be a field with char $F \neq 2$. Denote by $M^{r,s}$ the quadratic space

$$M^{r,s} = <-1> \perp \ldots \perp <-1> \perp <1> \perp \ldots \perp <1>$$

over F, where there are r summands of the form $<-1>$ and s of the form $<1>$. This section considers the relation $C(M^{r,s}) \cong \text{Mat}_{2^k}(D)$. We will see that only $D = F$ and $D = \left(\dfrac{-1,\ -1}{F}\right)$ can arise, and that they do so in a certain periodic way. In view of (11.4), our discussion will apply to any quadratic space if F is either $\mathbb{R}$ or $\mathbb{C}$. Incidentally, the $<-1>$ components are placed first so as to conform with notational practices in the application of Clifford algebras to topology and differential geometry. See Chapter 15.

Denote the Clifford algebra of $M^{r,s}$ by $C^{r,s}$. Since the existence of an isomorphism $C^{r,s} \cong \text{Mat}_{2^k}(D)$ implies by Theorems (9.11) and (9.7) that $n = \dim M^{r,s} = r + s$ is even, this is assumed throughout this section. Note that by Example 0 in Chapter 5A, $C^{0,0} = F$.

(11.9).   (1) *If* $r \geq s$ *and* $C^{r-s,0} \cong \text{Mat}_{2^k}(D)$, *then* $C^{r,s} \cong \text{Mat}_{2^{s+k}}(D)$.

(2) *If* $s \geq r$ *and* $C^{0,s-r} \cong \text{Mat}_{2^k}(D)$, *then* $C^{r,s} \cong \text{Mat}_{2^{r+k}}(D)$.

Proof. Suppose we are in case (1). Since $n = r + s = 2s + (r - s)$ is even, $r - s$ is even also. By changing the order of the summands, $M^{r,s}$ is isometric to

$$<1> \perp <-1> \perp \ldots \perp <1> \perp <-1> \perp <-1> \perp \ldots \perp <-1>$$

with s summands of the form $<-1> \perp <1>$ and $r - s$ summands of the form $<-1>$. Since $<-1> \perp <1>$ has isotropic vectors, it is a hyperbolic plane. Now apply (11.3). Case (2) is done in the same way.   QED.

In view of this proposition, we can restrict our attention to the cases $C^{n,0}$ and $C^{0,n}$ where $n = 2m$ is even.

(11.10). *Let* $m \geq 4$. *If* $C^{2(m-4),0} \cong \mathrm{Mat}_{2k}(D)$, *then* $C^{2m,0} \cong \mathrm{Mat}_{2k+4}(D)$, *and if* $C^{0,2(m-4)} \cong \mathrm{Mat}_{2k}(D)$, *then* $C^{0,2m} \cong \mathrm{Mat}_{2k+4}(D)$.

Proof: Let z be a special element of $C^{2m,0}$ or $C^{0,2m}$. Let $X^2 - aX - b$ be its polynomial and let $u = a^2 + 4b$. By an application of Theorem (10.12), $u \in (-1)^{m(2m-1)}(R^*)^2$ in either case. If $m = 2$, then $u \in (R^*)^2$, so that by (8.10) and (8.11), $C^{4,0} \cong C^{0,4}$. The use of Theorem (8.12, 2) and (11.9) shows that

$$C^{8,0} \cong C^{4,0} \otimes_F C^{4,0} \cong C^{4,0} \otimes_F C^{0,4} \cong C^{4,4} \cong \mathrm{Mat}_{16}(F) \text{ and}$$

$$C^{0,8} \cong C^{0,4} \otimes_F C^{0,4} \cong C^{4,0} \otimes_F C^{0,4} \cong \mathrm{Mat}_{16}(F).$$

Since $C^{0,0} = F$, this proves the proposition for $m = 4$. Applying (8.12, 2) once more, this time in combination with Exercise 2 of Chapter 2E, shows that

$$C^{2m,0} \cong C^{2m-8,0} \otimes_F C^{8,0} \cong C^{2m-8,0} \otimes_F \mathrm{Mat}_{16}(F) \cong \mathrm{Mat}_{24}(C^{2(m-4),0}).$$

This completes the proof for $C^{2m,0}$. The other case follows similarly.   QED.

We are left with $C^{2m,0}$ and $C^{0,2m}$ for $m = 1, 2$, or 3. By Section C, $C^{2,0}$ is the quaternion algebra $\left(\frac{-1, -1}{F}\right)$. Denote this algebra by Q. By (11.5), Q is a division algebra if and only if -1 is not a sum of two squares in F. Note that if -1 is a square in F, then it is obviously a sum of two squares in F.

(11.11). *Suppose that* $1 \leq m \leq 3$. *If* -1 *is the sum of two squares in* F, *then*

$$C^{2m,0} \cong C^{0,2m} \cong \mathrm{Mat}_{2m}(F).$$

If -1 is not the sum of two squares in F, then Q is a division algebra. In this case,

$$C^{0,2} \cong Mat_2(F), \quad C^{6,0} \cong Mat_8(F),$$

$$C^{2,0} \cong Q, \quad C^{4,0} \cong C^{0,4} \cong Mat_2(Q), \text{ and } \quad C^{0,6} \cong Mat_4(Q).$$

Proof. By (11.5), $C^{0,2} = \left(\frac{1,1}{F}\right) \cong Mat_2(F)$. Arguing as in the proof of (11.10), we get

$$C^{4,0} \cong C^{0,4} \cong C^{0,2} \otimes_F C^{2,0} \cong Mat_2(F) \otimes_F Q \cong Mat_2(Q),$$

$$C^{6,0} \cong C^{4,0} \otimes_F C^{2,0} \cong Mat_2(Q) \otimes_F Q \cong Mat_2(Q \otimes_F Q), \text{ and}$$

$$Q \otimes_F Q \cong C^{2,0} \otimes_F C^{2,0} \cong C^{2,2} \cong Mat_4(F).$$

So $C^{6,0} \cong Mat_8(F)$. Continuing in the same way,

$$C^{0,6} \cong C^{0,2} \otimes_F C^{4,0} \cong Mat_2(C^{4,0}) \cong Mat_2(Mat_2(Q)) \cong Mat_4(Q).$$

If -1 is not the sum of two squares in F, then Q is a division algebra over F and we are done. If -1 is the sum of two squares in F, then $Q \cong Mat_2(F)$ and the result follows by substituting this into the relations above.   QED.

E. Local and Global Fields

We begin by recalling the basics of valuation theory. The references Cassels-Fröhlich, Jacobson [1985, II], Lang, or O'Meara [1971], provide the details for the following discussion.

A *valuation* v on F is a map $v : F \to \mathbb{R}$, such that

$$v(a) > 0 \text{ if } a \neq 0 \text{ and } v(0) = 0,$$
$$v(ab) = v(a)v(b), \text{ and}$$
$$v(a + b) \leq v(a) + v(b) \quad \text{(Triangle Inequality)},$$

for all a and b in F. A valuation v is *non-Archimedean* if it satisfies

$$v(a + b) \leq \max(v(a), v(b)) \quad \text{(Strong triangle inequality)}$$

and it is *Archimedean* if it does not. Note that $v(F^*)$ is a subgroup of $\mathbb{R}^*$. If $v(F^*) = \{1\}$, v is the *trivial* valuation. Let v be nontrivial. The subgroup $v(F^*)$ can be shown to be either a discrete subset of $\mathbb{R}$ or an everywhere dense subset of $\{r \in \mathbb{R} \mid r > 0\}$. In the first instance, v is called *discrete*. If v is discrete, then it is non-Archimedean and $v(F^*)$ is an infinite cyclic group.

A valuation v defines a metric d on F by $d(a, b) = v(a - b)$ and provides F with the structure of a metric space. Two valuations v and v' are *equivalent* if they define the same metric topology on F. This is equivalent to the existence of a real number $\rho > 0$ such that $v'(a) = v(a)^\rho$ for all $a \in F^*$. If the metric space defined by v is complete, we say that v is *complete*. In general, the completion of F as a metric space is an extension field of F in a natural way. This field is denoted by F_v and is referred to as the *completion of* F *at* v.

Example 1. Let F be the field $\mathbb{Q}$ of rational numbers. The ordinary absolute value is an Archimedean valuation. It is denoted v_∞. The completion of $\mathbb{Q}$ at v_∞ is $\mathbb{R}$. Let p be any prime number. Define $v_p : \mathbb{Q} \to \mathbb{R}$ as follows: Let $a \in \mathbb{Q}$ be arbitrary, write a in the form $a = p^i(\frac{m}{n})$ with m and n both relatively prime to p, and set $v_p(a) = (\frac{1}{p})^i$. Check that v_p is discrete. This is the *p-adic valuation* on $\mathbb{Q}$. The completion of $\mathbb{Q}$ at v_p is the field $\mathbb{Q}_p$ of *p-adic numbers*. It can be shown that any nontrivial valuation on $\mathbb{Q}$ is equivalent to one of $v_\infty, v_2, v_3, \ldots$. See Chapter II of Cassels-Fröhlich.

Example 2. Let $F = K(X)$ be the field of fractions of a polynomial ring $K[X]$ over a finite field K. The situation here is very much analogous to that of the previous example. Fix a real number λ with $0 < \lambda < 1$. Define $v_0 : F \to \mathbb{R}$ as follows: Let $a \in F$ be arbitrary, write a in the form $a = (\frac{f}{g})$ with f and g in $K[X]$, and set $v_0(a) = \lambda^{-(\deg f - \deg g)}$. Check that v_0 is a discrete valuation on F. Next, let p be a monic irreducible polynomial in $K[X]$. Let $a \in F$, write a in the form $a = p^i(\frac{f}{g})$ with f and g both relatively prime to p, and set $v_p(a) = \lambda^i$. Again check that v_p is discrete. It can be shown that any nontrivial valuation on F is equivalent to v_0 or one of the v_p. Refer once more to Chapter II of Cassels-Fröhlich.

Suppose v is a non-Archimedean valuation on F. The set

$$\mathfrak{o} = \{a \in F \mid v(a) \le 1\}$$

is a subring of F, called the *valuation ring* of v. Observe that $\mathfrak{o}^* = \{a \in F \mid v(a) = 1\}$, and that $\mathfrak{p} = \mathfrak{o} - \mathfrak{o}^* = \{a \in F \mid v(a) < 1\}$ is the unique maximal ideal of $\mathfrak{o}$. So $\mathfrak{o}$ is a local ring. The field $\mathfrak{o}/\mathfrak{p}$ *is the residue class field* of v. If v is discrete and complete and if $\mathfrak{o}/\mathfrak{p}$ is finite, then F is called a *local field*. In this case, choose π in $\mathfrak{p}$ such that $v\pi$ is the largest value in $v(F^*)$ with $v\pi < 1$. Check that $\mathfrak{p}$ is the principal ideal $\pi\mathfrak{o}$, and that every element a in F is uniquely of the form $a = u\pi^i$, where $u \in \mathfrak{o}^*$ and i is an integer. Since every ideal of $\mathfrak{o}$ has the form $\mathfrak{p}^i = (\pi^i)$, $\mathfrak{o}$ is a principal ideal domain. Let $p = \mathrm{char}\ \mathfrak{o}/\mathfrak{p}$. If $\mathrm{char}\ F = 2$, then $p = 2$. If $\mathrm{char}\ F \neq 2$, then $p = 2$ and $p \neq 2$ are both possible; F, or v, is called *dyadic* in the first case, and *nondyadic* in the second.

Example 3. The field of p-adic numbers $\mathbb{Q}_p$ is an example of a local field. Its residue class field is the finite field F_p. The completion of K(X) for any of the valuations v_0 or v_p is also a local field. In the first case the residue class field is K. In the other cases it is the field of order $(\mathrm{card}\ K)^{\deg P}$.

We now assume that F is a local field and return to the study of quaternion algebras.

Suppose first that $\mathrm{char}\ F \neq 2$. Refer to §2 in Chapter 6 of Lam, or §63C of O'Meara [1971] for the following fact: There is an element $u \in \mathfrak{o}^*$ such that the quadratic space

$$<1> \perp <-\pi> \perp <-u> \perp <\pi u>$$

is the unique (up to isometry) four-dimensional anisotropic quadratic space over F. It follows from Section C that

$$\left(\frac{\pi, u}{F} \right)$$

is a quaternion division algebra. Since a quaternion algebra is determined (up to isomorphism) by its norm form (see Exercise 4) this is the unique quaternion division algebra over F (of course, up to isomorphism).

If $\mathrm{char}\ F = 2$, then there is a $u \in \mathfrak{o}^*$ such that

$$<1, u> \perp <\pi, u\pi^{-1}>$$

is the unique (again up to isometry) four-dimensional anisotropic quadratic space over F. For a complete proof of this fact, see Lemma 1.7 of Riehm [1965].

Therefore,

$$\left(\frac{\pi, u}{F}\right]$$

is a division algebra. Again, it is the unique such.

It is an important fact, see the references already provided that the unique four-dimensional anisotropic quadratic space over the local field F represents all non-zero elements of F. It follows directly that any quadratic space over F of dimension 5 or more is isotropic.

Now let M be an even dimensional quadratic space over F. We set $C(M) \cong Mat_k(D)$ for some divison algebra D over F, and turn to the problem of determining D in terms of M. In view of Theorem (9.11), M is even dimensional. By (11.3), it suffices to assume that M is anisotropic of dimension 2 or 4. While we will only consider the case char $F \neq 2$, the statement and proof of the following proposition are valid in the case char $F = 2$ with only trivial modifications. We denote $\left(\frac{\pi, u}{F}\right)$ by Q.

(11.12). *Let* F *be a local field with* char $F \neq 2$, *and let* M *be a two- or four-dimensional anisotropic quadratic space over* F.

(1) *Suppose* $M \cong <c> \perp <d>$. *If* $d \in (F^*)^2$, *or if* $d \notin (F^*)^2$ *and* c *is a norm from the quadratic extension* $F(\sqrt{d})$ *of* F, *then* $C(M) \cong Mat_2(F)$. *Otherwise,* $C(M) \cong Q$.

(2) *Suppose* M *is the unique four-dimensional anisotropic space over* F. *Then* $C(M) \cong Mat_2(Q)$.

Proof. Case (1) follows from (11.5) and the uniqueness of Q. So consider case (2). Put $N = <1> \perp <-\pi>$ and $N' = <-u> \perp <\pi u>$. By Example 2 in Chapter 7C, C(N) has a special element with polynomial $X^2 - \pi$. So by (8.12, 2), $C(M) \cong C(N) \otimes_F C(^{4\pi}N')$. By (1), $C(N) \cong Mat_2(F)$, and by (8.10), $C(^{4\pi}N') \cong C(^{\pi}N') = C(<-\pi u> \perp <u>)$. By Section C, the quadratic space $C(<-u> \perp <\pi u>)$ has the splitting $<1> \perp <\pi u> \perp <-u> \perp <-\pi>$. Since this is (the unique) anisotropic, $C(^{\pi}N')$ is a quaternion division algebra over F. Therefore, $C(^{\pi}N') \cong Q$. It now follows that $C(M) \cong Mat_2(Q)$.   QED.

With the facts about quadratic spaces over local fields already collected, it is a fairly routine exercise to classify such spaces in terms of "classical" invariants. See Exercise 15 and Chapter 13D.

For the rest of this section, we will assume that F is a *global field*. If char $F = 0$, this means that F is a finite extension field of $\mathbb{Q}$. If char $F = p$, it means that F is a finite extension field of the field of fractions $K(X)$ of a polynomial ring $K[X]$ over a finite field K of characteristic p. In the first case F is called an *algebraic number field* and in the second F is an *algebraic function field*. The valuation theory of global fields (see Examples 1 – 3 for instance) give rise to important local-global principles. These will not be needed at the moment, but they will play a role in Chapters 13 and 14. The fact that is important now is the following standard result in the theory of central simple algebras over a global field. Refer, for example, to the development on pages 238, 253, and 280 of Reiner for the proof.

Theorem. Let F *be a global field. If* D *is a central division algebra over* F *which has an anti-involution, then* D *is either* F *or a quaternion division algebra over* F.

When inserted into Theorem (11.8), this shows that if M is a quadratic space over a global field F and $C(M) \cong \mathrm{Mat}_k(D)$ for a central divison algebra D over F, then D is either F or a quaternion division algebra. This in turn has the following consequence.

(11.13). *Theorem. Let* F *be a global field and let* A *be a finite-dimensional central simple algebra over* F. *Then* $A \cong C(M)$ *for some quadratic space* M *if and only if* A *admits an anti-involution and* $\dim_F A = 2^n$ *with n even.*

Proof. If $A \cong C(M)$, then A has an anti-involution, since $C(M)$ does. By Theorem (9.11), dim $M = n$ is even, and by (5.3), $\dim_F A = 2^n$. To prove the converse, let $A \cong \mathrm{Mat}_k(D)$ for some central divison algebra D over F. If A has an anti-involution, then by Exercise 1, so does D. Therefore, D is either F or a quaternion division algebra over F. If $D = F$, set $M = H_1 \perp ... \perp H_m$ where each H_j is a hyperbolic plane and apply (11.3). In the other case, put $D = C(N)$ for a quadratic space N over F with dim $N = 2$, and set $M = H_1 \perp ... \perp H_{m-1} \perp N$. Now apply (11.3) again. QED.

Remark. The theorem does not hold for an arbitrary field F. If it did, it would follow (refer to the proof of Theorem (11.8)) that any central simple algebra A over F which admits an involution and has dimension 2^n with n even, is isomorphic to a tensor product of quaternion algebras. But this is false by Amitsur-Rowen-Tignol.

F. Exercises

Refer to §4.4 of Jacobson [1985, II], particularly to the theorem on page 206 and his Exercises 5, 6, and 9, for the proofs of Exercises 1 and 2.

1. Let D be a division algebra. Show that $\text{Mat}_k(D)$ has an anti-involution if and only if D has an anti-involution.

2. Let D and D' be division algebras over F. Then $\text{Mat}_k(D) \cong \text{Mat}_{k'}(D')$ if and only if $k = k'$ and $D \cong D'$.

3. Let M be a quadratic space over F with $\dim M$ even. Set $M = H_1 \perp ... \perp H_i \perp N$, where $H_1, ..., H_i$ are hyperbolic planes and N is the anisotropic part of M. Show that $\text{Arf} M = \text{Arf} N$.

4. Let M and N be two-dimensional quadratic spaces over a field F and consider the quaternion algebras $C(M)$ and $C(N)$. Prove that $C(M)$ and $C(N)$ are isomorphic if and only if the quadratic spaces $(C(M), nr)$ and $(C(N), nr)$ are isometric.

5. Let M be a two-dimensional quadratic space over a field and let N be the quadratic space $N = (C(M), nr)$. Show that $\text{Arf} N = 1$.

6. Show that $C^{r,s}$ with $r + s > 0$ ($r + s$ can be odd or even) has a special element with polynomial $X^2 - (-1)^{[r(r+1) + s(s-1)]/2}$.

7. Let $n \geq 1$ (n odd or even) and consider the Clifford algebras $C^{n,0}$ and $C^{0,n}$ of Section C. Let $C_0^{n,0}$ and $C_0^{0,n}$ be the respective even subalgebras and show that both $C_0^{n,0}$ and $C_0^{0,n}$ are isomorphic to $C^{n-1,0}$.

8. Let Q denote the quaternion algebra $\left(\frac{-1,-1}{F}\right)$. Show that $C^{3,0} \cong \text{Mat}_2(F) \oplus \text{Mat}_2(F)$ if -1 is the sum of two squares in F, and that $C^{3,0} \cong Q \oplus Q$ if not.

9. Show that $C^{5,0} \cong \text{Mat}_4(F) \oplus \text{Mat}_4(F)$ if $-1 \in (F^*)^2$, and that $C^{5,0} \cong \text{Mat}_4(F(\sqrt{-1}))$ otherwise.

10. Show that $C^{7,0} \cong \text{Mat}_8(F) \oplus \text{Mat}_8(F)$.

11. Show that $C^{0,3} \cong Mat_2(F) \oplus Mat_2(F)$ if $-1 \in (F^*)^2$, and that $C^{0,3} \cong Mat_4(F(\sqrt{-1}))$ otherwise.

12. Fill in all the information in the tables of Cases 1 – 3 on pages 128 and 129 of Lam.

In Exercises 13 – 15, F is a local field. Let Q denote the unique quaternion division algebra over F. Let M be a quadratic space over F with dim M even. Define the element $c(M) \in \mathbb{Z}^* = \{1, -1\}$ as follows: By a combination of (11.3) and (11.12), $C(M) \cong Mat_k(D)$, where D is isomorphic to either F or Q. If it is F, set $c(M) = 1$, and if it is Q, set $c(M) = -1$.

13. Use facts from Sections B, C and E to deduce that F and Q are the unique division algebras over F which admit an anti-involution.

14. Let M be a quadratic space over F with dim M even. Set M = $H_1 \perp ... \perp H_i \perp N$, where $H_1, ..., H_i$ are hyperbolic planes and N is the anisotropic part of M. Show that $c(M) = c(N)$.

15. Suppose that M and M' are even dimensional quadratic spaces over F. Prove that M is isometric to M' if and only if dim M = dim M', Arf M = Arf M', and $c(M) = c(M')$.

16. Give an example which shows that the conclusion of Exercise 15 fails for $F = \mathbb{R}$.

Hints:

1. One direction is easy, for if $\varphi : D \to D$ is an anti-involution, then the assignment $(x_{ij}) \to (\varphi x_{ij})$ defines an anti-involution of $Mat_k(D)$.

2. Again, one direction is easy. For, if $k = k'$ and $\varphi : D \to D'$ is an isomorphism, then the assignment $(x_{ij}) \to (\varphi x_{ij})$ defines an isomorphism $Mat_k(D) \to Mat_{k'}(D)$.

3. Use Example 3 of Chapter 7C and the multiplicativity of Arf.

4. Refer to O'Meara [1971] and Scharlau [1985].

5. Consider the cases char $F \neq 2$ and char $F = 2$ separately. If char $F \neq 2$, set $C(M) = \left(\frac{c, d}{F}\right)$. Then $N \cong <1> \perp <-c> \perp <-d> \perp <cd>$. Let

$L = <1> \perp <-c>$ and $L' = <-d> \perp <cd>$, and show that $\text{Arf } L = \text{Arf } L'$. The case $\text{char } F = 2$ is done in a similar way.

6. Use Exercise 6 of Chapter 7D.

7. Use Exercise 6, Theorem (8.12, 1), (8.10), and Example 2 of Chapter 5B.

8 – 12. Use (8.2), Exercise 7, Exercise 6, (11.11), and the lemma in Section A. Refer also to Theorem (11.1) and note that $-1 \in (Q^*)^2$.

13. Combine Theorem (11.8) with (11.3) and (11.12).

15. The difficult part is the demonstration of the fact that if the three invariants of M and M' are the same, then $M \cong M'$. Set $M = H_1 \perp \ldots \perp H_i \perp N$, where $H_1, \ldots, H_i$ are hyperbolic planes and N is anisotropic and set $M' = H_1 \perp \ldots \perp H_j \perp N'$, where $H_1, \ldots, H_j$ are hyperbolic planes and N' is anisotropic. It suffices to show that $i = j$ and $N \cong N'$. Note that $\dim N = 0, 2,$ or 4, and similarly for N'. By Exercises 3 and 14, $\text{Arf } N = \text{Arf } N'$ and $c(N) = c(N')$. Suppose first that $N = \{0\}$. Note that $\dim N' = 2$ is impossible by Example 3 of Chapter 7C, and $\dim N' = 4$ is impossible by (11.12, 2). That $i = j$ is clear. Consider next, the case $\dim N = 4$. That $\dim N' = 0$ is impossible we already know, and that $\dim N' = 2$ is impossible follows from Exercise 5 and Example 3 both of Chapter 7C. So $\dim N' = 4$. So by the uniqueness of four-dimensional anisotropic spaces over F, $N \cong N'$. Again, $i = j$ is clear. The case $\dim N = \dim N' = 2$ remains. Note that either $C(N) \cong C(N') \cong \text{Mat}_2(F)$, or $C(N) \cong C(N') \cong Q$. Assume $\text{char } F \neq 2$. Put $N \cong <c> \perp <d>$ and $N' \cong <c'> \perp <d'>$. Since $\text{Arf } N = \text{Arf } N'$, $cd \in c'd'(F^*)^2$ by (3.1). It follows by Exercise 4 and the Witt Cancellation Theorem that $<-c> \perp <-d> \cong <-c'> \perp <-d'>$, and hence that $N \cong N'$. Assume $\text{char } F = 2$. Here put $N \cong <c, d>$ and $N' \cong <c', d'>$. Since $\text{Arf } N = \text{Arf } N'$, $c'd' = cd + \gamma + \gamma^2$ by (3.1). To complete the proof, it suffices to show (again by Exercise 4 and the Witt Cancellation Theorem) that $W \cong <1, cd>$ and $W' \cong <1, c'd'>$ are isometric. Let $\{x, y\}$ be the basis of W that corresponds to $W \cong <1, cd>$, and show that $W \cong <1, c'd'>$ in $\{x, x\gamma + y\}$.

16. Consider the quadratic spaces $M^{4,0}$ and $M^{0,4}$ over $\mathbb{R}$.

12
Dis(R) and Qu(R)

Overview

In this chapter the set of isomorphism classes of separable quadratic algebras over a commutative ring R will be provided with the structure of an Abelian group. This is the "quadratic group" $Qu(R)$ of R. It contains the free quadratic group $Qu_f(R)$ as a subgroup and it is closely related to the group $Dis(R)$ of Chapter 4C. The group $Qu(R)$ will make it possible (in Chapter 13C) to extend the Arf invariant of Chapter 10D to nonsingular finitely generated projective quadratic modules over R. The connection between $Qu(R)$ and $Dis(R)$ will be used in Chapter 14E to measure the "difference" between quadratic forms and symmetric bilinear forms in arithmetic situations. In Sections A and B, R will be an arbitrary commutative ring. Thereafter it will be a domain which is integrally closed in its field of fractions F. In this case, see Section D, separable quadratic algebras can be characterized as the integral closures of R in quadratic Galois extensions of F in which all prime ideals of R are unramified.

A. The Quadratic Group Qu(R)

For the moment, S is any faithful commutative algebra over R equipped with an involution σ which satisfies $s + s^\sigma \in R$ for all $s \in S$.

Define $S \times S \to S \times S$ by $(s, s') \to (ss', s(s')^\sigma)$. Since the required linearity properties are met, there is an induced additive map $\psi_\sigma : S \otimes_R S \to S \times S$ which satisfies $s \otimes s' \to (ss', s(s')^\sigma)$. A check on the generators $s \otimes s'$ shows that

$$\psi_\sigma : S \otimes_R S \to S \oplus S$$

is a homomorphism of R-algebras.

(12.1). $\psi_\sigma : S \otimes_R S \to S \oplus S$ *is an isomorphism if and only if* S *is separable quadratic and* σ *is the conjugation of* S.

Proof. Suppose first that $\psi_\sigma : S \otimes_R S \to S \oplus S$ is an isomorphism. Since $(1, 0)$ is in the image, there are elements $x_1,\ldots, x_k, y_1,\ldots, y_k$ in S such that $\sum_i x_i y_i = 1$ and $\sum_i x_i y_i^\sigma = 0$. Put $e = \sum_i x_i \otimes y_i$. Let s in S be arbitrary. Since the elements $\sum_i (s x_i) \otimes y_i$ and $\sum_i x_i \otimes (y_i s)$ both have image $(s, 0)$, they are equal. So e is a separability idempotent for S. Now define $f_i \in \mathrm{Hom}_R(S, R)$ by $f_i(s) = s y_i + (s y_i)^\sigma$. Notice that

$$\sum_i x_i f_i(s) = \sum_i x_i(s y_i + (s y_i)^\sigma) = s(\sum_i x_i y_i) + s^\sigma(\sum_i x_i y_i^\sigma) = s$$

for any s in S. By basic properties of projective modules, e.g., see the corollary on page 153 of Jacobson [1985, II], S is finitely generated and projective over R. Let $\mathfrak{p}$ be any prime ideal of R. Since the local rank of $S \otimes_R S$ is $(\mathrm{rank}\ S_\mathfrak{p})^2$ and that of $S \oplus S$ is $2(\mathrm{rank}\ S_\mathfrak{p})$, it follows that $\mathrm{rank}\ S_\mathfrak{p} = 2$. The proof of the fact that S is a separable quadratic algebra is complete. To show that σ is the conjugation of S, it suffices in view of (10.4) to show that the set of fixed points is R. Suppose $s \in S$ satisfies $s^\sigma = s$. Since $\psi_\sigma(s \otimes 1) = (s, s)$ and $\psi_\sigma(1 \otimes s) = (s, s)$, $s \otimes 1 = 1 \otimes s$. Refer to the map $\varphi : S \to T$ in Chapter 10A. Since the image $(s \otimes 1)(1 - e)$ of s under φ is fixed by sw, it follows by (10.4) and the commutative diagram (*) which defines the conjugation of S that $s \in R$.

To establish the other implication, assume that S is separable quadratic and that σ is the conjugation of S. Refer to Chapter 2B and consider the map $S \oplus S \to S \otimes_R S$ given by $(s, s') \to \theta(s) + \varphi(s') = (s \otimes 1)e + (s' \otimes 1)(1 - e)$ for all s and s' in S. Since e is an idempotent, this is an R-algebra map. Since $\varphi : S \to T$ is an isomorphism, θ is injective, and $S \otimes_R S = T \oplus \theta S$, this is an algebra isomorphism. Since $(s \otimes 1)e = (1 \otimes s)e$ and $\varphi : S \to T$ takes σ to sw, it is easy to show that $S \oplus S \to S \otimes_R S$ takes $(ss', s(s')^\sigma)$ to $s \otimes s'$. We have therefore constructed an inverse of ψ_σ.   QED.

(12.2). *If* S *is finitely generated projective of rank 2 and if the ideal of* S *generated by* $\{s - s^\sigma \mid s \in S\}$ *is equal to* S, *then* S *is separable quadratic and* σ *is the conjugation.*

Proof. By hypothesis, there are elements $s_1, ..., s_k$ and $x_1, ..., x_k$ in S such

that $\sum\limits_{i=1}^{k} s_i(x_i - x_i^\sigma) = 1$. Now put $s_{k+1} = -\sum\limits_{i=1}^{k} s_i x_i^\sigma$ and $x_{k+1} = 1$. Then

$\sum\limits_{j=1}^{k+1} s_j x_j = 1$ and $\sum\limits_{j=1}^{k+1} s_j x_j^\sigma = 0$. It follows that $(1, 0)$ is in the image of the map

$\psi_\sigma : S \otimes_R S \rightarrow S \oplus S$. Since $s \otimes 1$ is sent to (s, s), it follows that the

algebra homomorphism ψ_σ is surjective. By (4.5), ψ_σ is an isomorphism.

Now apply (12.1).   QED.

We now construct the group Qu(R) of isomorphism classes of separable quadratic algebras. Let S and S' be two separable quadratic algebras, and let σ and σ' be the respective conjugations. Form the algebra $S \otimes_R S'$ and let

$$S * S' = \{c \in S \otimes_R S' \mid c^{\sigma \otimes \sigma'} = c\}$$

be the set of fixed points of the involution $\sigma \otimes \sigma'$ of $S \otimes_R S'$.

We show that $S * S'$ is separable quadratic over R with conjugation $\sigma \otimes id_{S'} = id_S \otimes \sigma'$. To do this it suffices — in view of (10.5) — to show that for any prime ideal $\mathfrak{p}$ of R, $(S * S')_\mathfrak{p}$ is separable quadratic over $R_\mathfrak{p}$ with conjugation $(\sigma \otimes id_{S'})_\mathfrak{p} = (id_S \otimes \sigma')_\mathfrak{p}$. Fix a prime ideal $\mathfrak{p}$ of R. Since $S_\mathfrak{p}$ and $S'_\mathfrak{p}$ are free separable quadratic algebras over $R_\mathfrak{p}$, we find by (10.1) and the

analysis of the algebra $(a, b)^\varepsilon * (c, d)^\eta$ in Chapter 3D that $S_\mathfrak{p} * S'_\mathfrak{p}$ is a separable quadratic algebra over $R_\mathfrak{p}$ and $\sigma_\mathfrak{p} \otimes id_{S'_\mathfrak{p}} = id_{S_\mathfrak{p}} \otimes \sigma'_\mathfrak{p}$ is its

conjugation. As in Chapter 10A (where the special case $S = S'$ is considered), there is an isomorphism of $R_\mathfrak{p}$-algebras

$$(S \otimes_R S')_\mathfrak{p} \rightarrow S_\mathfrak{p} \otimes_{R_\mathfrak{p}} S'_\mathfrak{p}$$

which takes $(s \otimes s')_\mathfrak{p}$ to $s_\mathfrak{p} \otimes s'_\mathfrak{p}$ for any s in S and s' in S'. The involution $(\sigma \otimes \sigma')_\mathfrak{p}$ corresponds to $\sigma_\mathfrak{p} \otimes \sigma'_\mathfrak{p}$ under this isomorphism. The isomorphism carries the fixed point set $(S * S')_\mathfrak{p}$ of $(\sigma \otimes \sigma')_\mathfrak{p}$ onto the fixed point set $S_\mathfrak{p} * S'_\mathfrak{p}$ of $\sigma_\mathfrak{p} \otimes \sigma'_\mathfrak{p}$. It follows that $(S * S')_\mathfrak{p}$ is a separable quadratic algebra. Since the isomorphism carries the involution $(\sigma \otimes id_{S'})_\mathfrak{p} = (id_S \otimes \sigma')_\mathfrak{p}$ to $\sigma_\mathfrak{p} \otimes id_{S'_\mathfrak{p}} = id_{S_\mathfrak{p}} \otimes \sigma'_\mathfrak{p}$, $(\sigma \otimes id_{S'})_\mathfrak{p} = (id_S \otimes \sigma')_\mathfrak{p}$ is the conjugation of $(S * S')_\mathfrak{p}$.

Let [S] denote the isomorphism class of a quadratic R-algebra S and let Qu(R) be the set of all isomorphism classes. The product $[S][S'] = [S * S']$ makes Qu(R) into an Abelian group. The group laws are tedious to verify. Localization arguments of the type already illustrated on a number of occasions in this text reduce things to the free case, where they were already established. See Chapter 3A. The details are left to the reader. The group

$$Qu(R)$$

is called the *quadratic group* of R. The discussion in Chapter 3D, in particular (3.9), when specialized to $\varepsilon = \eta = 0$, shows that the free quadratic group $Qu_f(R)$ is a subgroup of Qu(R). Recall that $Qu_f(R)$ is an elementary Abelian 2-group. We now check that the same is true for Qu(R). Let [S] in Qu(R) be arbitrary. We must show that $[S]^2$ is equal to the identity element $[1, 0]$ of $Qu_f(R)$, i.e., that $S * S$ is isomorphic to the algebra $(1, 0) \cong R \oplus R$. Consider the isomorphism $\psi_\sigma : S \otimes_R S \to S \oplus S$ given by (12.1). It takes the typical element $\sum_i s_i \otimes s_i'$ to $(\sum_i (s_i s_i'), \sum_i s_i(s_i')^\sigma)$. The involution $\sigma \otimes \sigma$ corresponds to (σ, σ) under this map. Since R is the fixed point set of σ, it is easy to see that the restriction of this isomorphism takes $S * S$ onto $R \oplus R$. Therefore, $[S]^2 = 1$ as asserted.

In the general situation, Qu(R) is subtle and difficult to compute. In number theoretic situations it is determined by the arithmetic in the Hilbert class field. See Chapter 14D. In the special case char R = 2, Qu(R) provides nothing new:

(12.3). *If* char R = 2, *then* $Qu(R) = Qu_f(R) \cong R/\wp(R)$, *where* $\wp(R) = \{r + r^2 \mid r \in R\}$.

Proof. For the isomorphism, refer to Exercises 7 – 10 in Chapter 3E. To verify that $Qu(R) = Qu_f(R)$, we must show − see (10.1) − that any quadratic R-algebra is free. So let S be any quadratic algebra over R and let σ be its conjugation. By (10.7), $S = Rt \oplus$ dis S. Since dis $S = \{s \in S \mid s^\sigma = -s = s\}$, it follows by the defining property of σ that dis S = R.   QED.

B. More about Dis(R)

Let P be a finitely generated projective module of rank 1 over R. Denote the isomorphism class of P by [P]. The set of all [P] is a group known as the *Picard group* Pic(R) of R. The product of Pic(R) is given by the tensor product, the identity element is [R], and the inverse of [P] is $[P^*]$ where P^*

is the dual of P. See Jacobson [1985, II], for example, for the details. Following our general notational convention,

$$Pic(R)_2$$

denotes the subgroup of $Pic(R)$ consisting of elements of order 1 or 2.

Now refer to Chapter 4C, in particular to the group $Dis(R)$. The assignment $[P, f] \rightarrow [P]$ defines a homomorphism $Dis(R) \rightarrow Pic(R)_2$. For $[P]$ in $Pic(R)_2$, there is an isomorphism $\varphi : P \rightarrow P^*$. So $f : P \times P \rightarrow R$ defined by $f(x, y) = \varphi(x)(y)$ provides P with a nonsingular bilinear form. It follows that $Dis(R) \rightarrow Pic(R)_2$ is surjective. Since both groups are elementary Abelian 2-groups, and therefore vector spaces over $\mathbb{Z}_2$, this map splits. If $[P, f]$ is in the kernel, then P is free, so that $[P, f] = [R, f_u]$ for some unit u in R. By (4.12), the homomorphism $R^*/(R^*)^2 \rightarrow Dis(R)$ given by $u(R^*)^2 \rightarrow [R, f_u]$ is injective. We have proved:

(12.4). *The sequence* $1 \rightarrow R^*/(R^*)^2 \rightarrow Dis(R) \rightarrow Pic(R)_2 \rightarrow 1$ *is split exact. In particular,* $Dis(R) \cong R^*/(R^*)^2 \times Pic(R)_2$.

For the rest of this section we assume that R is a domain with field of fractions F. Refer to Chapter 4C for the concepts used. Let $Inv(R)$ be the set of invertible fractional ideals in F. Let $\mathfrak{a}$ and $\mathfrak{b}$ be in $Inv(R)$. With the operations

$$\mathfrak{a}\mathfrak{b} = \{\sum_{\text{fin}} ab \mid a \in \mathfrak{a}, b \in \mathfrak{b}\} \quad \text{and} \quad \mathfrak{a}^{-1} = \{d \in F \mid d\mathfrak{a} \subseteq R\}$$

$Inv(R)$ is an Abelian group with identity element R. The set of nonzero principal ideals, i.e., those of the form cR with c a nonzero element in F, constitute a subgroup $Pr(R)$ of $Inv(R)$. The quotient $Inv(R)/Pr(R)$ is denoted

$$Cl(R)$$

and is called the *ideal class group* of R. Let $\mathfrak{a} \in Inv(R)$. Using observations made in Chapter 4C, we find that $\mathfrak{a}$ is a finitely generated projective R-module of rank 1 and that $Inv(R) \rightarrow Pic(R)$ given by $\mathfrak{a} \rightarrow [\mathfrak{a}]$ is a surjective homomorphism. Check that it induces an isomorphism $Cl(R) \cong Pic(R)$. The restriction of this isomorphism provides the special case

$$1 \rightarrow R^*/(R^*)^2 \rightarrow Dis(R) \rightarrow Cl(R)_2 \rightarrow 1$$

of the earlier split exact sequence. Note that the map $\text{Dis}(R) \to \text{Cl}(R)_2$ is given by $[\mathfrak{a}, c] \to [\mathfrak{a}]$.

If every fractional ideal is invertible, i.e., if it is in $\text{Inv}(R)$, then R is a *Dedekind domain*. Refer, for example, to Jacobson [1985, II] for the basic properties of such domains, in particular for the fact that the following conditions on R are equivalent:

(1) R is Dedekind.

(2) R is Noetherian, integrally closed in F, and every nonzero prime ideal of R is maximal.

(3)   Every proper (ordinary) ideal of R can be written in a unique way as a product of prime ideals.

If R is Dedekind and F is a global field, then R is called an *arithmetic Dedekind domain*. In this situation, the preceding groups are important concerns in the theory of numbers. For example, the Dirichlet Unit Theorem (see Chapter 14A) provides information about the group $R^*/(R^*)^2$; the ideal class group $\text{Cl}(R)$ is a finite group whose order is the *class number* of F; and as already remarked, $\text{Qu}(R)$ is related to the arithmetic in the Hilbert class field of F.

C. Connecting Qu(R) with Dis(R)

The goal of this section is the comparison of $\text{Qu}(R)$ with $\text{Dis}(R)$. Define the map

$$\text{dis} : \text{Qu}(R) \to \text{Dis}(R)$$

by assigning to any $[S]$ in $\text{Qu}(R)$ the isomorphism class of the discriminant module $(\text{dis } S, p)$ of Chapter 10B. So $\text{dis } [S] = [\text{dis } S, p]$. If $R = \{0\}$, the assertions made in this section are trivial. So assume that $R \neq \{0\}$.

(12.5). $\text{dis} : \text{Qu}(R) \to \text{Dis}(R)$ *is a homomorphism*.

Proof. Let S and S' be separable quadratic algebras over R with conjugations σ and σ' respectively. The discriminant modules $(\text{dis } S, p)$ and $(\text{dis } S', p)$ are denoted $\text{dis } S$ and $\text{dis } S'$. We show first that dis is independent of the choice of S in $[S]$. If $\phi : S \to S'$ is an isomorphism, then the diagram

$$S \xrightarrow{\phi} S'$$

$$\sigma \downarrow \qquad \downarrow \sigma'$$

$$S \xrightarrow{\phi} S'$$

commutes, by an application of (10.4). This implies directly that ϕ takes dis S onto dis S'. It follows that dis S $\cong$ dis S'.

Consider $S * S' = \{c \in S \otimes_R S' \mid c^{(\sigma \otimes \sigma')} = c\}$. To prove that dis is a homomorphism, we must show that the discriminant modules dis S $\otimes_R$ dis S' and

$$\text{dis } (S * S') = \{d \in S * S' \mid d + d^{(\sigma \otimes \text{id}_{S'})} = 0\}$$

are isomorphic. For s in dis S and s' in dis S', the element $s \otimes s'$ in $S \otimes_R S'$ satisfies $(s \otimes s')^{(\sigma \otimes \sigma')} = (-s \otimes -s') = (s \otimes s')$. So $s \otimes s'$ is in $S * S'$. Also $(s \otimes s')^{(\sigma \otimes \text{id}_{S'})} = (-s \otimes s') = -(s \otimes s')$, and hence $s \otimes s'$ is in dis $(S * S')$. Defining dis $S \times$ dis $S' \to$ dis $(S * S')$ by $(s, s') \to s \otimes s'$, provides an R-module homomorphism dis $S \otimes_R$ dis $S' \to$ dis $(S * S')$. We show that it is an isomorphism of discriminant modules. A typical localization argument reduces the proof to the case where R is local, and therefore to the case where S and S' are free. So by (10.1), we may put $S = (a, b)$ with $a^2 + 4b \in R^*$ and $S' = (c, d)$ with $c^2 + 4d \in R^*$. Set $u = a^2 + 4b$ and $u' = c^2 + 4d$. By Chapter 3A,

$$(a, b) * (c, d) \cong (ac, a^2d + c^2b + 4bd).$$

By the Example of Chapter 10B, dis $S \cong (R, f_u)$, dis $S' \cong (R, f_{u'})$, and dis $(S * S') \cong (R, f_{uu'})$. Since the homomorphism in question corresponds to the isomorphism $(R, f_u) \otimes_R (R, f_{u'}) \cong (R, f_{uu'})$, the proof is complete.   QED.

Consider the R-algebra R/4R and the change-of-scalars homomorphism Dis(R) $\to$ Dis(R/4R). Observe that R/4R = $\{0\}$ if and only if $2 \in R^*$, and that Dis(R/4R) is trivial in this case.

(12.6). *The sequence* Qu(R) $\xrightarrow{\text{dis}}$ Dis(R) $\longrightarrow$ Dis(R/4R) *is exact.*

Proof. We prove first that the image of Qu(R) under dis is contained in the kernel of Dis(R) $\to$ Dis(R/4R). We can assume that R/4R $\neq \{0\}$. We need to take a typical [S] in Qu(R) and show that the discriminant module (dis S) $\otimes_R$ (R/4R) over R/4R is isomorphic to (R/4R, f_1). So let [S] in Qu(R) and let σ be the conjugation of S. By (10.7), S = Rt $\oplus$ dis S, with $t + t^\sigma = 1$. Put $1 = rt + d$ with r in R and d in dis S. Then $2 = 1 + 1^\sigma = (rt + d) + (rt^\sigma - d) = r$. So $d = 1 - 2t$ and $d^2 = 1 - 4t(1 - t) = 1 - 4tt^\sigma$. Now let x in dis S be arbitrary, and observe that $x = (xd)d + 4tt^\sigma x$. Since both xd and tt^σ are fixed by σ, they are in R. It follows that

$$\text{dis } S = Rd + 4(\text{dis } S).$$

We claim that $d \otimes (1 + 4R)$ is a basis of (dis S) $\otimes_R$ (R/4R). Let $z = \sum_i x_i \otimes (r_i + 4R)$ with x_i in dis S and r_i in R be arbitrary. Put $x_i = u_i d + 4y_i$ with u_i in R and y_i in dis S. So $z = \sum_i (u_i d + 4y_i) \otimes (r_i + 4R)$
$= \sum_i (u_i r_i d \otimes (1 + 4R)) = (\sum_i u_i r_i)(d \otimes (1 + 4R))$. So $d \otimes (1 + 4R)$ spans (dis S) $\otimes_R$ (R/4R) over R/4R. Since (dis S) $\otimes_R$ (R/4R) is a discriminant module, it is faithful, and hence $d \otimes (1 + 4R)$ is a basis. Since the form p of dis S is given by product, it follows that the form $P_{(R/4R)}$ of (dis S) $\otimes_R$ (R/4R) satisfies

$$P_{(R/4R)}(d \otimes (1 + 4R), d \otimes (1 + 4R)) = d^2 + 4R = 1 + 4R.$$

We have proved that $((\text{dis } S) \otimes_R (R/4R), P_{(R/4R)})$ is isomorphic to (R/4R, f_1) as required.

To complete the proof, we let (P, f) be any discriminant module over R, assume that [P, f] is in the kernel, and construct [S] $\in$ Qu(R) such that [dis S, p] = [P, f]. We will require an element x in P with certain properties. If $2 \in R^*$, take x = 0. Suppose that $2 \notin R^*$. Since $(P \otimes_R (R/4R), f_{(R/4R)}) \cong$ (R/4R, 1) and the R/4R modules P/4P and $P \otimes_R$ (R/4R) are isomorphic under the assignment $p + 4P \to p \otimes (1 + 4R)$, it follows that P/4P has a basis, say x + 4P, such that $f(x, x) = 1 - 4r_0$ with r_0 in R. Observe that P $= Rx + 4P$ and $f(x, x) = 1 - 4r_0$ with r_0 in R, whether $2 \in R^*$, or not.

Take a free R-module of rank 1 with basis, say t, and consider the R-module S = Rt $\oplus$ P. It is clear that S is a finitely generated projective module of rank 2. Define a product on S by:

$$tt = (1 - 2r_0)t - r_0 x; \quad tp = pt = -f(x, p)t + 2r_0 p, \ p \in P; \text{ and}$$
$$pq = 2f(p, q)t + f(p, q)x,$$

for all p and q in P; and by R-linear extension to all of S. Observe that this product is commutative. Note also (use (4.10) for the second series of equalities), that

$$(2t + x)t = 2(1 - 2r_0)t - 2r_0x - f(x, x)t + 2r_0x = (2 - 4r_0 - f(x, x))t = t, \text{ and}$$

$$(2t + x)p = -2f(x, p)t + 4r_0p + 2f(x, p)t + f(x, p)x = 4r_0p + f(x, x)p = p.$$

So $1_S = 2t + x$ is an identity element. Notice that $p{\cdot}q = f(p, q)1_S$. To verify that the product is associative, it suffices to check the associativity on the generators $\{t\} \cup P$. These computations again make use of (4.10).

i) For $p, q,$ and q' in P,

$$p(qq') = p(f(q, q')1_S) = f(q, q')p) = f(p, q)q' = (pq)q'.$$

ii) For p and q in P,

$$
\begin{aligned}
t(pq) &= f(p, q)t = (f(x, p)f(x, q) - f(x, x)f(p, q) + f(p, q))t \\
&= (f(x, p)f(x, q) + (1 - f(x, x))f(p, q))t + r_0(-2f(x, p)q) + 2f(p, q)x) \\
&= (f(x, p)f(x, q) + 4r_0f(p, q))t + (-2r_0f(x, p)q + 2r_0f(p, q)x) \\
&= -f(x, p)(-f(x, q)t + 2r_0q) + 2r_0(2f(p, q)t + f(p,q)x) \\
&= -f(x, p)tq + 2r_0pq = (-f(x, p)t + 2r_0p)q = (tp)q.
\end{aligned}
$$

The two other associativity equalities involving $t, p,$ and q, are consequences of the commutativity and the identity just verified. For example,

$$p(tq) = (tq)p = t(qp) = t(pq) = (tp)q = (pt)q.$$

iii) For any p in P,

$$
\begin{aligned}
t(tp) &= t(-f(x, p)t + 2r_0p) = -f(x, p)((1 - 2r_0)t - r_0x) + 2r_0(-f(x, p)t + 2r_0p) \\
&= -f(x, p)(1 - 2r_0)t - 2r_0f(x, p)t + r_0(f(x, p)x + 4r_0p) \\
&= -f(x, p)(1 - 2r_0)t - 2r_0f(x, p)t + r_0(f(x, x) + 4r_0)p \\
&= -f(x, p)(1 - 2r_0)t - 2r_0f(x, p)t + r_0(2 - 4r_0 - f(x, x))p \\
&= (1 - 2r_0)(-f(x, p)t) - 2r_0f(x, p)t + (1 - 2r_0)2r_0p - r_0f(x, p)x \\
&= (1 - 2r_0)(-f(x, p)t + 2r_0p) - r_0(2f(x, p)t + f(x, p)x) \\
&= (1 - 2r_0)tp - r_0xp = (tt)p.
\end{aligned}
$$

The other associativity equalities involving t and p follow from this one and the commutativity. We have now proved that S is an R-algebra.

Define a linear map σ on S by $t^\sigma = t + x$ and $p^\sigma = -p$ for all p in P. Observe that $(1_S)^\sigma = (2t + x)^\sigma = 2(t + x) - x = 2t + x = 1_S$. Also, $t^{\sigma^2} = (t + x)^\sigma = t + x - x = t$. If we can show that σ preserves the product structure of S, then σ is an involution on S. It suffices to check this on pairs of the generators $\{t\} \cup P$:

$$(tt)^\sigma = (1 - 2r_0)t^\sigma - r_0 x^\sigma = (1 - 2r_0)(t + x) + r_0 x = (1 - 2r_0)t + (1 - r_0)x$$
$$= (1 - 2r_0)t + (4r_0 - r_0 + f(x, x))x$$
$$= (1 - 2r_0)t - r_0 x + 2(-f(x, x)t + 2r_0 x) + 2f(x, x)t + f(x, x)x$$
$$= t^2 + 2tx + x^2 = (t^\sigma)^2,$$

$$(tp)^\sigma = -f(x, p)t^\sigma - 2r_0 p = -f(x, p)t - f(x, p)x - 2r_0 p$$
$$= f(x, p)t - 2r_0 p - 2f(x, p)t - f(x, p)x = -(t + x)p$$
$$= t^\sigma p^\sigma, \text{ and finally}$$

$$(pq)^\sigma = f(p, q)1_S = f(-p, -q)1_S = (-p)(-q).$$

We have proved that σ is an involution on S. The separability of S is next. Since $t + t^\sigma = 2t + x = 1_S$ and $p^\sigma = -p$, it follows that $s + s^\sigma \in R$ for all s in S. By (4.8), $\sum_i f(x_i, y_i) = 1$ for some x_i and y_i in P. So $\sum_i x_i y_i = 1_S$ in S. Put $x_i = r_i x + 4p_i$ with p_i in P, and set $s_i = -r_i t + 2p_i$. Since $s_i - s_i^\sigma = -r_i t + 2p_i + r_i(t + x) + 2p_i = x_i$, $\sum_i (s_i - s_i^\sigma)y_i = 1_S$. Since S is faithful, it is separable by (12.2). We have produced an element [S] in Qu(R).

It remains to prove that dis [S] = [P, f]. As an important first step we establish that σ is the conjugation of S. In view of (10.4), it must be established that $R1_S = R(2t + x)$ is the fixed point set of σ. Let rt + p in S be arbitrary and assume that $(rt + p)^\sigma = rt + p$. So rx = 2p. Since

$$(rx)t = -rf(x, x)t + 2rr_0 x = -rf(x, x)t + 4r_0 p \quad \text{and} \quad 2pt = -2f(x, p)t + 4r_0 p,$$

$rf(x, x) = 2f(x, p)$. So $r(1 - 4r_0) = 2f(x, p)$, and therefore, r = 2r' for some $r' \in R$. Put p' = p - r'x. Note that 2p' = 2p - rx = 0. So

$$p' = p'1_S = p'(2t + x) = p'x = 2f(p', x)t + f(p', x)x = f(p', x)x.$$

Put r'' = f(p', x). Clearly, 2r'' = 0. Now,

$rt + p = 2r't + (p' + r'x) = 2r't + (r'' + r')x = 2(r'' + r')t + (r'' + r')x = (r'' + r')1_S.$

So, $R = R1_S$ is the fixed point set of σ, and σ is the conjugation of S as asserted. We check that the discriminant modules dis S and (P, f) are isomorphic. Let s in S with $s^\sigma = -s$ be arbitrary. Put $s = rt + p$. So $rt + rx - p = -rt - p$, and $rx = -2rt$. Thus, $r = 0$ and $s \in P$. Therefore, dis S = $\{s \in S \mid s^\sigma = -s\} = P$. Since $pq = f(p, q)1_S$, the assignment $p \to p1_S$ defines an isomorphism (P, f) $\to$ dis S. Therefore, dis [S] = [P, f] and the proof is now complete.   QED.

Consider the exact sequences of (3.6), (12.6), and (12.4) and let $Qu_f(R) \to Qu(R)$ be the inclusion map.

(12.7). *The diagram*

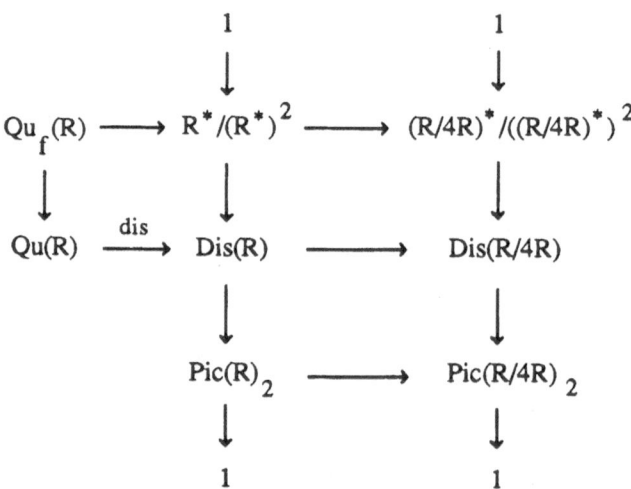

commutes.

Proof. We begin with the square on the left. Let [a, b] in $Qu_f(R)$ with (a, b) $= R[X]/(X^2 - aX - b)$, be arbitrary. In view of the Example of Chapter 10B, the discriminant modules (R, f_u), where $u = a^2 + 4b$, and dis (a, b) $= R(a - 2v)$, where $v = X + (X^2 - aX - b)$, are isomorphic. So the square commutes. The square next to it commutes, since $R^*/(R^*)^2 \longrightarrow (R/4R)^*/((R/4R)^*)^2$ is defined by $u(R^*)^2 \dashrightarrow (u + R/4R) + ((R/4R)^*)^2$, and [R, f_u] goes to [R/4R, $f_{u + 4R}$] under Dis(R) $\to$ Dis(R/4R). The last square is left to the reader.
    QED.

(12.8). *Corollary. If* char $R = 0$ *in* R, *then* dis : $Qu(R) \to Dis(R)$ *is the trivial map. If* $2 \in R^*$, *then* dis : $Qu(R) \to Dis(R)$ *is an isomorphism.*

Proof. If $2 = 0$, note that $4R = \{0\}$, and $Dis(R) \to Dis(R/4R)$ is injective. Assume that $2 \in R^*$. In this case, $Dis(R) \to Dis(R/4R)$ is the trivial map. So dis : $Qu(R) \to Dis(R)$ is surjective. The injectivity remains. Suppose $[S]$ in $Qu(R)$ has trivial image. So dis S is free, and hence by (10.8), S is free. So by (10.1), $[S] \in Qu_f(R)$. A combination of (3.4) and (12.7) shows that $[S] = 1$.

QED.

D. The Case of an Integrally Closed Domain

Let R be a domain with field of fractions F. We begin with the notation that will be in effect throughout this section.

Let S be a separable quadratic algebra over R. By (9.2), $S \otimes_R F$ is separable quadratic over F. Set

$$E = S \otimes_R F$$

and let σ be the conjugation of E. Since S is finitely generated projective, the map $S \to E$, given by $s \to s \otimes 1$, is injective. We will identify S with its image in E. By an application of (10.5) to the prime ideal $\mathfrak{p} = 0$, we find that the conjugation of S is the restriction of σ to S. Let t be an element in S with $t + t^\sigma = 1$. By (10.7),

$$S = Rt \oplus \text{dis } S \quad \text{and} \quad E = Ft \oplus \text{dis } E.$$

Let d be a basis of dis E. Since $(d^2)^\sigma = (d^\sigma)^2 = (-d)^2 = d^2$, $d^2 \in F$. Since dis $S \subseteq$ dis $E = Fd$, dis $S = \mathfrak{a}d$ with $\mathfrak{a} \subseteq F$. By (10.7), $\mathfrak{a}$ is a finitely generated projective R-module. It is therefore an invertible fractional ideal of F. By (10.9) and (4.8), $\mathfrak{a}^2 d^2 = R$. Let $\beta = d^2$ and observe that

$$[\text{dis } S] = [\mathfrak{a}, \beta] \in Dis(R).$$

Suppose that char $F \neq 2$. Since $d^\sigma = -d$, $d \notin F$. Use the basis 1 and d of E to show that

$$E \cong F[X]/(X^2 - \beta),$$

as F-algebras with conjugation. If char $F = 2$, then dis $E = F$. Since $t + t^\sigma = 1$, $t^2 + t + tt^\sigma = 0$. Since $\gamma = tt^\sigma$ is fixed by σ, $\gamma \in F$. Using the basis 1 and t of E,

$$E \cong F[X]/(X^2 + X + \gamma),$$

again as F-algebras with conjugation.

We proceed next to a commutative diagram that will be relevant in the number theoretic situations considered in Chapter 14C. By (10.1), $Qu(F) = Qu_f(F)$, and by basic properties of the tensor product, $[S] \to [S \otimes_R F]$ defines a homomorphism $Qu(R) \to Qu_f(F)$. The inclusion $Qu_f(R) \to Qu(R)$ provides the composite

$$Qu_f(R) \to Qu(R) \to Qu_f(F).$$

It is easy to see that the assignment $[\mathfrak{a}, c] \to c(F^*)^2$ defines a homomorphism

$$Dis(R) \to F^*/(F^*)^2.$$

It is, see Chapter 4C, the composite of the homomorphism $Dis(R) \to Dis(F)$ given by change of scalars with the isomorphism $Dis(F) \to F^*/(F^*)^2$ defined by $[F, f_c] \to c(F^*)^2$. Composition with the injection $R^*/(R^*)^2 \to Dis(R)$ of (4.12), gives

$$R^*/(R^*)^2 \to Dis(R) \to F^*/(F^*)^2.$$

The maps just defined, together with dis and the homomorphism δ of Chapter 3B, combine into the diagram

$$(\#) \qquad \begin{array}{ccccc} Qu_f(R) & \longrightarrow & Qu(R) & \longrightarrow & Qu_f(F) \\ \delta \downarrow & & \downarrow \text{dis} & & \downarrow \delta \\ R^*/(R^*)^2 & \longrightarrow & Dis(R) & \longrightarrow & F^*/(F^*)^2 \end{array}.$$

We show that it commutes. Let S be a separable quadratic algebra over R. If $S = R[X]/(X^2 - aX - b)$ with $a^2 + 4b \in R^*$, then by the Example of Chapter 10B, the discriminant module (dis S, p) is isomorphic to (R, f_u), where $u = a^2 + 4b$. Since $\delta[S] = u(R^*)^2$, this shows that the square on the left commutes. As to the square on the right, observe that if char $R = 2$, then by (12.8) and

Remark 2 in Chapter 3B, both vertical maps are trivial and the commutativity is obvious. So assume that char $R \neq 2$. Since $[\text{dis } S] = [\mathfrak{a}, \beta] \in \text{Dis}(R)$ and $E \cong F[X]/(X^2 - \beta)$, $\delta[E] = 4\beta(F^*)^2 = \beta(F^*)^2$ and the commutativity follows.

(12.9). *Suppose R is integrally closed in F. If $E = S \otimes_R F$ is a field, then it is a quadratic Galois extension of F and the field of fractions of S. If E is not a field, then $E \cong F \oplus F$ and $S \cong R \oplus R$ both as quadratic algebras with conjugation. In either case, S is the integral closure of R in E.*

Proof. Denote by C the integral closure of R in E.

(1) We begin with the properties of E. Suppose that E is a field. Since E is separable as an algebra over F, it is separable as a field extension of F, by the remarks that conclude Chapter 2C. Since F is the field of fixed points of σ, E is Galois over F. The field of fractions of S inside E contains both F and S. Since any element of E is a finite sum of terms rs with $r \in F$ and $s \in S$, it follows that this field of fractions is E. Suppose that E is not a field. Then $X^2 - \beta$ (if char $F \neq 2$) or $X^2 + X + \gamma$ (if char $F = 2$) has a root. So by (3.3) and (10.4), there is an F-algebra isomorphism $\varphi : E \to F \oplus F$ such that σ corresponds to the "switch" $(\alpha, \alpha') \to (\alpha', \alpha)$. Note that $\varphi\alpha = (\alpha, \alpha)$ for $\alpha \in F$. Let $c \in C$ and put $\varphi c = (c_1, c_2)$. Since (c_1, c_2) is integral over R,

$$(c_1, c_2)^n + r_{n-1}(c_1, c_2)^{n-1} + \ldots + r_0 = 0$$

with $r_0, \ldots, r_{n-1}$ in R. So $(c_i)^n + r_{n-1}(c_i)^{n-1} + \ldots + r_0 = 0$ for i either 1 or 2. Since R is integrally closed in F, c_1 and c_2 are both in R. Since C is closed under addition, $\varphi C = R \oplus R$.

(2) We show that $C \supseteq S$. Since $t + t^\sigma = 1$, $t^2 - t + tt^\sigma = 0$. Since $tt^\sigma \in S$ is fixed by σ, $tt^\sigma \in R$. So t satisfies a monic integral polynomial, and thus $t \in C$. Now let $s \in \text{dis } S$. Again, $ss^\sigma \in R$. Since $s^2 + ss^\sigma = s(s + s^\sigma) = 0$, $s \in C$. Since C contains both Rt and dis S, C contains S.

(3) Let $I = \{e \in \text{dis } E \mid e \text{ is integral over } R\}$. We will prove that $C = Rt \oplus I$. Since C is closed under addition and $C \supseteq S \supseteq Rt$, C contains this sum. Now let $c \in C$ be arbitrary. Put $c = ut + v$ with $u \in F$ and $v \in \text{dis } E$. Note that

$$c^2 = u^2t^2 + 2utv + v^2 = u^2t - u^2tt^\sigma + 2utv - vv^\sigma, \text{ and}$$
$$c^2 - uc = -u^2tt^\sigma + 2utv - vv^\sigma - uv =$$
$$= -u^2tt^\sigma - vv^\sigma + uv(2t - 1) = -u^2tt^\sigma - vv^\sigma + uv(t^\sigma - t).$$

Since tt^σ, vv^σ, and $v(t^\sigma - t)$ are all fixed by σ, all are in F. Note that tt^σ in R, since it is in S. So c satisfies the monic quadratic polynomial

$$c^2 - uc + (u^2 tt^\sigma + vv^\sigma - uv(t^\sigma - t))$$

over F.

(3a) Suppose E is a field. Since E has degree 2 over F, this polynomial is the characteristic polynomial of c over F. By page 611 in Jacobson [1985, II], both u and $u^2 tt^\sigma + vv^\sigma - uv(t^\sigma - t)$ are in R. So $u^2 tt^\sigma$ is in R and $r = vv^\sigma - uv(t^\sigma - t)$ is in R. Since, $v^2 + u(t^\sigma - t)v + r = 0$, v is integral over R. Therefore, $v \in I$. Thus c is in $Rt \oplus I$ and $C = Rt \oplus I$ as required.

(3b) Suppose E is not a field. Return to the isomorphism $\varphi : E \to F \oplus F$. Note that $\varphi I = \{(r, -r) \mid r \in R\}$. Put $\varphi t = (t_1, t_2)$. Since $t + t^\sigma = 1$, $(t_1, t_2) + (t_2, t_1) = (1, 1)$. So $t_1 + t_2 = 1$, and $\varphi t = (t_1, 1 - t_1)$. Since $t \in C$, $t_1 \in R$. Put $\varphi c = (r_1, r_2)$ with $r_i \in R$, and $\varphi v = (\alpha, -\alpha)$ with $\alpha \in F$. Since $c = ut + v$, $(r_1, r_2) = (t_1 u, u - t_1 u) + (\alpha, -\alpha)$. Therefore, $r_1 = t_1 u + \alpha$ and $r_2 = u - t_1 u - \alpha$. So $u \in R$ and hence $\alpha \in R$. Therefore, $\varphi v \in \varphi I$ and thus $v \in I$. Again, $C = Rt \oplus I$ as required.

(4) Recall that dis $E = Fd$. Put $I = \mathfrak{c}d$. By page 612 in Jacobson [1985, II], C is a finitely generated projective R-module. It follows that $\mathfrak{c} \cong C/Rt$ is a finitely generated projective R-module. Therefore, by page 604 in Jacobson [1985, II], $\mathfrak{c}$ is an invertible fractional ideal of F. Let γ and γ' be in $\mathfrak{c}$. Since $(\gamma d)(\gamma' d) = \gamma\gamma' d^2$ is in F and in C, $\gamma\gamma' d^2$ is in R. It follows that $Rd^{-2} \supseteq \mathfrak{c}^2$. It was already observed at the beginning of this section that dis $S = \mathfrak{a}d$ where $\mathfrak{a}$ is an invertible fractional ideal of F with $\mathfrak{a}^2\beta = R$, where $\beta = d^2$. Since $C \supseteq S$, $I \supseteq$ dis S, and therefore, $\mathfrak{c} \supseteq \mathfrak{a}$. It follows that $R\beta^{-1} \supseteq \mathfrak{c}^2 \supseteq \mathfrak{a}^2 = R\beta^{-1}$. So $\mathfrak{c}^2 = \mathfrak{a}^2$, and $(\mathfrak{c}\mathfrak{a}^{-1})^2 = \mathfrak{c}^2\mathfrak{a}^{-2} = R$. This implies by Exercise 4 that $\mathfrak{c} = \mathfrak{a}$. So

$$S = Rt \oplus \text{dis } S = Rt \oplus \mathfrak{a}d = Rt \oplus \mathfrak{c}d = C.$$

Therefore S is the integral closure of R in E. The proof of the proposition is now complete.   QED.

(12.10). *Suppose that R is integrally closed. Then the maps of the two horizontal rows of diagram (#) are all injective. If char R = 2, all the vertical maps of the diagram are trivial; if char R ≠ 2, all the vertical maps are injective; if 2 ∈ R*, all the vertical maps are isomorphisms.*

Proof. That $R^*/(R^*)^2 \to \text{Dis}(R)$ is injective is already known. Consider $\text{Dis}(R) \to F^*/(F^*)^2$ next. Suppose that $[\mathfrak{c}, c]$ is in the kernel. So $c \in (F^*)^2$

and $[\mathfrak{c}, c] = [\mathfrak{a}, 1]$ for some fractional ideal $\mathfrak{a}$ with $\mathfrak{a}^2 = R$. Let $b = a + r$, with $a \in \mathfrak{a}$ and $r \in R$, be any element in the finitely generated R-submodule $\mathfrak{a} + R$ of F. Observe that $b(\mathfrak{a} + R) \subseteq \mathfrak{a} + R$, so that by the lemma on page 409 in Jacobson [1985, II], b is R-integral. Since R is integrally closed, $b \in R$. So $\mathfrak{a} + R = R$. Thus $\mathfrak{a} \subseteq R$ and hence $\mathfrak{a} = R$ by Exercise 4. So $[\mathfrak{c}, c] = [\mathfrak{a}, 1] = [R, 1]$, and $\mathrm{Dis}(R) \to F^*/(F^*)^2$ is injective. The first map on the top row of (#) is an inclusion. We prove the injectivity of $Q(R) \to Q(F)$. Let $[S]$ be in the kernel of $Q(R) \to Q(F)$. So $E = S \otimes_R F$ is isomorphic to $F \oplus F$. Therefore by (12.9), $S \cong R \oplus R$. Since $R \oplus R \cong (1, 0)$, $[S] = 1$. The proof of the first statement is complete.

If char $R = 2$, then both δ are trivial by Remark 2 in Chapter 3B, and dis is trivial by (12.8). If $2 \in R^*$, then both δ are isomorphisms by (3.5) and dis is an isomorphism by (12.8). If char $R \neq 2$, then both δ are injective by (3.4). This in turn implies that dis is injective by the commutativity of the diagram.

QED.

Remarks. 1) If R is Dedekind domain of char $R \neq 2$, then the isomorphism

$$\mathrm{Dis}(R) \cong R^*/(R^*)^2 \times \mathrm{Cl}(R)_2$$

and the fact that dis $: Qu(R) \to \mathrm{Dis}(R)$ is injective together put a "bound" on $Qu(R)$. If char $R = 2$, then there is no apparent connection between $Qu(R)$ and $\mathrm{Dis}(R)$. However, in this case we know by (12.3), that $Qu(R) \cong R/\wp(R)$.

2) If $2 \notin R^*$ and $2 \neq 0$, the matter of the image of dis $: Q(R) \to \mathrm{Dis}(R)$ is rather subtle. We will investigate the situation in the case of an arithmetic Dedekind domain by examining the images of these groups in $F^*/(F^*)^2$ under the injections provided by the commutative diagram (#).

E. The Classical Discriminant

We continue the assumption that R is a domain which is integrally closed in its field of fractions F.

Let E be a finite Galois extension of F of degree n. Let $\{1 = \sigma_1, ..., \sigma_n\}$ be the Galois group of E/F. For any basis $\{x_1, ..., x_n\}$ of E over F, the matrix $C = (\sigma_i(x_j))$ in $\mathrm{Mat}_n(E)$ is known to be invertible. See page 281 in Jacobson [1985, II]. For $x \in E$, let $T_{E/F}(x) = \sigma_1(x) + ... + \sigma_n(x)$. Since it is fixed by all σ_i, $T_{E/F}(x)$ is an element of F. Check that $(C^t)C = (T_{E/F}(x_i x_j))$. It follows that $\det (T_{E/F}(x_i x_j)) = d$ is a nonzero element in F.

Let S be the integral closure of R in E. Observe that $\sigma_i S = S$ for all i. If all of the basis elements $\{x_1, ..., x_n\}$ fall into S, then $d \in S \cap F = R$.

Define $\mathfrak{d}_{E/F}$ to be the ideal of R that is generated by all elements d that arise in this way. This is the *classical discriminant* of number theory.

Suppose from now on that E is a quadratic extension of F and let $\{1, \sigma\}$ be the Galois group. So F is the fixed field of σ and $\sigma S = S$. The restriction of σ to S is an involution with fixed point set R. By basic facts, in Jacobson [1985, II], for example, S is a finitely generated projective module of rank 2 over R. For s_1 and s_2 in S, consider the 2×2 matrix $(T_{E/F}(s_i s_j))$ and check that

$$\det (T_{E/F}(s_i s_j)) = (s_1 s_2^\sigma - s_1^\sigma s_2)^2.$$

If $\{s_1, s_2\}$ is dependent over F, then $(s_1 s_2^\sigma - s_1^\sigma s_2)^2 = 0$. It follows that $\mathfrak{d}_{E/F}$ is the ideal of R generated by all elements of the form $(s_1 s_2^\sigma - s_1^\sigma s_2)^2$.

(12.11). *S is a separable quadratic R-algebra if and only if* $\mathfrak{d}_{E/F} = R$.

Proof. Suppose S is a separable quadratic algebra. By (10.4), σ is the conjugation of S. By (12.1), ψ_σ is surjective. So there exist elements $x_1,..., x_k, y_1,..., y_k$ in S such that $\sum_i x_i y_i = 1$ and $\sum_i x_i y_i^\sigma = 0$. Therefore, $\sum_i x_i(y_i - y_i^\sigma) = 1$. By applying σ, $\sum_i x_i^\sigma(y_i^\sigma - y_i) = 1$. By multiplying the two terms, $\sum_{i,j} x_i(y_i - y_i^\sigma) x_j^\sigma(y_j^\sigma - y_j) = 1$. It follows that

$$\sum_i x_i x_i^\sigma(y_i - y_i^\sigma)(y_i^\sigma - y_i)) - \sum_{i<j} (x_i x_j^\sigma + x_j x_i^\sigma)(y_i - y_i^\sigma)(y_j - y_j^\sigma)) = 1.$$

Note that the elements $x_i x_i^\sigma$ and $x_i x_j^\sigma + x_j x_i^\sigma$ are in R. To show that $\mathfrak{d}_{S/R} = R$, it suffices to show that $(x - x^\sigma)(y - y^\sigma)$ is in $\mathfrak{d}_{E/F}$ for any x and y in S. Let x and y in S be arbitrary. By (10.7), there is a t in S with $t + t^\sigma = 1$, such that $x = rt + d$ and $y = r't + d'$ with r and r' in R and d and d' in dis S. Note that $c = t - t^\sigma$ is in dis S. Since $x - x^\sigma = rt + d - (rt^\sigma - d) = rc + 2d$ and similarly $y - y^\sigma = r'c + 2d'$,

$$(x - x^\sigma)(y - y^\sigma) = rr'c^2 + 2rcd' + 2r'dc + 4dd'.$$

If a^2 is in $\mathfrak{d}_{E/F}$ for any a in dis S, then $2ab = (a + b)^2 - a^2 - b^2$ is in $\mathfrak{d}_{E/F}$ for any a and b in dis S. This implies that $(x - x^\sigma)(y - y^\sigma)$ is in

$\mathfrak{d}_{E/F}$ as required. For any $a \in$ dis S, $a^2 = (a(t + t^\sigma))^2 = (at^\sigma - a^\sigma t)^2 \in \mathfrak{d}_{E/F}$, so that this requirement is met.

Assume conversely that $\mathfrak{d}_{E/F} = R$. Thus 1 is an R-linear combination of elements of the form $(s - s^\sigma)^2$ with s in S. So $\sum_{i=1}^{k} x_i(s_i - s_i^\sigma) = 1$ for elements $x_1,..., x_k$ in S and $s_1,..., s_k$ in S. Therefore, by (12.2), S is separable quadratic. QED.

Putting the facts of Section D together with those above provides us with the following characterization of separable quadratic algebras over R.

(12.12). *Theorem. Suppose that* R *is an integrally closed domain with field of quotients* F. *Let* S *be any separable quadratic algebra over* R. *Then either* $S \cong R \oplus R$, *or else there exists a quadratic Galois extension* E *of* F *with* $\mathfrak{d}_{E/F} = R$, *such that* S *is isomorphic to the integral closure of* R *in* E. *Moreover, any such* S *is a separable quadratic algebra over* R.

Proof. Suppose a separable quadratic algebra S is given and that S is not isomorphic to $R \oplus R$. By Section D, especially (12.9), S is contained in a quadratic Galois extension E of F and S is the integral closure of R in E. By (12.11), $\mathfrak{d}_{E/F} = R$. The converse follows from the facts in this section, except in the case $S \cong R \oplus R \cong R[X]/(X^2 - X)$ where (2.2) applies. QED.

Assume that R is a Dedekind domain in F. Let E be a finite Galois extension of F and let S be the integral closure of R in E. Then S is a Dedekind domain in E. A prime ideal $\mathfrak{p}$ of R is *unramified* in S if all prime ideals in the factorization of the ideal $\mathfrak{p}S$ of S into prime ideals of S occur to the first power only, i.e., if the factorization has the form $\mathfrak{p}S = \mathcal{P}_1 \cdots \mathcal{P}_k$ with all the prime ideals $\mathcal{P}_i$ of S distinct. If this is not the case, $\mathfrak{p}$ is *ramified* in S. A basic fact asserts (refer to Chapter 10 of Jacobson [1985, II], for example) that a prime $\mathfrak{p}$ of R is *ramified* in S if and only if $\mathfrak{p}$ appears in the factorization of $\mathfrak{d}_{S/R}$ into primes of R. Now let E/F be quadratic. In view of Theorem (12.11), we have proved that S is a separable quadratic algebra over R if and only if all prime ideals of R are unramified in S.

F. Exercises

1. Show that if R is a principal ideal domain, then $Qu(R) = Qu_f(R)$. Deduce that $Qu(\mathbb{Z}) = 1$ and that this in turn implies that $\mathbb{Q}$ has no unramified quadratic extensions. This, see Samuel, is a special case of the more general fact that $\mathbb{Q}$ has no nontrivial unramified extensions of any degree.

2. Let F be a local field and R its valuation ring. Then $Qu(F) = Qu_f(F)$ and $Qu(R) = Qu_f(R)$. Suppose char $F \neq 2$. If F is nondyadic, then $Qu_f(F) \cong \mathbb{Z}_2 \times \mathbb{Z}_2$ and $Qu_f(R) \cong \mathbb{Z}_2$. If F is dyadic, then $Qu_f(F) \cong (\mathbb{Z}_2)^{t+1}$ and $Qu_f(R) \cong (\mathbb{Z}_2)^t$, with $t = 1 + ki$, where k is the cardinality of the residue class field and i is defined by $2 = u\pi^i$, $u \in R^*$ (see Chapter 11E). If char $F = 2$, then $Qu_f(F)$ is infinite and $Qu_f(R) \cong \mathbb{Z}_2$.

3. Specialize the commutative diagrams (#) and that of (12.7) to the case $R = \mathbb{Z}$ and to the case of a valuation ring of a local field.

4. Let R be a domain with field of fractions F. If $\mathfrak{a}$ is an invertible fractional ideal of F such that $\mathfrak{a} \subseteq R$ and $\mathfrak{a}^2 = R$, then $\mathfrak{a} = R$.

5. Suppose R is a domain which is finitely generated as a ring, e.g., R is a Noetherian domain. Then by Roquette, R^* is a finitely generated Abelian group. Show that $R^*/(R^*)^2 \cong (\mathbb{Z}_2)^k$ for some nonnegative integer k.

6. Let A be any R-algebra which is free of finite rank. Let $a \in A$ and consider the R-module map $m_a : A \to A$ defined by $m_a(x) = ax$. Let $\text{Tr}(a)$ be the trace m_a in some basis of A. Show that $\text{Tr}(ab) = \text{Tr}(ba)$. Conclude that $\text{Tr}(a)$ is independent of the choice of the basis, and that $T_A : A \times A \to R$ defined by $T_A(a, b) = \text{Tr}(ab)$ is a symmetric R-bilinear form on A.

7. Let A be an R-algebra which is free of finite rank. In §9.5, Chapter III of Bourbaki [1975, Algebra], the *discriminant* $\Delta_{A/R}$ of A is defined to be the principal ideal $(\det_{\mathcal{X}} T_A)R$, where $\mathcal{X}$ is some basis of A. By facts in Chapter 4B, this ideal is independent of the choice of $\mathcal{X}$. Study the properties of $\Delta_{A/R}$ which Bourbaki develops.

8. Let S be a free quadratic algebra. So $S \cong R[X]/(X^2 - aX - b)$ for some a and b in R. Show that $\Delta_{S/R} = (a^2 + 4b)R$ and deduce that S is

separable if and only if $\Delta_{S/R} = R$. Now suppose that R is an integrally closed domain with field of fractions F, and that S is the integral closure of R in a quadratic Galois extension E of F. Show that $\Delta_{S/R} = \mathfrak{d}_{S/R}$.

9. Let $A = R[X]/(X^3 - aX - b)$ with a and b in R. Then $\mathcal{X} = \{1, v, v^2\}$, where $v = X + (X^3 - aX - b)$ is a basis of A. Show that $\det_{\mathcal{X}} T_A = 4a^3 - 27b^2$. So $\Delta_{A/R} = (4a^3 - 27b^2)R$.

10. Let (M, q) be a free quadratic module over R of rank 2 with basis $\{x_1, x_2\}$. Put $q(x_1) = s_1$, $q(x_2) = -s_2$ and $h(x_1, x_2) = t$. Consider the Clifford algebra $A = C(M)$ of M and its basis $\mathcal{X} = \{1, x_1x_2, x_1, x_2\}$. Show that $\det_{\mathcal{X}} T_A = -4^2(t^2 + 4s_1s_2)^2$. So $\Delta_{A/R} = 4^2(t^2 + 4s_1s_2)^2R$. If $2 \in R^*$ prove that $C(M)$ is separable if and only if $\Delta_{C(M)/R} = R$.

11. Suppose that R is a field and that A is a commutative and finite-dimensional algebra over R. Prove that A is separable if and only if T_A is nonsingular, i.e., if and only if $\Delta_{A/R} = R$.

12. Let $R = \mathbb{Z}[\sqrt{-3}]$ and $A = R[X]/(X^3 - 1)$. Show that the conclusion of Exercise 11 does not hold for R.

13. Let $A = \text{Mat}_n(R)$ and consider the basis $\mathcal{X} = \{\varepsilon_{ij} \mid 1 \le i, j \le n\}$, where ε_{ij} is the $n \times n$ matrix which has a 1 in the (i,j) entry and 0 elsewhere. Show that $\det_{\mathcal{X}} T_A = \pm(n1_R)^{(n^2)}$ and hence that $\Delta_{A/R} = n^{(n^2)}R$. In particular, $\Delta_{A/R} = R$ if and only if $n1_R \in R^*$.

14. Does the conclusion of Exercise 11 hold if the assumption that A is commutative is omitted ?

15. Let M be a quadratic module over R which is an orthogonal sum of hyperbolic planes. Show that $\Delta_{C(M)/R} = R$ if and only if $2 \in R^*$. Does this remain true for the Clifford algebra $C(M)$ of any free quadratic module M of finite rank ?

16. In number theoretic situations there is a discriminant defined for a maximal order A via the "reduced trace," which is trivial if and only if A is separable. Refer to the Λ-Construction in Chapter 14B for details and references to the literature. It would be interesting to know whether this can be generalized to an algebra A over a commutative ring, i.e., is it possible

to define (under suitable assumptions, e.g., A finitely generated projective, faithful) a "discriminant" for A which is trivial if and only if A is separable ?

Hints:

1. Finitely generated projective modules over R are free.

2. Assume char $F \neq 2$. If F is nondyadic use 63:9 in O'Meara [1971] and (3.4). If F is dyadic, $Qu_f(F)$ has order $2^{(2+ki)}$ by the same argument. As to $Qu_f(R)$, by page 160 of O'Meara, in particular by 63:2, any ε in R^* has the form $\varepsilon = \delta^2 + \alpha$, with $\alpha \in 4R$. So by (3.6), $Qu_f(R) \cong R^*/(R^*)^2$. Now apply 63:9 of O'Meara once more. If char $F = 2$, then by Exercise 10 in Chapter 3E, $Qu_f(F) \cong F/\wp(F)$ and $Qu_f(R) \cong R/\wp(R)$. That $Qu_f(F)$ is infinite follows from Lemma 1 in Sah [1960]. For the last statement, show that $\pi R \subseteq \wp(R)$ and use the fact that for any finite field K with char $K = 2$, $K/\wp(K)$ has order 2.

3. Use Example 3 in Chapter 3B. Also, if R is a semilocal ring (i.e., one with only finitely many maximal ideals), then Pic(R) is trivial.

6. Routine linear algebra.

8. Use Exercise 3 in Chapter 1E. Show that $\det_{\chi} T_A = a^2 + 4b$, where χ is the basis $\{1, v = X + (X^2 - aX - b)\}$ of S. In the second part, let $\{1, \sigma\}$ be the Galois group of E/F. Let $\{x_1, x_2\}$ be any basis of E over F contained in S, express it in terms of χ, and show that $(x_1 x_2^\sigma - x_1^\sigma x_2)^2$ is a multiple of $a^2 + 4b$.

10. Make use of Chapter 6B. Show that $\det_{\chi} f = (t^2 + 4s_1 s_2)^2$. Apply (6.4).

11. See the Lemma of Chapter 10.5 of Jacobson [1985, II] and Proposition 4.1 on page 85 of the lecture notes Knus-Ojanguren [1974].

12. By Exercise 6 of Chapter 2E, A is a separable R-algebra. Since 27 is not a unit in R, see Samuel, T_A is not nonsingular by Exercise 9.

13. Since $\varepsilon_{ij}\varepsilon_{km} = \delta_{jk}\varepsilon_{im}$, where δ_{jk} is the Kronecker delta, $\text{Tr}(\varepsilon_{ij}\varepsilon_{ji}) = n$ and $\text{Tr}(\varepsilon_{ij}\varepsilon_{km}) = 0$ otherwise. It follows that the matrix of T in the basis $\mathcal{X}$ is of the form nP, where P is a $n^2 \times n^2$ permutation matrix.

14. By Exercise 3 of Chapter 2E, the R-algebra $A = \text{Mat}_n(R)$ is always separable. However, by Exercise 13, $\Delta_{A/R} = R$ if and only if $n \in R^*$.

15. Note that rank $M = 2m$. Use Exercise 2 in Chapter 5D and induction to show that $C(M) \cong \text{Mat}_{2m}(R)$. Now apply Exercise 13. I don't know the answer to the second question (a localization argument in combination with (4.16), and Exercise 10 might provide the answer).

13
Brauer Groups and Witt Groups

Overview

The final three chapters of this book will focus on connections and applications of the theory presented so far to some important concerns in mathematics especially the theory of algebras and quadratic forms. This chapter recalls the Brauer group, the Brauer-Wall group, and the Witt groups. We shall see that these concepts are closely related to each other and to the themes of this book. The links are provided by the invariants of quadratic forms, and in particular by the Arf invariant (already defined in Chapters 7C and 10D in the free case). In view of the extensive literature on these topics, proofs or sketches of proofs are provided only when they seemed not available in appropriately explicit form. The constructions are presented in the generality of commutative rings and then illustrated in the classical situations: over the real and complex numbers, and local and global fields.

A. Brauer and Brauer-Wall Groups

For the details of the subjects outlined here, refer to Auslander-Goldman, Wall [1964], Bass [1967], Small [1971], DeMeyer-Ingraham, the lecture notes Knus-Ojanguren [1974], Orzech-Small, and Knus [1991]. For the number theoretical aspects, see Deuring, Cassels-Fröhlich, Weil, and Reiner, and for connections with algebraic geometry, refer to Grothendieck and Manin.

We begin by defining an equivalence relation on the collection of Azumaya algebras over R. Let P be an R-module which is faithful and finitely generated projective. For such a P, $\mathrm{End}_R P$ is an Azumaya algebra over R (see Exercise 5 in Chapter 2E). While $\mathrm{End}_R P$ is hardly a trivial algebra — even in the special case of a finite-dimensional vector space over a field — it will be considered a "trivial" Azumaya algebra for purposes of the current discussion.

Let A and B be two Azumaya algebras over R. We say that A and B are *Brauer equivalent* if there are two faithful finitely generated projective R-modules P and Q such that

$$A \otimes_R \text{End}_R P \cong B \otimes_R \text{End}_R Q.$$

So A and B are Brauer equivalent if they are isomorphic up to tensor product by trivial algebras. Note that if A and B are isomorphic, then they are equivalent (just take P = Q = R). There are a number of equivalent ways of defining this equivalence relation. One of these asserts that A and B are equivalent if and only if the respective categories of finitely generated projective modules (either left or right) over A and B are equivalent categories.

Let A be an Azumaya algebra over R. Denote its equivalence class by [A]. With the product operation $[A][B] = [A \otimes_R B]$, the set of equivalence classes is an Abelian group. This is the *Brauer group*

$$Br(R)$$

of R. It measures the complexity of the theory of Azumaya algebras over R. It is clear that $[R] = 1$ and in turn that $[\text{End}_R P] = 1$ for any faithful finitely generated projective P. Let A^o be the opposite algebra of A. The assignment which sends a pair (a, b^o) with a in A and b^o in A^o to the endomorphism given by $x \rightarrow axb$, defines an isomorphism

$$A \otimes_R A^o \cong \text{End}_R A.$$

Since A is faithful and finitely generated projective as an R-module, it follows that $[A][A^o] = 1$. Therefore, $[A]^{-1} = [A^o]$ for any [A] in Br(R). Observe that if A has an antiautomorphism, then $A^o \cong A$. In this case, $[A]^2 = 1$ and hence, $[A] \in Br(R)_2$.

When R = F is specialized to be a field, then Br(F) is the classical Brauer group. Its elements are in one-to-one correspondence with the isomorphism classes of finite-dimensional central division algebras over F. The fact that $\mathbb{C}$ is algebraically closed implies, for example, that

$$Br(\mathbb{C}) = 1.$$

Frobenius' theorem asserts that $\mathbb{R}$ and the quaternion algebra $\left(\frac{-1,-1}{\mathbb{R}}\right)$ are up to isomorphism the only finite-dimensional central simple algebras over $\mathbb{R}$. Consequently,

$$Br(\mathbb{R}) \cong \mathbb{Z}_2.$$

Let F be a local field. The *Hasse invariant* defines an isomorphism

$$\text{Br}(F) \cong \mathbb{Q}/\mathbb{Z}.$$

The equivalence class of the quaternion division algebra $\left(\frac{u, \pi}{F}\right)$ (or $\left[\frac{u, \pi}{F}\right]$ in case char $F = 2$) — refer to Chapter 11E — corresponds to $\frac{1}{2} + \mathbb{Z}$ under this map. The fact that $\frac{1}{2} + \mathbb{Z}$ is the unique element of order 2 in $\mathbb{Q}/\mathbb{Z}$ is confirmation of the fact that this algebra is the unique (again of course up to isomorphism) nontrivial finite-dimensional division algebra over F with anti-automorphism. See Exercise 13 in Chapter 11F.

Notice that if F is the complex numbers, the real numbers, or a local field, then $\text{Br}(F)$ is isomorphic to a subgroup of $\mathbb{Q}/\mathbb{Z}$. In the first case this is $\{0\}$, in the second case it is $\{0, \frac{1}{2} + \mathbb{Z}\}$, and in the last it is the entire group.

Now let F be a global field. So if char $F = 0$, F is an algebraic number field, i.e., a finite extension of $\mathbb{Q}$, and if char $F = p \neq 0$, F is an algebraic function field, i.e., a finite extension of the field of fractions $K(X)$ of $K[X]$ for some finite field K with char $K = p$. Let Ω be a *complete* set of valuations of F. This means that all valuations in Ω are nontrivial and that any non-trivial valuation on F is equivalent to exactly one in Ω. Archimedean valuations occur in Ω if and only if F is an algebraic number field. Let $v \in \Omega$. If v is Archimedean, then its restriction to $\mathbb{Q}$ is equivalent to the ordinary absolute value, and the completion F_v is isomorphic (as field and metric space) to either $\mathbb{R}$ or $\mathbb{C}$; v is called *real* in the first case and *complex* in the second. Let v be non-Archimedean. Then v is discrete and the completion F_v is a local field. If F is an algebraic number field, then the restriction of v to $\mathbb{Q}$ is equivalent to one of the p-adic valuations. If F is an algebraic function field, the restriction of v to $K(X)$ is equivalent to v_0 or v_p for some p; see Examples 1 and 2 of Chapter 11E. A non-Archimedean valuation is also called *finite*. Note finally that only finitely many valuations in Ω restrict to a given valuation of $\mathbb{Q}$ or $K(X)$, and every valuation on $\mathbb{Q}$ or $K(X)$ is equivalent to such a restriction.

The following notation will be in effect. If $\{G_i\}_{i \in I}$, for some finite or countably infinite index set I, is a family of multiplicative groups, we denote by $\prod_{i \in I} G_i$ the *restricted Cartesian product* of the G_i. This is the subgroup of the full Cartesian product consisting of all tuples with all but finitely many components equal to 1. In order to be consistent with the notation used for the direct sum of modules, $\prod_{i \in I} G_i$ will be written $\bigoplus_{i \in I} G_i$ if the G_i are additive.

For $v \in \Omega$, consider the completion F_v. The map $[A] \to [A \otimes_F F_v]$ given by change of scalars provides a homomorphism

$$\beta_v : \text{Br}(F) \to \text{Br}(F_v).$$

The image of a given [A] is trivial for all but finitely many v, and we can put these maps together component-wise to get $Br(F) \xrightarrow{\beta} \prod_{v \in \Omega} Br(F_v)$. Recall that F_v is isomorphic to $\mathbb{C}$ if v is complex, to $\mathbb{R}$ if v is real, and that F_v is a local field in all other cases, i.e., for all finite v. We therefore obtain a homomorphism $\prod_{v \in \Omega} Br(F_v) \longrightarrow \bigoplus_{v \in \Omega} \mathbb{Q}/\mathbb{Z}$ and by composition with $\bigoplus_{v \in \Omega} \mathbb{Q}/\mathbb{Z} \xrightarrow{sum} \mathbb{Q}/\mathbb{Z}$, the *(global) Hasse invariant*

$$\text{inv} : \prod_{v \in \Omega} Br(F_v) \longrightarrow \mathbb{Q}/\mathbb{Z}.$$

The spectacular classical theory of Hasse-Brauer-Noether-Albert provides the following characterization of Br(F).

Theorem. The sequence

$$1 \to Br(F) \xrightarrow{\beta} \prod_{v \in \Omega} Br(F_v) \xrightarrow{\text{inv}} \mathbb{Q}/\mathbb{Z} \to 0$$

is exact.

We return to an arbitrary commutative ring R and proceed to the graded version of the Brauer group.

Let $A = A_0 \oplus A_1$ be a graded R-algebra. Since $A_i A_j \subseteq A_{i+j}$, notice that if a and b are in A_{hom}, then $ab \in A_{hom}$. In terms of the degree, $\partial(ab) = \partial a + \partial b$. Define the *graded center* CEN A of A to be the R-linear span of

$$\{a \in A_{hom} \mid ab = (-1)^{\partial a \partial b} ba \text{ for all } b \in A_{hom} \}$$

in A. We say that a graded and separable algebra A is a *graded Azumaya algebra* if $CEN A = R1_A$. Notice that an ordinary Azumaya algebra provided with the trivial grading is a graded Azumaya algebra.

(13.1). *Let M be a faithful, finitely generated projective nonsingular quadratic module over R. Then C(M) is a graded Azumaya algebra over R.*

Proof. The grading on C(M) is the grading $C(M) = C_0(M) \oplus C_1(M)$ given in Chapter 5B. By Theorem (9.9), C(M) is separable. The requirement CEN C(M) = R remains. The verification of this will make use of the localization results of Chapter 9D, especially Theorem (9.11). Observe first that CEN C(M) is contained in the Arf algebra $A(M)$. Let $\mathfrak{p}$ be a prime ideal of R. If rank $M_{\mathfrak{p}}$ is odd, then CEN $C(M_{\mathfrak{p}})$ is contained in $A(M_{\mathfrak{p}}) = \text{Cen } C(M_{\mathfrak{p}})$. Let $c \in \text{CEN } C(M_{\mathfrak{p}})$. If $c \in C_1(M_{\mathfrak{p}})$, then for all x in $M_{\mathfrak{p}}$, $cx = xc = -cx$. So by (4.18), $cx = 0$, and by (5.2), $c = 0$. Therefore, CEN $C(M_{\mathfrak{p}}) \subseteq C_0(M_{\mathfrak{p}})$. So, CEN $C(M_{\mathfrak{p}}) \subseteq \text{Cen } C_0(M_{\mathfrak{p}}) = R_{\mathfrak{p}}$, and CEN $C(M_{\mathfrak{p}}) = R_{\mathfrak{p}}$. Assume rank $M_{\mathfrak{p}}$ is even. Then $A(M_{\mathfrak{p}}) \subseteq C_0(M_{\mathfrak{p}})$, and it follows that CEN $C(M_{\mathfrak{p}}) = \text{Cen } C(M_{\mathfrak{p}}) = R_{\mathfrak{p}}$. An application of the change-of-scalars construction of Chapter 5C now shows that $R_{\mathfrak{p}} \to (\text{CEN } C(M))_{\mathfrak{p}}$ is surjective. Since this is so for any $\mathfrak{p}$, the inclusion $R \to \text{CEN } C(M)$ is surjective and the proof is complete.   QED.

If P is a faithful finitely generated projective graded R-module, then $\text{END}_R P$ is a graded Azumaya algebra (refer to Chapter 8C). These are the "trivial" graded Azumaya algebras. We say that two graded Azumaya algebras A and B are *Brauer-Wall equivalent* if there exist two faithful finitely generated projective graded R-modules P and Q such that

$$A \overset{\wedge}{\otimes}_R \text{END}_R P \cong B \overset{\wedge}{\otimes}_R \text{END}_R Q$$

as graded algebras. Denote by $[A]_g$ the equivalence class of A. With $[A]_g [B]_g = [A \overset{\wedge}{\otimes}_R B]_g$, the set of equivalence classes is a group. This is the *Brauer-Wall group*

$$BW(R)$$

of R. The identity element is $[R]_g$, where R has the trivial grading. The inverse of $[A]_g$ is given by a "twisted" opposite algebra of A, which is defined as follows: Start with the additive group A and define a multiplication for elements a and b in A_{hom} by $a \cdot b = (-1)^{\partial a \partial b} ba$. Extending $\cdot$ to all of A by use of the distributive law defines the R-algebra A^{go}. The grading of A provides one for A^{go}, and in this way A^{go} is a graded Azumaya algebra. The isomorphism $A \otimes_R A^o \to \text{End}_R A$ discussed earlier is a graded algebra isomorphism $A \overset{\wedge}{\otimes}_R A^{go} \to \text{END}_R A$. So $[A]_g^{-1} = [A^{go}]_g$.

It turns out that the natural homomorphism

$$Br(R) \to BW(R)$$

given by $[A] \to [A]_g$, where A is supplied with the trivial grading, is injective. In this way, we consider $Br(R) \subseteq BW(R)$. The structure of the quotient $BW(R)/Br(R)$ is described by a graded version of the group $Qu(R)$, which is developed next.

B. The Graded Quadratic Group QU(R)

A separable quadratic algebra over R which is graded is called a *graded separable quadratic algebra over* R. In view of Chapter 7A and Theorem (9.11), the Arf algebra $A(M)$ of a nonsingular faithful finitely generated projective quadratic module M over R is a graded separable quadratic algebra. The graded quadratic algebra $(0, b)^1$ of Chapter 3C is an example of a free separable quadratic algebra with nontrivial grading.

(13.2). *Let S be a free separable quadratic algebra. Assume that S has a non-trivial grading* $S = S_0 \oplus S_1$ *where both S_0 and S_1 are free. Then*

(1) $S_0 = R$ *and* $S_1 = $ dis S, *and*

(2) $2 \in R^*$ *and* S *is isomorphic to a graded algebra* $(0, b)^1$.

Proof. In view of Exercise 3 in Chapter 1E and (2.2), we may assume that $S = R[X]/(X^2 - cX - d)$ with $c^2 + 4d$ in R^*. Note that both S_0 and S_1 are free of rank 1. Consider the basis $\{1, v = X + (X^2 - cX - d)\}$ of S. Let $r + r'v$ be a basis of S_0. Since $S_0 \supseteq R$, $t(r + r'v) = 1$ for some t in R. It follows that t is a unit in R and that $r' = 0$. Therefore, $S_0 = R$. Now let $r + r'v$ be a basis of S_1. Since $\{1, r + r'v\}$ is a basis of S, r' must be a unit. After multiplying $r + r'v$ by $-(r')^{-1}$, the basis of S_1 has the form $r - v$. Since $S_1 S_1 \subseteq S_0$,

$$(r - v)^2 = r^2 - 2rv + (d + cv) = (r^2 + d) - (2r - c)v \in R.$$

So $2r - c = 0$. Since $4(r^2 + d) = 4r^2 + 4d = c^2 + 4d$ is in R^*, $2 \in R^*$. Since $2(r - v) = c - 2v$, we find by the Example in Chapter 10B, that $S_1 = R(c - 2v)$ $ = $ dis S. It remains to prove that S is isomorphic to some $(0, b)^1$. Put $b = c^2 + 4d$ and consider the graded algebra $A = (0, b)^1 = R[X]/(X^2 - b)^{odd}$. Put $w = X + (X^2 - b)$. Check that $1 \to 1$ and $w \to -c + 2v$ defines an isomorphism from A onto S. Since A_1 has w as basis, this isomorphism is graded.

QED.

(13.3). *Let* S *be a separable quadratic algebra with conjugation* σ. *Let* S = $S_0 \oplus S_1$ *be any grading of* S. *Then* $\sigma S_0 = S_0$ *and* $\sigma S_1 = S_1$.

Proof. It suffices to show that σ maps S_0 into S_0 and S_1 into S_1. To do this for S_0, we show that the composite $S_0 \xrightarrow{\sigma} S \to S_0$, where the second map is the projection, is injective. For this, it suffices to prove for any prime ideal $\mathfrak{p}$ of R, that the localization $(S_0)_\mathfrak{p} \xrightarrow{\sigma_\mathfrak{p}} S_\mathfrak{p} \to (S_0)_\mathfrak{p}$ of this composite is injective. If the grading $S_\mathfrak{p} = (S_0)_\mathfrak{p} \oplus (S_1)_\mathfrak{p}$ of the quadratic algebra $S_\mathfrak{p}$ is trivial, this is clear. If it is nontrivial, then by (13.2,1), $\sigma_\mathfrak{p}(S_0)_\mathfrak{p} = (S_0)_\mathfrak{p}$ and $\sigma_\mathfrak{p}(S_1)_\mathfrak{p} = (S_1)_\mathfrak{p}$. So the localization of the composite is injective in this case also. The proof for S_1 is the same.   QED.

The fact that σ has the preceding property is crucial in the definition of the graded version of the quadratic group Qu(R). Let S and S' be graded separable quadratic algebras over R and let σ and σ' be the respective conjugations. Consider the algebra $S \overset{\wedge}{\otimes}_R S'$ and define

$$S * S' = \{c \in S \overset{\wedge}{\otimes}_R S' \mid c^{\sigma \otimes \sigma'} = c\}.$$

It can be shown by a localization argument analogous to that in the ungraded case, that S * S' is a graded separable quadratic algebra whose conjugation is $\sigma \otimes \mathrm{id}_{S'} = \mathrm{id}_S \otimes \sigma'$. The details are left to the reader.

Denote by [S] and [S'] the graded isomorphism classes of S and S' respectively. As in the ungraded case, the set

QU(R)

of graded isomorphism classes can be made into a group – the *graded quadratic group of* R – by defining [S][S'] = [S * S']. Reference to Sections C and D of Chapter 3, particularly to (3.9), shows that the group $QU_f(R)$ is a subgroup of QU(R). If R is a field, in fact more generally, if R is a principal ideal domain or a local ring (finitely generated projective modules are free for both), then $QU(R) = QU_f(R)$. Let S be any separable quadratic algebra and let [S] be the corresponding element in Qu(R). Supplying S with the trivial grading gives rise to an element [S] in QU(R). A moment's reflection shows that we have embedded Qu(R) into QU(R) and that we can regard Qu(R) as a subgroup of QU(R).

The existence of graded separable quadratic algebras over R is directly related to the existence of idempotents in R. We will see, more precisely, that the structure of the quotient $QU(R)/Qu(R)$ is determined by the structure of the set of idempotents of R.

Example. Let e be an idempotent of R. The principal ideal Re is a ring with identity element e. Assume that 2e is an invertible element in this ring, i.e., that $2r'e = e$ with r' in R. Consider the quadratic algebra $S = R[X]/(X^2 - X)$. Set $v = X + (X^2 - X)$ and consider the submodules

$$S_0 = \{r1 + tv \mid et = 0\} \quad \text{and} \quad S_1 = \{r1 + 2rv \mid (1 - e)r = 0\}$$

of S. We claim that $S = S_0 \oplus S_1$ is a grading of S. Since $S_0 \supseteq R$ and $v = (-r'e1 + (1 - e)v) + (r'e1 + 2r'ev)$, $S = S_0 + S_1$. If $r1 + tv \in S_0 \cap S_1$, then $t = 2r$, $2re = 0$, and $(1 - e)r = 0$. So $t = 2r = 0$ and $r = er = 2r'er = 0$. Thus $S = S_0 \oplus S_1$. It remains to be shown that $S_i S_j \subseteq S_{i+j}$ for i and j in $\mathbb{Z}_2$. But this is routine. For example, for $r1 + tv$ in S_0 and $r_1 1 + 2r_1 v$ in S_1,

$$(r1 + tv)(r_1 1 + 2r_1 v) = rr_1 1 + (2rr_1 + tr_1 + 2tr_1)v = rr_1 1 + 2rr_1 v \in S_1,$$

since $tr_1 = tr_1(e + (1 - e)) = 0$. Observe that the grading $S = S_0 \oplus S_1$ is trivial if $e = 0$, and that it is nontrivial otherwise.

Let $S = S_0 \oplus S_1$ be a graded separable quadratic algebra over R. Let $\mathfrak{p}$ be any prime ideal of R and consider the localization $S_\mathfrak{p} = (S_0)_\mathfrak{p} \oplus (S_1)_\mathfrak{p}$. Since rank $S_\mathfrak{p} = 2$ and $(S_0)_\mathfrak{p} \supseteq R_\mathfrak{p}$, it follows that rank $(S_1)_\mathfrak{p}$ is either 0 or 1. Since the map Spec $R \to \mathbb{Z}$ given by $\mathfrak{p} \to$ rank $(S_1)_\mathfrak{p}$ is continuous, the inverse images of 0 and 1 in Spec R are both open and hence both closed. So by the discussion preceding the proof of Theorem (9.10), there exists a unique idempotent e in R such that

$$\text{rank } (S_1)_\mathfrak{p} = 0 \Leftrightarrow \mathfrak{p} \supseteq Re \quad \text{and} \quad \text{rank } (S_1)_\mathfrak{p} = 1 \Leftrightarrow \mathfrak{p} \supseteq R(1 - e).$$

We call e the *idempotent of* S. For any r in R let $r_\mathfrak{p}$ be the image of r under $R \to R_\mathfrak{p}$. If $e \in \mathfrak{p}$, then $1 - e \notin \mathfrak{p}$ and hence $(1 - e)_\mathfrak{p} \in R_\mathfrak{p}^*$. Since $e(1 - e) = 0$, $e_\mathfrak{p} = 0$ and $(1 - e)_\mathfrak{p} = 1$. These considerations show that

$$\text{rank } (S_1)_\mathfrak{p} = 0 \Leftrightarrow e_\mathfrak{p} = 0 \quad \text{and} \quad \text{rank } (S_1)_\mathfrak{p} = 1 \Leftrightarrow e_\mathfrak{p} = 1.$$

Let S' be another graded separable quadratic algebra. Let e' be the idempotent of S' and let e" be that of $S * S'$. How is e" related to e and e'? Again let $\mathfrak{p}$ be a prime of R. Since $S_{\mathfrak{p}}$ and $S'_{\mathfrak{p}}$ are free quadratic over $R_{\mathfrak{p}}$, we can put $S_{\mathfrak{p}} = R_{\mathfrak{p}}[X]/(X^2 - aX - b)$ and $S'_{\mathfrak{p}} = R_{\mathfrak{p}}[X]/(X^2 - cX - d)$ with a, b, c, and d in $R_{\mathfrak{p}}$ and $a^2 + 4b$ and $c^2 + 4d$ in $R^*_{\mathfrak{p}}$. Assume, for example, that $e_{\mathfrak{p}} = e'_{\mathfrak{p}} = 1$. So rank $(S_1)_{\mathfrak{p}} = 1$ and rank $(S'_1)_{\mathfrak{p}} = 1$. By (13.2), $S_{\mathfrak{p}} \cong (0, b')^1$ and $S'_{\mathfrak{p}} \cong (0, d')^1$. Since

$$(S * S')_{\mathfrak{p}} \cong S_{\mathfrak{p}} * S'_{\mathfrak{p}} \cong (0, b')^1 * (0, d')^1,$$

we find by an application of (3.9) that $(S * S')_{\mathfrak{p}}$ has the trivial grading, i.e., that $e''_{\mathfrak{p}} = 0$. Arguing in the same way in the other cases shows that

$$e''_{\mathfrak{p}} = 0 \iff \left(e_{\mathfrak{p}} = 0 \text{ and } e'_{\mathfrak{p}} = 0 \right) \text{ or } \left(e_{\mathfrak{p}} = 1 \text{ and } e'_{\mathfrak{p}} = 1 \right) .$$

Consider the idempotent $e * e' = e + e' - 2ee'$ and observe that $(e * e')_{\mathfrak{p}}$ and $e''_{\mathfrak{p}}$ are equal at all primes $\mathfrak{p}$ (i.e., if $e''_{\mathfrak{p}} = 0$, then $(e * e')_{\mathfrak{p}} = 0$, and if $e''_{\mathfrak{p}} = 1$ then $(e * e')_{\mathfrak{p}} = 1$). Therefore by (4.1), $e'' = e * e'$.

Denote the set of idempotents of R by

$$\mathbb{Z}_2(R) .$$

With the operation $e * e' = e + e' - 2ee'$, $\mathbb{Z}_2(R)$ is a group. Its identity is 0, and since $e * e = 0$, it is an elementary Abelian 2-group. It follows from the preceding discussion that there is a homomorphism

$$QU(R) \to \mathbb{Z}_2(R)$$

defined by sending [S] to the idempotent of S.

(13.4). *The sequence* $1 \to Qu(R) \to QU(R) \to \mathbb{Z}_2(R)$ *is exact. The image of* $QU(R) \to \mathbb{Z}_2(R)$ *is the subgroup* $\{e \in \mathbb{Z}_2(R) \mid e = 2re \text{ for some } r \in R\}$.

Proof. Observe first that $[S] \to 0$ if and only if S has the trivial grading. It remains to check the assertion about the image. Let e be the idempotent of a graded separable quadratic algebra S over R. Let $\mathfrak{p}$ be a prime ideal of R. If rank $(S_1)_{\mathfrak{p}} = 1$, then by (13.2), 2 is in $R^*_{\mathfrak{p}}$. So either $e_{\mathfrak{p}} = 0$ or $2 \in R^*_{\mathfrak{p}}$. Therefore, $(Re/2Re)_{\mathfrak{p}} \cong (Re)_{\mathfrak{p}}/(2Re)_{\mathfrak{p}}$ is the zero module. Since this is so for any $\mathfrak{p}$, Re/2Re = {0}. Therefore, e = 2re for some r in R. This shows in combination with the earlier example that the image of $QU(R) \to \mathbb{Z}_2(R)$ is as stated.   QED.

Observe that if $2 \in R^*$, then $QU(R) \to \mathbb{Z}_2(R)$ is surjective. Conversely, if 1 is in the image, then $2 \in R^*$. If for some e, e and $1 - e$ are both in the image, then $e * (1 - e) = 1$ is in the image and the map is surjective.

(13.5). *Suppose that either*

(1) *the only idempotents of* R *are* 0 *and* 1 *and that* $2 \notin R^*$, *or*

(2) $2 \in$ Rad R.

Then $QU(R) = Qu(R)$.

Proof. In case (1), $\mathbb{Z}_2(R) = \{0, 1\}$. Since $2 \notin R^*$, 1 is not in the image of $QU(R) \to \mathbb{Z}_2(R)$. So this map is trivial. Therefore, $Qu(R) \to QU(R)$ is onto. In case (2), any e in the image of $QU(R) \to \mathbb{Z}_2(R)$ is in the Jacobson radical of R. So $1 - e$ is a unit of R. Since $1 - e$ is an idempotent, $1 - e = 1$, so $e = 0$. Again, the map is trivial.   QED.

Remark. Condition (1) is satisfied for any domain or local ring in which 2 is not invertible.

We now return to the injection

$$Br(R) \to BW(R)$$

given by $[A] \to [A]_g$. Let $[A]_g \in BW(R)$ be arbitrary. By replacing A with the graded tensor product of A with $END_R(R \oplus R)$ we can assume that both components of the grading $A = A_0 \oplus A_1$ are faithful. It turns out that the centralizer $Cen_A A_0$ of A_0 in A is a graded separable quadratic algebra. If A is the Clifford algebra of a nonsingular faithful finitely generated projective module over R, this fact is provided by Theorem (9.11). Refer to Small [1971] or Knus [1991] for the proof of the theorem that follows.

(13.6). *Theorem. The assignment* $[A]_g \to [Cen_A A_0]$ *defines a homomorphism* $BW(R) \to QU(R)$ *and the sequence*

$$1 \to Br(R) \to BW(R) \to QU(R) \to 1$$

is exact.

Remark. If $R = F$ is a field with char $F \neq 2$, then $QU(F) \cong \mathbb{Z}_2 \times F^*/(F^*)^2$ by (3.8). In this case, Lam (see pages 118 and 119) constructs $BW(F)$ by "glueing" $Br(F)$, $\mathbb{Z}_2$ and $F^*/(F^*)^2$ together.

Examples. Since $Br(\mathbb{C}) = 1$, $BW(\mathbb{C}) \cong QU(\mathbb{C}) \cong \mathbb{Z}_2$. Since $Br(\mathbb{R}) \cong \mathbb{Z}_2$, card $BW(\mathbb{R}) = 8$. By Lam's construction, $BW(\mathbb{R})$ has an element of order 8. So $BW(\mathbb{R}) \cong \mathbb{Z}_8$. Since $Br(\mathbb{Z}) = 1$ (see Chapter 14B, for example) and $QU(\mathbb{Z}) = Qu(\mathbb{Z}) = Qu_f(\mathbb{Z}) = 1$ (by (13.5) and Exercise 1 in Chapter 12F), $BW(\mathbb{Z}) = 1$.

C. The Witt Group of Quadratic Forms

Let R be a commutative ring. A finitely generated projective nonsingular quadratic module over R will from now on be called a *quadratic space over* R. An equivalence relation can be defined on the collection of quadratic spaces over R in a natural way and the resulting set of equivalence classes can be made into a group, the quadratic Witt group of R. This group is an analogue of the Brauer group of R; indeed there is a direct link with the Brauer group (actually the Brauer-Wall group) via the Clifford algebra. The construction of the quadratic Witt group and the related invariants of quadratic forms are the focal points of this section. The presentation is mostly in outline form. The reader is referred to Bass [1967] and [1974], Milnor-Husemoller, Baeza [1978], Micali-Revoy and Knus [1991] for the details of the theory over R; to Serre [1973] for the theory over $\mathbb{Z}$; to Lam, Knebusch-Kolster, and Scharlau [1985] for the (by now) classical theory over fields; and finally to Knebusch [1977] and Knus [1991] for the connections with algebraic geometry.

A quadratic space (M, q) is *hyperbolic* if it contains a direct summand P, i.e., a submodule P such that $M = P \oplus Q$ for some submodule Q, such that $P = P^{\perp}$ and q is identically 0 on P. It is clear that this extends the definition of Chapter 11B from fields to commutative rings. Examples of hyperbolic spaces are constructed as follows. Let P be a finitely generated projective R-module. Its dual $P^* = \mathrm{Hom}_R(P, R)$ is also finitely generated projective, and

$$q : P \oplus P^* \to R$$

defined by $q(x, \rho) = \rho(x)$ is a quadratic form on $P \oplus P^*$. The associated symmetric bilinear form $h : (P \oplus P^*) \times (P \oplus P^*) \to R$ is given by $h((x, \rho), (x', \rho')) = \rho(x') + \rho'(x)$. In this way, $P \oplus P^*$ is a quadratic space over R. Denote it by $H(P)$ and observe that $H(P)$ is hyperbolic. It can be shown that any hyperbolic space is isometric to an $H(P)$. Hyperbolic spaces will be considered trivial in the discussion that follows. In a hyperbolic space $q(x) = 0$

holds often, and it is generally the case that the "more zeros" a quadratic form has, the easier it is to handle.

Now let M and N be quadratic spaces over R. We say that M and N are *Witt equivalent* if there exist hyperbolic spaces H(P) and H(Q) such that

$$M \perp H(P) \cong N \perp H(Q) .$$

So M and N are Witt equivalent if they are isometric up to hyperbolics. One can check that Witt equivalence is an equivalence relation in the category of quadratic spaces over R. Denote the equivalence class of M by [M]. Defining [M] + [N] = [M ⊥ N] makes the set of equivalence classes into an Abelian group. This is the *quadratic Witt group*

$$Wq(R) .$$

Suppose for a moment that R is a field. A change of basis shows that a quadratic space is hyperbolic if and only if it has the form $H_1 \perp ... \perp H_i$ for hyperbolic planes $H_1, \ldots, H_i$. Therefore by observations in Chapter 11B, any quadratic space M has a splitting

$$M = H(P) \perp N,$$

where N is the anisotropic part of M. Observe that dim H(P) = 2i, where i is the Witt index ind M of M. It follows that the statements: M is a hyperbolic space, the anisotropic part of M is {0}, and dim M = 2(ind M) are equivalent to each other. As a consequence of the Witt Cancellation Theorem, two quadratic spaces M and M' are isometric if and only if their anisotropic parts are isometric and ind M = ind M' (or dim M = dim M'). Further, the anisotropic parts of M and M' are isometric if and only if [M] = [M'] in Wq(R). Therefore, the isometry class of M is determined by [M] together with ind M (or dim M). In this sense, [M] carries the essential information about M. This is far from being true if R is a commutative ring, where a class [M] can contain spaces that are fundamentally different. For example, in the case of an arithmetic domain, spaces with completely different properties can become isometric after the (orthogonal) addition of a single hyperbolic plane to both. Such examples can be constructed over ℤ by applying the results in §1 of Chapter V of Serre [1973].

We now turn to the important invariants of quadratic forms. Let M be a faithful quadratic space over R. Let w be any element Wq(R). It follows by (4.3) and the (orthogonal) addition of a hyperbolic plane to a representative of w, that w is represented by such an M. By (13.1), the Clifford algebra C(M) is a graded Azumaya algebra over R. Define the *Clifford invariant* of M by

$$\text{Cliff } M = [C(M)]_g \in BW(R)$$

and define

$$Cliff: Wq(R) \rightarrow BW(R)$$

by $Cliff$ [M] = Cliff M. That this is a well-defined homomorphism is a consequence of Theorem (5.5) and the fact that for a hyperbolic space H(P), $C(H(P)) \cong END_R (\wedge P)$, where the grading of $\wedge P$ is given by $\wedge P = (\bigoplus_{i \geq 0} \wedge^{2i}(P)) \oplus (\bigoplus_{i \geq 0} \wedge^{2i+1}(P))$.

By Theorem (9.11), the Arf algebra A(M) is a graded separable quadratic algebra. Define the *Arf invariant*

$$Arf\ M \in QU(R)$$

of M to be the graded isomorphism class of A(M) (compare this with the definition given in Chapters 7C and 10D). Let

$$Arf: Wq(R) \rightarrow QU(R)$$

be given by Arf [M] = Arf M. It follows from the definitions that Arf is the composite of $Cliff: Wq(R) \rightarrow BW(R)$ and the map $BW(R) \rightarrow QU(R)$ of Section B. In particular, Arf is a homomorphism.

The *rank* invariant is considered next. Let $g : Spec(R) \rightarrow \mathbb{Z}$ be a continuous function. The sets

$$\{g(\mathfrak{p}) \mid g(\mathfrak{p}) \text{ is even or } 0\} \quad \text{and} \quad \{g(\mathfrak{p}) \mid g(\mathfrak{p}) \text{ is odd}\}$$

partition Spec(R) into two disjoint open and closed sets. Therefore as in the discussion preceding Theorem (9.10), there is a unique idempotent e such that $e \in \mathfrak{p}$ if and only if $g(\mathfrak{p})$ is even and $1 - e \in \mathfrak{p}$ if and only if $g(\mathfrak{p})$ is odd. The set $\{g : Spec(R) \rightarrow \mathbb{Z} \mid g \text{ is continuous}\}$ inherits an additive group structure from $\mathbb{Z}$, and $g \rightarrow e$ defines a homomorphism

$$\{g : Spec(R) \rightarrow \mathbb{Z} \mid g \text{ is continuous}\} \rightarrow \mathbb{Z}_2(R) .$$

Applying it to rank M : Spec(R) $\rightarrow \mathbb{Z}$ provides an idempotent e such that rank $M_{\mathfrak{p}}$ is even $\Leftrightarrow e \in \mathfrak{p}$, and rank $M_{\mathfrak{p}}$ is odd $\Leftrightarrow 1 - e \in \mathfrak{p}$. We refer to e as the *rank idempotent* of M. Notice that the rank idempotent of M is 0 if and only if the rank of M − in the sense of Chapter 4A − is even, and that it is 1 if and only if the rank of M is odd.

It is not hard to check that

$$rank : Wq(R) \rightarrow \mathbb{Z}_2(R)$$

defined by $rank\ [M] = e$ is a homomorphism. By Chapter 9D, $A(M_{\mathfrak{p}}) \cong A(M)_{\mathfrak{p}}$ for any $\mathfrak{p} \in \mathrm{Spec}(R)$. Therefore by Chapter 10C, the grading of $A(M)_{\mathfrak{p}}$ is trivial $\Leftrightarrow$ rank $M_{\mathfrak{p}}$ is even $\Leftrightarrow$ $e \in \mathfrak{p}$. This shows that $rank$ is the composite of Arf and $QU(R) \to \mathbb{Z}_2(R)$. The kernel

$$Wq_{ev}(R)$$

of $rank$ is the set of $[M]$ in $Wq(R)$ with rank M even.

Write the kernel of Arf as $\mathrm{Ker}\ Arf$ and that of $Cliff$ as $\mathrm{Ker}\ Cliff$. Observe that

$$\mathrm{Ker}\ Cliff \subseteq \mathrm{Ker}\ Arf \subseteq Wq_{ev}(R) \subseteq Wq(R) .$$

It follows by (13.6) that the image of $\mathrm{Ker}\ Arf$ under $Cliff$ is contained in $\mathrm{Br}(R)$. Since a Clifford algebra comes equipped with an anti-involution, this image is in fact contained in $\mathrm{Br}(R)_2$. Since $rank$ is the composite of Arf and $QU(R) \to \mathbb{Z}_2(R)$, we find by (13.4) that the image of $Wq_{ev}(R)$ under Arf is contained in $\mathrm{Qu}(R)$. Therefore the restrictions of Arf and $Cliff$ provide the homomorphisms

$$Cliff : \mathrm{Ker}\ Arf \to \mathrm{Br}(R)_2 \quad \text{and} \quad Arf : Wq_{ev}(R) \to \mathrm{Qu}(R).$$

It is clear that their kernels are $\mathrm{Ker}\ Cliff$ and $\mathrm{Ker}\ Arf$ respectively. These kernels are described in number theoretic situations in Chapter 14.

Remark 1. Let $[M] \in \mathrm{Ker}\ Arf$. So rank M is even. By Theorem (9.11), $C(M)$ is an Azumaya algebra and we can consider $[C(M)] \in \mathrm{Br}(R)_2$. It is a fact that $Cliff\ [M] = [C(M)]$. Restated, this says that $C(M)$ with the usual grading is Brauer-Wall equivalent to $C(M)$ with the trivial grading.

A discussion of the images of the various homomorphisms defined earlier follows next.

By Exercise 12 of Chapter 9E, $rank : Wq(R) \to \mathbb{Z}_2(R)$ has image

$$\{e \in \mathbb{Z}_2(R) \mid e = 2re \text{ for some } r \in R\}.$$

The maps $Arf : Wq(R) \to QU(R)$ and $Arf : Wq_{ev}(R) \to \mathrm{Qu}(R)$ are both surjective. Since $rank$ and $QU(R) \to \mathbb{Z}_2(R)$ have the same image, the surjectivity of the second map implies that of the first. Since this will be important in Chapter 14, we supply a proof (that is much more explicit than that of Micali-Revoy.

(13.7). *Theorem.* $Arf: Wq_{ev}(R) \to Qu(R)$ *is surjective.*

Proof. Let A be a separable quadratic algebra over R and let $A = Rt \oplus dis\, A$ be the decomposition given by (10.7). Denote the conjugation of A by α. Let $nr(x) = xx^{\alpha}$ be the norm form of A and h the associated bilinear form. By Exercise 2 of Chapter 10E, h is nonsingular. Denote the quadratic space (A, nr) by M. By Theorem (9.11) and Exercise 14 of Chapter 6E, the Arf algebra of M is $A(M) = Cen\, C_0(M) = C_0(M)$. To prove the theorem, it suffices to show that the separable quadratic algebras A and $A(M)$ are isomorphic. Put $1 = rt + d$ with $r \in R$ and $d \in dis\, A$. Taking *tr* shows that $r = 2$. Write the product of $C(M)$ as $\cdot$. Define $\varphi : A \to C_0(M)$ by $\varphi(t) = 2nr(t)1_C + t \cdot d$ and $\varphi(x) = h(t, x)1_C - 2t \cdot x$ and by extending linearly to A. To prove that φ is an isomorphism of R-algebras, it suffices (by localizing) to assume that $dis\, A$ is free. So let s be a basis of $dis\, A$. At this point, insert the proof of (10.11), (part (4) $\Rightarrow$ (1) for rank M even). Observe that $d = as$. By routine computations, see Exercise 5, $h(t, s) = au$. It follows that $\varphi(1) = (4ub + a^2u)1_C = 1_C$ and $\varphi(uz) = t \cdot s$. In A, $(uz)^2 = u^2(az + b) = au(uz) + u^2b$, and in $C(M)$,

$$(t \cdot s)^2 = t \cdot (au - t \cdot s) \cdot s = au(t \cdot s) - nr(t)nr(s) = au(t \cdot s) + u^2b.$$

It follows that φ is an algebra isomorphism.   QED.

The maps *Cliff* : $Ker\, Arf \to Br(R)_2$ and *Cliff* : $Wq(R) \to BW(R)$ remain to be considered. Due to Hasse's deep and far-reaching contributions to both quadratic forms and the Brauer group, the designations of these images as *Hasse group* and *Hasse-Wall group*, reespectively, and the corresponding notation

$$Hs(R) \quad \text{and} \quad HW(R)$$

seem appropriate (see Laborde). If R is a field, then $Hs(R) = Br(R)_2$. This is the theorem of Merkurjev and Albert (see Theorem (11.8) and its proof). The same is true if R is a number theoretic domain; see Theorem (14.6) in Chapter 14B. However, there seems little doubt that there are examples of rings R for which $Hs(R) \neq Br(R)_2$ (several colleagues have assured me of this, but none has supplied a concrete example[1]). The following proposition shows that $HW(R)$ is in essence determined by $Hs(R)$ and $QU(R)$.

(13.8). *The restriction of the map* $BW(R) \to QU(R)$ *determines an exact sequence* $1 \to Hs(R) \to HW(R) \to QU(R) \to 1$.

Proof. In view of (13.6) and the surjectivity of $Arf: Wq(R) \to QU(R)$, it only needs to be shown that the image of $Hs(R) \to HW(R)$ contains the kernel of

---

[1] In their preprint *Non-surjectivity of the Clifford Invariant Map*, Parimala and Sridharan show that affine algebras over the 3-adic numbers yield such examples.

$HW(R) \rightarrow QU(R)$. Let $h \in HW(R)$ be in the kernel and let $[M] \in Wq(R)$ be a preimage of h under *Cliff*. Note that $[M] \in Ker\ Arf$ and that h is the image of $[M]$ under *Cliff* : $Ker\ Arf \rightarrow Hs(R)$.   QED.

Remark 2. Combining the exact sequences of (13.8) and (13.4) with the fact that $Qu(R)$ is an elementary Abelian 2-group shows that $HW(R)$ is a torsion group of exponent 8, i.e., that $h^8 = 1$ for all $h \in HW(R)$. If R is a domain or local ring with $2 \notin R^*$, then by (13.5), $HW(R)$ has exponent 4.

There is an invariant which can be "interposed" between *Arf* and *rank*. This is the discriminant already encountered in Chapter 10D in the free case. For $Wq_{ev}(R)$ it can be defined as the composite $Wq_{ev}(R) \xrightarrow{Arf} Qu(R) \xrightarrow{dis} Dis(R)$. It can be lifted to $Wq(R)$ as follows. The map dis : $Qu(R) \rightarrow Dis(R)$ has a "graded" analogue $QU(R) \rightarrow DIS(R)$. See Bass [1974], Micali-Revoy, or §6.3 in Chapter III of Knus [1991] (observe that the notation differs from one reference to the other). Composing it with *Arf* provides the invariant $Wq(R) \xrightarrow{Arf} QU(R) \rightarrow DIS(R)$. We will consider a different version of the discriminant momentarily. It will be formulated in the context of the Witt group of nonsingular symmetric bilinear forms.

D. The Witt Group of Symmetric Bilinear Forms

A nonsingular finitely generated projective symmetric bilinear module M over R will be called a *symmetric inner product space*. We call M *metabolic* or *split* if it contains a direct summand P such that $P = P^{\perp}$. The construction of the Witt group

$$W(R)$$

of inner product spaces over R now proceeds in a way completely analogous to that of the Witt group $Wq(R)$. Simply replace the hyperbolic spaces of the earlier construction by metabolic spaces. Refer to Milnor-Husemoller, Baeza [1978], or Scharlau [1985] for the details. The tensor product provides $W(R)$ with a multiplication, which gives it the structure of a ring. The primary focus here, however, will be on $W(R)$ as an Abelian group. For an inner product space M with bilinear form β, the corresponding element of $W(R)$ is denoted by $[M]$ or $[M, \beta]$.

Let M with quadratic form q and associated bilinear form h be a quadratic space over R. So M with h is a symmetric inner product space. Sending $[M, q]$ to $[M, h]$ defines a group homomorphism

$$Wq(R) \rightarrow W(R).$$

Suppose for a moment that $2 \in R^*$. For an inner product space (M, β) over R, define $q : M \to R$ by $q(x) = \frac{1}{2}\beta(x, x)$. Check that M with q is a quadratic space over R. The assignment $[M, \beta] \to [M, q]$ provides a map $Wq(R) \to Wq(R)$ which is an inverse of the earlier homomorphism. So if $2 \in R^*$, in particular if R is a field of characteristic not 2, then $Wq(R)$ and $W(R)$ are isomorphic. If, on the other hand, $2 = 0$ in R, then M with h is metabolic, so that $Wq(R) \to W(R)$ is the trivial map.

Define

$$rank : W(R) \to \mathbb{Z}_2(R)$$

by $rank\ [M] = e$, where e is the rank idempotent of M. This is a homomorphism whose kernel

$$W_{ev}(R)$$

is the subgroup of $W(R)$ consisting of all classes $[M]$ with rank M even.

We will assume for the remainder of this section that R is a domain. This insures that finitely generated projective modules have constant rank (which is needed in the following discussion), and it is also the case – in a number theoretic setting – which will have our attention in Chapter 14. Let M with β be a symmetric inner product space over R.

Let n be the rank of M. Consider the n-th exterior power $\Lambda^n(M)$ of M. Recall from Chapter 5D that $\Lambda^n(M)$ is finitely generated projective of rank 1. By page 30 in Bourbaki [1959], there is a unique symmetric bilinear form $\hat{\beta}$ on $\Lambda^n(M)$ with the property that

$$\hat{\beta}(x_1 \wedge x_2 \wedge ... \wedge x_n, y_1 \wedge y_2 \wedge ... \wedge y_n) = \det (\beta(x_i, y_j))$$

for all $x_1,..., x_n$ and $y_1,..., y_n$ in M. If M is free with basis $\{x_1,..., x_n\}$, then $\Lambda^n M \cong R$ is free with basis $x_1 \wedge ... \wedge x_n$. Since β is nonsingular, $\det (\beta(x_i, x_j)) \in R^*$. It follows that $\hat{\beta}$ is nonsingular. Therefore, the scaled form $(-1)^{n(n-1)/2} \hat{\beta}$ is nonsingular. So $(\Lambda^n M, (-1)^{n(n-1)/2} \hat{\beta})$ is a free discriminant module over R. By a typical localization argument, the pair $(\Lambda^n M, (-1)^{n(n-1)/2} \hat{\beta})$ is also a discriminant module in the general case. Let M and M' be symmetric inner product spaces of ranks n and n' respectively. The preceding construction supplies both $\Lambda^n(M) \otimes_R \Lambda^{n'}(M')$ and $\Lambda^{n+n'}(M \perp M')$ with the structure of discriminant modules. By basic properties of the exterior algebra, see Chapter 5D, there is an isomorphism $\Lambda^n(M) \otimes_R \Lambda^{n'}(M')$ onto $\Lambda^{n+n'}(M \perp M')$ which sends

$$(x_1 \wedge ... \wedge x_n) \otimes (y_1 \wedge ... \wedge y_{n'}) \quad \text{to} \quad (x_1 \wedge ... \wedge x_n \wedge y_1 \wedge ... \wedge y_{n'}).$$

If n and n' are both even, this is an isometry of discriminant modules; see Exercise 7. The *discriminant*

$$\text{dis } M \in \text{Dis}(R)$$

of M is defined to be the isometry class of the discriminant module $(\Lambda^n M, (-1)^{n(n-1)/2}\hat{\beta})$. Compare this with the discriminant of Chapter 10D. The isometry just referred to provides, via the assignment $[M, \beta] \to \text{dis } M$, the homomorphism

$$\textit{dis} : W_{ev}(R) \to \text{Dis}(R).$$

Let (P, f) be a discriminant module over R. The symmetric inner product space $M = (R, f_{-1}) \perp (P, f)$ satisfies dis $M \cong (P, f)$; see Exercise 8. It follows that *dis* is surjective. Evidently, the homomorphism $Wq(R) \to W(R)$ restricts to $Wq_{ev}(R) \to W_{ev}(R)$.

Let F be the field of fractions of R. For a symmetric inner product space M over R of rank n with form β, let $V = M \otimes_R F$ be the inner product space obtained by change of scalars and denote its form by β_F. The assignment $[M] \to [V]$ defines a homomorphism

$$W(R) \to W(F)$$

and by restriction, $W_{ev}(R) \to W_{ev}(F)$. Consider the discriminant modules $(\text{dis } M) \otimes_R F$ and dis $V = (\Lambda^n V, (-1)^{n(n-1)/2}\hat{\beta}_F)$ over F. Verify that

$$(x_1 \wedge ... \wedge x_n) \otimes 1 \; \to \; (x_1 \otimes 1) \wedge ... \wedge (x_n \otimes 1)$$

defines an isometry from $(\text{dis } M) \otimes_R F$ onto dis V. So $[(\text{dis } M) \otimes_R F] = [\text{dis } V]$ in $\text{Dis}(F)$. Let $\mathcal{Y} = \{y_1,..., y_n\}$ be a basis of V and set $c = (-1)^{n(n-1)/2}\det_{\mathcal{Y}}\beta_F$. Note that the assignment $y_1 \wedge ... \wedge y_n \to 1$ defines an isometry from dis V onto (F, f_c). Therefore, the image of $[\text{dis } M]$ under the composite $\text{Dis}(R) \to \text{Dis}(F) \xrightarrow{\sim} F^*/(F^*)^2$ of Chapter 12D is $c(F^*)^2$. It is easy to check that the discriminant of Chapter 10D defines a homomorphism

$$W_{ev}(F) \to F^*/(F^*)^2.$$

Assume that n is even and note that [V] goes to $c(F^*)^2$ under this map. We have established the commutativity of the top square of the following diagram.

(13.9). *The diagram*

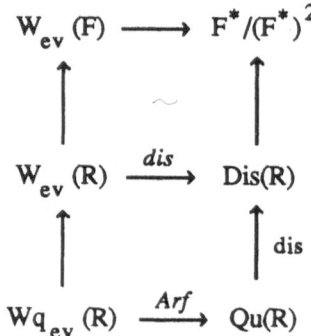

commutes.

Proof. The square on the bottom remains. Let $[M] \in Wq_{ev}(R)$, where M is a nonzero quadratic space of rank n over R. Let h be the underlying bilinear form and let A = A(M) be the Arf algebra. It must be shown that the discriminant modules (dis A, product) and $(\wedge^n M, (-1)^{n(n-1)/2} \hat{h})$ are isometric. If M is free, this is given by Theorem (10.12). In the general case, localize and apply (4.13).   QED.

Remark 3. The bottom square of the diagram commutes for any commutative ring R. This follows from Proposition (4.6.3) in Chapter IV of Knus [1991].

The rest of this section will specialize the preceding discussion to the case of a field F. In the process, other invariants traditionally encountered in the literature will be related to those discussed earlier. Let V be an n-dimensional quadratic space over F with underlying forms q and h.

Assume that char F ≠ 2 and refer to the notation and concepts in Lam's book. In Lam, q is defined by a symmetric bilinear form B on V via the equation $q(x) = B(x, x)$. It follows easily that $B(x, y) = \frac{1}{2} h(x, y)$. Recall the isomorphism $Wq(F) \cong W(F)$. It is clear that the homomorphism

$$\dim_0 : W(F) \to \mathbb{Z}_2$$

corresponds to *rank*. So $Wq_{ev}(F)$ is isomorphic to the fundamental ideal IF. Let $\mathcal{X}$ be a basis of V and put $d = (-1)^{n(n-1)/2} \det_{\mathcal{X}} h$. So dis $V = d(F^*)^2$. It is easy to see that Lam's signed determinant $d_{\pm}$ satisfies $d_{\pm} V = $ dis V if n is even, and $d_{\pm} V = 2d(F^*)^2$ if n is odd. Recall from (3.8) that $QU(F) = QU_f(F) \cong \mathbb{Z}_2 \times F^*/(F^*)^2$. So Lam's $Q(F)$ is isomorphic to $QU(F)$. By Theorem (10.12) and Corollary (12.8),

$$(\dim_0, d_{\pm}) : W(F) \to Q(F)$$

is equivalent to *Arf* (except for the factor 2 in odd dimension). It follows that Ker *Arf* $\cong$ Ker $(\dim_0, d_{\pm}) = I^2 F$, the square of the fundamental ideal.

The Clifford invariant is closely related to the classical Hasse invariant or Hasse symbol. The connection is as follows. If dim V is even, then $C(V)$ is an Azumaya algebra, and if dim V is odd, then $C_0(V)$ is Azumaya. Note that both $C(V)$ and $C_0(V)$ are equipped with antiautomorphisms, namely, the involution $^{-}$ and its restriction. Define the *Witt invariant* $c(V)$ of V by

$$c(V) = \begin{cases} [C(V)] \in \text{Br}(F)_2, \text{ if } \dim V \text{ is even} \\ [C_0(V)] \in \text{Br}(F)_2, \text{ if } \dim V \text{ is odd} . \end{cases}$$

Observe that this definition is also valid in characteristic 2 (where the odd case does not arise). The invariants $c(V)$, $\dim_0$, and $d_{\pm}$ taken together are equivalent to the Clifford invariant Cliff V. Indeed, $c(V)$, $\dim_0$, and $d_{\pm}$ can be "glued together" to give Cliff V. See page 120 in Lam and Exercise 11. Since the Witt invariant is equivalent to the Hasse invariant, see page 123 in Lam, this establishes the aforementioned connection. We point out, finally, that it is a consequence of the theorem of Merkurjev – see Theorem (11.8) – that Ker *Cliff* is isomorphic to the third power $I^3 F$ of the fundamental ideal.

Suppose next that char $F = 2$. This theory is different, but analogous. Refer to Sah [1972], or to Appendix 1 in Milnor-Husemoller for an overview. The map $Wq(F) \to W(F)$ is now trivial. The tensor product provides $Wq(F)$ with the structure of a module over the ring $W(F)$, but it does not make $Wq(F)$ into a ring (there is no 1). The kernel $W_{ev}(F) = IF$ of *rank* $: W(F) \to \mathbb{Z}_2$ contains $Wq(F)$ by (4.18).

By (13.5), $QU(F) = Qu(F) = Qu_f(F)$. By Exercises 6 – 10 of Chapter 3E, any separable quadratic F-algebra is isomorphic to $A_b = F[X]/(X^2 + X + b)$ for some b in F, and the assignment $[A_b] \to b + \wp(F)$ defines an isomorphism

$Qu_f(F) \to F/\wp(F)$. Theorem (9.11) in combination with Theorem (4.8) in Chapter 9 of Scharlau [1985] shows that $\text{Arf } V \in Qu(F)$ corresponds to the classical Arf invariant of V under this isomorphism. The kernel of $Arf : Wq(F) \to Qu(F)$ is equal to $(IF)Wq(F)$. Note finally, that $\text{Arf } V$ and $c(V)$ together are equivalent to $\text{Cliff } V$ (see Exercise 11) and that $\text{Ker } Cliff = (I^2 F)Wq(F)$.

E. The Classical Situations

This section will discuss quadratic spaces, the Witt group, and the invariants over the complex numbers, the reals, and local and global fields. Refer to O'Meara [1971], Lam, Milnor-Husemoller, and Scharlau [1985] for the details.

Let F be a field and let V and V' be two finite-dimensional quadratic spaces over F. Set $\dim V = n$ and $\dim V' = n'$.

Suppose first that $F = \mathbb{C}$. Then by (11.4),

$$V \cong <1> \perp \ldots \perp <1> \quad \text{(n copies)}.$$

It follows that V and V' are isometric if and only if $n = n'$. Since $\{0\}$ and $<1>$ are the only anisotropic spaces over $\mathbb{C}$, $Wq(\mathbb{C}) \cong \mathbb{Z}_2$.

Suppose that $F = \mathbb{R}$. Again by (11.4),

$$V \cong <1> \perp \ldots \perp <1> \perp <-1> \perp \ldots \perp <-1>.$$

Such a splitting can be achieved in different ways for a given V, but the number of $<1>$, say p, and the number of $<-1>$, say p', that occur are invariants. The *signature* sig V of V is defined by

$$\text{sig } V = p - p' \in \mathbb{Z}.$$

Since $\dim V = p + p'$, $\text{sig } V = 2p - \dim V$. The spaces V and V' are isometric if and only if $\dim V = \dim V'$ and $\text{sig } V = \text{sig } V'$, since this is equivalent to the condition that the respective p and p' for V and V' are equal. Since V is hyperbolic if and only if $\text{sig } V = 0$, there is an isomorphism (in fact a ring isomorphism)

$$\text{sig} : Wq(\mathbb{R}) \to \mathbb{Z}.$$

Assume next that F is a local field. In view of the isomorphism (see Section A), $Br(F)_2 \to \{0, \frac{1}{2} + \mathbb{Z}\} \cong \{1, -1\} = \mathbb{Z}^*$, we regard the Witt invariant $c(V)$ as an element in $\mathbb{Z}^*$.

(13.10). *Local Classification Theorem. Let* F *be a local field. Two quadratic spaces* V *and* V' *over* F *are isometric if and only if*

$$\dim V = \dim V', \quad \text{Arf } V = \text{Arf } V' \in QU(F), \quad \text{and} \quad c(V) = c(V') \in Z^*.$$

If char $F \neq 2$, this is a reformulation of Theorem 63:20 in O'Meara [1971]. To see this, use the connection between *Arf* and $d_\pm$ as well as the "dictionary" between $c(V)$ and the Hasse invariant. In the even dimensional situation, refer to Exercise 15 in Chapter 11F. This proves the classification theorem for char F = 2, since dim V and dim V' are necessarily even in this case (see (4.18)).

The Local Classification Theorem implies that *Cliff* : $Wq(F) \to HW(F)$ is an isomorphism. Refer to Exercise 12 for the proof and the consequences. See Lam, Scharlau [1985], and Sah [1972] for more about the structure of $Wq(F)$.

Suppose finally, that F is a global field. Let Ω be a complete set of valuations of F. Let $v \in \Omega$ be real. Since $F_v \cong \mathbb{R}$, define

$$\text{sig}_v V = \text{sig} (V \otimes_F F_v) \in \mathbb{Z}.$$

Let $v_1,...., v_r$ be all of the real valuations in Ω and define the *total signature* of V by

$$\text{sig } V = (\text{sig}_{v_1} V, \ldots, \text{sig}_{v_r} V) \in \mathbb{Z} \times ... \times \mathbb{Z} \quad (r \text{ copies}).$$

The assignment *sig* [V] = sig V, defines a homomorphism (it is actually a ring homomorphism)

$$sig : Wq(F) \to \mathbb{Z}^r.$$

Notice that [V] is in the kernel of *sig* if and only if V is hyperbolic at all real valuations in Ω. If F is an algebraic number field that has no real valuations or if F is an algebraic function field, then r = 0. Here by definition, sig V = 0 and *sig* : $Wq(F) \to \mathbb{Z}^r = \{0\}$ is the trivial map. The signature can be defined for general fields, indeed for commutative rings. See Milnor-Husemoller and Knebusch [1980].

Denote the subset of finite, i.e., non-Archimedean, valuations in Ω by Ω_{fin}. For $v \in \Omega_{\text{fin}}$, define $c_v(V) \in \mathbb{Z}^*$ by $c_v(V) = c(V \otimes_F F_v)$. The *total Witt invariant* of V is defined to be the element

$$w(V) \in \prod_{v \in \Omega_{\text{fin}}} \mathbb{Z}^*,$$

whose v-th coordinate is $c_v(V)$.

The following theorem is a very deep result that lies at the core of the theory of quadratic forms over global fields.

(13.11). *Hasse-Minkowski Theorem. Let* F *be a global field. Two quadratic spaces* V *and* V' *over* F *are isometric if and only if*

(1) $\dim V = \dim V'$,

(2) Arf V = Arf $V' \in QU(F)$,

(3) $w(V) = w(V') \in \prod_{v \in \Omega_{\text{fin}}} \mathbb{Z}^*$, *and*

(4) sig V = sig $V' \in \mathbb{Z}^r$.

If char $F \neq 2$, this is a reformulation of the theorem in 66:5 of O'Meara [1971]. The translation from O'Meara's version to the present one uses Lam's dictionary relating the Hasse symbol and $c(V)$ already referred to, and the connection between *Arf* and d_+. The bilinear form B in both O'Meara and Lam is equal to our $\frac{1}{2}$ h, but this is of no consequence since char $F \neq 2$. If char $F = 2$, the theorem follows from the "local-global" theory in Pollak in combination with the Local Classification Theorem. Note that (4) is obviously vacuous if $r = 0$.

It should not come as a surprise that this result has important consequences for the structure of $Wq(F)$.

(13.12). *Let* F *be a global field. Then the intersection of the kernels of* Cliff : $Wq(F) \to HW(F)$ *and* sig : $Wq(F) \to \mathbb{Z}^r$ *is trivial.*

Proof. Let $[V]$ be in the intersection. Since *Cliff* $[V] = 0$, it follows that both *rank* $[V] = 0$ and *Arf* $[V] = 0$. So $\dim V$ is even. By Remark 1 in Section C, the condition *Cliff* $[V] = 0$ implies that $C(V) \cong \text{Mat}_k(F)$ for some k. In particular, $c(V) = (1,1,\ldots)$. Choose a hyperbolic space H of the same dimension as V and observe that V and H have the same invariants (1) – (4) of (13.11). So $[V] = [H] = 0$. QED.

(13.13). *Theorem. The torsion subgroup* $Wq(F)_{\text{tor}}$ *of* $Wq(R)$ *is equal to the kernel of* sig : $Wq(F) \to \mathbb{Z}^r$ *and* Cliff : $Wq(F) \to HW(F)$ *is injective on* $Wq(F)_{\text{tor}}$. *In particular, if* $r = 0$, *then* $Wq(F) = Wq(F)_{\text{tor}}$.

Proof. Let $[V] \in Wq(F)_{tor}$. Since it is a submodule of a free $\mathbb{Z}$-module, $sig\ Wq(F)$ is torsion free. So $[V]$ is in the kernel of sig. Suppose, conversely, that $[V]$ is in the kernel of sig. Consider the element $Cliff\ [V]$ in $HW(F)$. By Remark 2 of Section C, $Cliff\ (8[V]) = (Cliff\ [V])^8 = 1$. But $8[V]$ is also in the kernel of sig, and therefore $8[V] = 0$ by (13.12).   QED.

Let $[V]$ be in $Wq(F)$ and put $sig\ [V] = (k_1,..., k_r)$. Since $sig_v\ V = 2p_v - \dim V$, $sig_v\ [V] - sig_{v'}\ [V] = 2(p_v - p_{v'})$ for any real valuations v and v' in Ω. Therefore, the image of sig is in the subset

$$\mathbb{Z}^r\ (mod\ 2) = \{(k_1,...., k_r) \mid k_i \equiv k_j\ (mod\ 2)\ all\ i, j\}$$

of $\mathbb{Z}^r$. See Lam for a verification of the fact that $sig : Wq(F) \to \mathbb{Z}^r\ (mod\ 2)$ is surjective. Only the case char $F \neq 2$ is relevant, for if char $F = 2$, then F is a function field and sig is trivial. This provides the exact sequence

$$0 \to Wq(F)_{tor} \to Wq(F) \xrightarrow{sig} \mathbb{Z}^r\ (mod\ 2) \to 0.$$

A basis of $\mathbb{Z}^r\ (mod\ 2)$ lets a splitting map be defined, and it follows that

$$Wq(F) \cong Wq(F)_{tor} \oplus \mathbb{Z}^r\ (mod\ 2).$$

Given the facts already available, it is not hard to prove, again see Lam for the details, that

$$sig\ (Ker\ Arf) = sig\ (I^2F) = (4\mathbb{Z})^r,$$

and that sig maps $Ker\ Cliff = I^3F$ isomorphically onto $(8\mathbb{Z})^r$.

Lam and Scharlau [1985] provide more information about $Wq(F)$. However, we have collected all that will be needed later, with the exception of the following proposition from §4 in Chapter IV of Milnor-Husemoller. The first part is a consequence of the Hasse-Minkowski Theorem; the second gives some indication as to the interdependence of the invariants.

(13.14). *Theorem. Let* F *be an algebraic number field. An element* $[V]$ *in* $Ker\ Arf$ *is uniquely determined by the total Witt invariant* $w(V)$ *and by the total signature* $sig\ [V]$. *These are subject only to the condition that the image of* $w(V)$ *under* $\prod_{v \in \Omega_{fin}} \mathbb{Z}^* \xrightarrow{product} \mathbb{Z}^*$ *is equal to* $\prod_{1 \leq i \leq r} (-1)^{k_i}$, *where* $sig\ [V] = (4k_1, \ldots, 4k_r)$.

F. Exercises

1. Let F be a field with char $F \neq 2$. Use basic properties of the quaternion algebras $\left(\frac{a, b}{F}\right)$ to show that there is a surjective homomorphism

$$F^* /(F^*)^2 \otimes_{\mathbb{Z}} F^*/(F^*)^2 \to Br(F)_2.$$

2. Let F be a field with char $F = 2$. Use the basic properties of the quaternion algebra $\left[\frac{a, b}{F}\right]$ to show that there is a surjective homomorphism

$$F^*/(F^*)^2 \otimes_{\mathbb{Z}} F/\wp(F) \to Br(F)_2.$$

3. Let F be a global field with char $F \neq 2$. Let a and b in F^* be arbitrary and consider the equivalence class of the quaternion algebra $\left(\frac{a, b}{F}\right)$ in $Br(F)_2$. Recall from Section A that it maps to either 0 or $\frac{1}{2} + \mathbb{Z}$ under $Br(F)_2 \to Br(F_v)_2 \to \mathbb{Q}/\mathbb{Z}$ for any $v \in \Omega$. Define $\left(\frac{a, b}{v}\right) \in \mathbb{Z}^*$ to be 1 in the first instance and -1 in the second. Show that $\left(\frac{a, b}{v}\right)$ is the *Hilbert symbol* of number theory (see O'Meara [1971]). Show that the *Hilbert Reciprocity Law* – this asserts that $\left(\frac{a, b}{v}\right) = 1$ for all but finitely many $v \in \Omega$ and that $\prod_{v \in \Omega} \left(\frac{a, b}{v}\right) = 1$ – is a consequence of the Hasse-Noether-Brauer-Albert Theorem. Is there an analogue if char $F = 2$?

4. Let S be a separable quadratic algebra over R. Let $S = Rt \oplus dis\ S$ be the splitting provided by (10.7). Show that the assignment $rt \wedge d \to rd$, for $r \in R$ and $d \in dis\ S$, defines an isomorphism from $\Lambda^2 S$ onto $dis\ S$. Let h be the symmetric bilinear form on S (as given by Exercise 2 of Chapter 10E) and check that this map is an isomorphism of the discriminant modules $(\Lambda^2 S, \hat{h})$ and $(dis\ S, p)$.

5. Supply the missing details in the proof of Theorem (13.7), i.e., carry out the localization step and show that $h(t, s) = au$.

6. Let $[M] \in Wq_{ev}(R)$. Show that there is a separable quadratic algebra A such that $[M] = [A] \perp [N]$, where A is the quadratic space of rank 2 over R determined by A and its norm form, and $[N] \in Ker\ Arf$.

7. Let (M, β) and (M', β') be symmetric inner product spaces of respective ranks n and n' over a domain R. Denote the form of $M \perp M'$ by γ. Assume that n and n' are both even and show that the isomorphism $\Lambda^n(M) \otimes_R \Lambda^{n'}(M') \to \Lambda^{n+n'}(M \perp M')$ is an isomorphism

$$(\Lambda^n(M), (-1)^{n(n-1)/2}\hat{\beta}) \otimes_R (\Lambda^{n'}(M'), (-1)^{n'(n'-1)/2}\hat{\beta'}) \to$$
$$(\Lambda^{n+n'}(M \perp M'), (-1)^{(n+n')(n+n'-1)/2}\hat{\gamma})$$

of discriminant modules.

8. Let R be a domain and consider the symmetric inner product space $(P, f) \perp (R, f_{-1})$. Let g be the form of $P \perp R$ and show that $p \wedge r \to pr$ defines an isomorphism $(\Lambda^2(P \perp R), (-1)^{n(n-1)/2}\hat{g}) \to (P, f)$ of discriminant modules. Deduce that $dis : W_{ev}(R) \to Dis(R)$ is surjective.

9. Let R be a domain and let $[M] \in W_{ev}(R)$. Show that $[M] = [(P, f) \perp (R, f_{-1})] + [N]$ for some discriminant module (P, f) and some $[N]$ in the kernel of $dis : W_{ev}(R) \to Dis(R)$.

10. Let R be a commutative ring and M a symmetric inner product space over R. Refer to Bass [1967] and [1974] and denote Bass' DISC(R) by DIS(R). Can $dis\, M \in DIS(R)$ be defined so that $dis : W(R) \to DIS(R)$ given by $[M] \to dis\, M$ is a surjective homomorphism such that dis, $Arf : Wq(R) \to QU(R)$, $Wq(R) \to W(R)$, and $DIS(R) \to QU(R)$ together provide an extension of the bottom square of the commutative diagram of (13.9) ?

11. Let F be a field and let V and V' be quadratic spaces over F. Prove that $Cliff\, V = Cliff\, V'$ if and only if $Arf\, V = Arf\, V'$ and $c(V) = c(V')$.

12. Let F be a local field. Prove that $Cliff : Wq(F) \to HW(F)$ is an isomorphism. Deduce that $Wq(F)$ is a torsion group of exponent 8. If $-1 \in (F^*)^2$, show that $Wq(F)$ is a torsion group of exponent 4. Suppose char $F \neq 2$. If F is nondyadic, then $Wq(F)$ has order 2^4; and if F is dyadic, then $Wq(F)$ has order 2^{4+ki} where k is the cardinality of the residue class field and i is determined by $2 = u\pi^i$ with u a unit in the valuation ring of F. If char $F = 2$, show that $Wq(F)$ is infinite.

13. Let F be a global field and let V and V' be two quadratic spaces over F. Show that $c(V) = c(V') \Rightarrow w(V) = w(V')$. Give an example which shows that the converse is false.

14. Let F be an algebraic number field with $r > 0$. For any even $n \geq 4$ give an example of two nonisometric quadratic spaces over F of dimension n such that Arf V = Arf V' and $w(V) = w(V')$. Is $r > 0$ necessary ?

In Exercises 15 – 19, R is a Dedekind domain with field of fractions F and M is a finitely generated projective module over R of rank n. By the structure theory of such modules there is an isomorphism $M \cong R \oplus \cdots \oplus R \oplus \mathfrak{a}$, where $\mathfrak{a}$ is an invertible fractional ideal. By Chapter 5D, the R-modules $\Lambda^n(M)$ and $\mathfrak{a}$ are isomorphic.

15. Show: there is a nonsingular symmetric bilinear form on M $\Leftrightarrow$ $\mathfrak{a}$ has the structure of a discriminant module $\Leftrightarrow$ $[\mathfrak{a}]^2 = 1$ in Cl(R).

16. There is a nonsingular quadratic form $q : M \to R$ if and only if $\mathfrak{a}$ has the structure of a discriminant module whose isomorphism class is in the kernel of Dis(R) $\to$ Dis(R/4R) and n is even if $2 \notin R^*$.

17. Suppose M is a symmetric inner product space. Then M is free if and only if dis M $\in R^*/(R^*)^2$. In particular, if $Cl(R)_2$ is trivial, then any such M is free.

18. Suppose M is a quadratic space. Then: M is free $\Leftrightarrow$ A(M) is free $\Leftrightarrow$ Arf M $\in QU_f(R)$.

Let $W_f(R)$ be the subset of all $w \in W(R)$ such that $w = [M]$ with M free. It is clear that $W_f(R)$ is a subgroup (indeed subring) of W(R). Define the subgroup $Wq_f(R)$ of Wq(R) in the analogous way.

19. $W(R)/W_f(R) \cong Cl(R)_2$ and $Wq(R)/Wq_f(R) \cong QU(R)/QU_f(R)$. The last group is isomorphic to $Qu(R)/Qu_f(R)$ if $2 \notin R^*$.

Hints:

1. Refer to §57B in O'Meara [1971] or page 89 in Scharlau [1985], or to Lam for the basic properties. Apply the Merkurjev-Albert Theorem for the surjectivity.

2. See page 314 in Scharlau [1985] and Sah [1972] and apply the Merkurjev-Albert Theorem.

4. The typical localization argument allows the assumption that S is free. Now make use of the example in Chapter 4D.

5. $h(t, s) = ts^{\alpha} + st^{\alpha} = s - 2ts = s - 2(-uat + 2bus) = u(u^{-1}s + 2at - 4bs) =$
   $u((a^2 + 4b)s + 2at - 4bs) = u(a^2s + 2at) = au(2t + as) = au.$

7. For $n \times n$ and $n' \times n'$ matrices A and B over R, $\det\begin{bmatrix} A & 0 \\ 0 & B \end{bmatrix} =$
   $(\det A)(\det B)$ and for n and n' both even, $(-1)^{n(n-1)/2}(-1)^{n'(n'-1)/2} =$
   $(-1)^{(n+n')(n+n'-1)/2}$.

8. $\hat{g}(p \wedge r, p' \wedge r') = (-1) \det\begin{bmatrix} f(p, p') & 0 \\ 0 & -rr' \end{bmatrix} = (-1)(-rr')f(p, p') = rr'f(p, p').$

10. I don't know whether this can be done but one would expect so. It may well exist in the literature.

11. Assume that Arf $V = $ Arf V' and $c(V) = c(V')$. Since the grading of $A(V)$ is trivial if and only if dim V is even, it follows that dim V and dim V' are either both even or both odd. Take the even case first. The scaled space $^{-1}V'$ satisfies $[^{-1}V'] = -[V']$. So Arf $[V \perp {}^{-1}V'] = 1$. By Remark 1 of Section C, $(\text{Cliff } V)(\text{Cliff } V')^{-1} = \text{Cliff } (V \perp {}^{-1}V') = [C(V \perp {}^{-1}V')] = [C(V) \hat{\otimes}_F C(^{-1}V')]$. Put Arf $V = $ Arf $V' = [a, b]$ with $u = a^2 + 4b \neq 0$. By (8.12), $[C(V) \hat{\otimes}_F C(^{-1}V')] = [C(V) \otimes_F C(^{-u}V')] = [C(V)][C(^{-u}V')] = [C(V')][C(^{-u}V')] = [C(^{-u}V')][C(^{-u}V')]$ by (8.11). In the odd case, apply Theorem (8.2). The other implication is routine.

12. To establish the isomorphism, the injectivity suffices, and for this use the Local Classification Theorem and Exercise 11. Refer to the proof of (13.12) if necessary. Now consider (13.8), use the fact that $Br(F)_2 = Hs(F) \cong \mathbb{Z}^*$, and refer to Chapter 3C and Exercise 2 in Chapter 12F.

13. For the example use the fact that $Hs(F) = Br(F)_2$. Restrict the exact sequence of the Hasse-Brauer-Noether-Albert Theorem to $Br(F)_2$ (see the version of this sequence in Chapter 14B if necessary) and consider a situation where $r \geq 2$.

14. Consider the quadratic spaces $V = M^{4,0}$ and $V' = M^{0,4}$ of Chapter 11D. By (10.12) and (12.8), Arf $V = $ Arf V', and by (11.11), $c(V) = c(V')$. So $w(V) = w(V')$. Add hyperbolic planes to V and V' to get examples in higher dimensions. Do $V = M^{2,0}$ and $V' = M^{0,2}$ provide an example in dimension 2 ?

15. If M is a symmetric inner product space, then $\mathfrak{a}$ has the structure of a discriminant module by the definition of dis M. This implies, by Chapter 4C, that $[\mathfrak{a}]^2 = 1$ in $\text{Cl}(R)$. If $[\mathfrak{a}]^2 = 1$, then $\mathfrak{a}^2 = c^{-1}R$ with $c \in F^*$. So $(\mathfrak{a}, f_c)$ along with $M \cong <1> \perp ... \perp <1> \perp (\mathfrak{a}, f_c)$ provide M with a nonsingular symmetric bilinear form.

16. Suppose M has a q. As in the proof of (13.9), (dis A, product) and $(\Lambda^n M, (-1)^{n(n-1)/2} \hat{h})$ are isomorphic discriminant modules. Since A is separable quadratic, it follows by (12.6) that its class is in the kernel of $\text{Dis}(R) \to \text{Dis}(R/4R)$. By (4.18), n is even if $2 \notin R^*$. Suppose conversely that $\mathfrak{a}$ has the required structure. Then by (12.6) and (10.7) there is a separable quadratic algebra S over R such that $S \cong R \oplus \mathfrak{a}$ as R-modules. So $R \oplus \mathfrak{a}$ has the structure of a quadratic space over R.

17. Refer to the hints for Exercise 15.

19. Let $W_{fev}(R)$ be the subgroup of $W(R)$ consisting of all elements which have a free even dimensional representative. Show that the composite $W_{ev}(R) \to \text{Dis}(R) \to \text{Dis}(R)/(R^*/(R^*)^2)$ is surjective. By Exercise 17, the kernel is $W_{fev}(R)$. Now consider $W_{ev}(R) \to W(R) \to W(R)/W_f(R)$. Since $[M, f] + W_f(R) = [M, f] + [<1>] + W_f(R) = [M \perp <1>] + W_f(R)$, any element of $W(R)/W_f(R)$ is of the form $[M, f] + W_f(R)$ with rank M even. It follows that this composite is onto and hence that $W(R)/W_f(R) \cong W_{ev}(R)/W_{fev}(R) \cong \text{Dis}(R)/(R^*/(R^*)^2)$. For the second part use Arf and Exercise 18.

14
The Arithmetic of Wq(R)

Overview

The primary purpose of this chapter is the analysis of the group $Wq(R)$ in the case of an arithmetic Dedekind domain R. The characterization of the elements of $Br(R)_2$ as the Brauer classes of Clifford algebras of quadratic spaces over R of rank 4 and trivial Arf invariant is an important step along the way. The quadratic spaces which it provides make it possible to describe $Ker\ Arf$ by use of the total signature. In the special case where R is the ring of integers in a number field, the quadratic group $Qu(R)$ is closely related to the ideal class group, and this connection together with the description of $Ker\ Arf$ implies that $Wq(R) \cong Cl(R)_2 \oplus G$, where $Cl(R)$ is the ideal class group of R and G is a free Abelian group of rank with r the number of real embeddings of the number field. An additional focus is the comparison of the number theory of $Wq(R)$ with that of $W(R)$ and the structure of the quotient $W(R)/Wq(R)$.

A. Arithmetic Dedekind Domains

This section contains an overview of some basic properties of arithmetic Dedekind domains. For the details see Borevich-Shafarevich, Cassels-Fröhlich, Jacobson [1985,II], and O'Meara [1971], for instance.

Fix a global field F. Recall that if char $F = 0$, then F is an algebraic number field, i.e., a finite extension of $\mathbb{Q}$, and if char $F = p \neq 0$, then F is an algebraic function field, i.e., a finite extension of the field of fractions of $K[X]$ for some finite field K with char $K = p$. Let Ω be a complete set of valuations on F. Let $v \in \Omega$ and consider the completion F_v. If v is finite, then F_v is a local field. If not, then v is either real or complex, i.e., F_v is either $\mathbb{R}$ or $\mathbb{C}$. We will assume that each $v \in \Omega$ is *normalized*. This (can be achieved by replacing each v by an equivalent valuation and) means: If v is finite, then the extension v' of v to F_v has the property that the largest value $v'a$ with $v'a < 1$ is $\frac{1}{k}$, where k is the cardinality of the residue class field of v'; and if v is Archimedean, then the restriction of v' to $\mathbb{Q}$ is the ordinary absolute value of $\mathbb{Q}$ if v is real, and the square of the ordinary absolute value if v is complex. Denote the sets of finite, real, and complex valuations in Ω, respectively, by

$$\Omega_{fin}, \ \Omega_{re}, \text{ and } \ \Omega_{co}.$$

The last two sets are empty if F is an algebraic function field.

Now let R be a Dedekind domain with field of fractions F. So R is an arithmetic Dedekind domain. An important fact asserts that there exists a subset S of Ω which is nonempty and contains both Ω_{re} and Ω_{co}, such that

$$R = R_S = \{r \in F \mid v(r) \leq 1 \ \text{for all} \ v \in \Omega - S\},$$

where $\Omega - S = \{v \in \Omega \mid v \notin S\}$. Conversely, any such S and this equality defines a Dedekind domain with field of fractions F.

The domain R satisfies a number of finiteness properties. For example, the ideal class group $Cl(R)$ is finite. For any nonzero ideal $\mathfrak{a}$ in R, the quotient ring $R/\mathfrak{a}$ is finite. It follows that $Pic(R/\mathfrak{a}) = 1$ (since a finite ring is semilocal, i.e., has only finitely many ideals). Note, in particular, that the term $Pic(R/4R)_2$ of the diagram in Chapter 12C is trivial if char $F \neq 2$.

The nonzero prime ideals of R correspond bijectively to the valuations in $\Omega - S$ as follows: Fix a nonzero prime ideal $\mathfrak{p}$ of R. Since $\mathfrak{p} \cap \mathbb{Z}$ is a prime ideal of $\mathbb{Z}$, $\mathfrak{p} \cap \mathbb{Z} = p\mathbb{Z}$ for some prime p of $\mathbb{Z}$. The local ring $R_\mathfrak{p}$ has unique maximal ideal $\mathfrak{p}R_\mathfrak{p}$ and the field $R_\mathfrak{p}/\mathfrak{p}R_\mathfrak{p}$ is a finite extension of the field of p elements. Let $N\mathfrak{p}$ be its cardinality. Let $a \in F$ and consider the fractional ideal aR. Since R is Dedekind, we may put $aR = \mathfrak{p}^j\mathfrak{q}$, where j is an integer and $\mathfrak{q}$ is a product of prime ideals (and inverses), none of which equals $\mathfrak{p}$ or $\mathfrak{p}^{-1}$. The equation $v(a) = (\frac{1}{N\mathfrak{p}})^j$ defines a valuation in $\Omega - S$. In the other direction, let $v \in \Omega - S$ and let $\mathfrak{p} = R \cap \{r \in F \mid v(r) < 1\}$.

The group R^* is given by

$$R^* = \{r \in F \mid v(r) = 1 \ \text{for all} \ v \in \Omega - S\}.$$

Let $\mu = \{a \in F^* \mid a^k = 1, \text{ for some } k\}$ be the group of all roots of 1 in F. This is a finite cyclic group contained in R^*. If char $F \neq 2$, then μ contains the involution -1, and it follows that card μ is even; if char $F = 2$, then μ has no nontrivial involutions, so that card μ is odd. If F is an algebraic function field, then μ contains the nonzero elements of the finite coefficient field of the defining polynomial ring. In particular, if char $F = p \neq 2$, then $2 \in R^*$. If F is an algebraic number field then $2 \in R^*$ if and only if $v(2) = 1$

for all $v \in \Omega - S$. This is equivalent to saying that the *dyadic* valuations in Ω, i.e., those that restrict to the 2-adic valuation of $\mathbb{Q}$, lie in S.

Let $S_{fin} = S \cap \Omega_{fin}$, and observe that

$$S = \Omega_{re} \cup \Omega_{co} \cup S_{fin}.$$

Let

$$\text{card } S = s, \quad \text{card } \Omega_{re} = r, \quad \text{card } \Omega_{co} = c, \quad \text{and card } S_{fin} = f.$$

If F is an algebraic number field, then $r + 2c =$ degree $F/\mathbb{Q}$. While r and c are finite, f can be infinite. If f, or equivalently s, is finite, then R_S is called a *Hasse domain*. If F is an algebraic number field and $S = \Omega_{re} \cup \Omega_{co}$, then the Hasse domain R is denoted by R_∞. This is the integral closure of $\mathbb{Z}$ in F, i.e., the *ring of algebraic integers* of F. If $F = \mathbb{Q}$, then $R_\infty = \mathbb{Z}$.

Dirichlet Unit Theorem. If R is a Hasse domain, then the group R^* contains $s - 1$ infinite cyclic groups $U_1, \ldots, U_{s-1}$ such that R^* is equal to the direct product $\mu U_1 \ldots U_{s-1}$.

The homomorphism $\mu \rightarrow \mu^2$ has kernel $\{\pm 1\}$. So $\mu/\mu^2 \cong \mathbb{Z}^* = \{\pm 1\}$ if char $F \neq 2$, and $\mu/\mu^2 = \{1\}$ if char $F = 2$. Since $U_i/U_i^2 \cong \mathbb{Z}^*$ for all i, it follows that

$$R^*/(R^*)^2 \cong \mathbb{Z}^* \times \ldots \times \mathbb{Z}^*,$$

where there are s copies if char $F \neq 2$ and $s - 1$ copies if char $F = 2$. In particular, $R^*/(R^*)^2$ has order 2^s in the first case and 2^{s-1} in the second.

B. The Arithmetic of $Br(R)_2$

Throughout this section, $R = R_S$ is a Dedekind domain with a global field of fractions F. The goal is the study of the 2-torsion subgroup of the Brauer group $Br(R)$. The facts about $Br(R)$ which are used here can be found in Auslander-Goldman, DeMeyer-Ingraham, Fossum, the lecture notes Knus-Ojanguren [1974], or Orzech-Small.

Refer to Chapter 13A. Since $Br(F_v) \cong (\mathbb{Q}/\mathbb{Z})_2 = \{1, \frac{1}{2} + \mathbb{Z}\}$ for v real and $Br(F_v) = 1$ for v complex, the Hasse-Noether-Brauer-Albert Theorem provides the exact sequence

$$1 \to Br(F) \xrightarrow{\beta} \prod_{v \in \Omega} Br(F_v) \xrightarrow{\sim}$$

$$\left(\bigoplus_{v \in \Omega_{re}} (\mathbb{Q}/\mathbb{Z})_2 \right) \oplus \left(\bigoplus_{v \in \Omega_{fin}} \mathbb{Q}/\mathbb{Z} \right) \xrightarrow{sum} \mathbb{Q}/\mathbb{Z} \to 0.$$

Let $Br(R) \to Br(F)$ be the homomorphism defined by $[A] \to [A \otimes_R F]$ for an Azumaya algebra A over R. This map is injective, and an element $[B]$ of $Br(F)$ is in the image if and only if $\beta_v[B] = 1$ for all $v \in \Omega - S$, i.e., if and only if $B_v = B \otimes_F F_v$ is isomorphic to $Mat_m(F_v)$, for some m, for all $v \in \Omega - S$. Identifying $Br(R)$ with its image in $Br(F)$ provides the exact sequence

$$1 \to Br(R) \longrightarrow Br(F) \xrightarrow{\gamma} \prod_{v \in \Omega - S} Br(F_v),$$

where $\gamma : Br(F) \longrightarrow \prod_{v \in \Omega - S} Br(F_v)$ comes from those $\beta_v : Br(F) \to Br(F_v)$ with $v \in \Omega - S$. The injection $\beta : Br(F) \longrightarrow \prod_{v \in \Omega} Br(F_v)$ sends $Br(R)$ into $\prod_{v \in S} Br(F_v)$ and hence into $\prod_{v \in \Omega_{re} \cup S_{fin}} Br(F_v)$. This gives the exact sequence

$$1 \to Br(R) \longrightarrow \prod_{v \in \Omega_{re} \cup S_{fin}} Br(F_v) \xrightarrow{\sim}$$

$$\left(\bigoplus_{v \in \Omega_{re}} (\mathbb{Q}/\mathbb{Z})_2 \right) \oplus \left(\bigoplus_{v \in S_{fin}} \mathbb{Q}/\mathbb{Z} \right) \xrightarrow{sum} \mathbb{Q}/\mathbb{Z}.$$

The group $Br(F_v)_2$ is isomorphic to $(\mathbb{Q}/\mathbb{Z})_2$ for any finite or real v. Replacing $(\mathbb{Q}/\mathbb{Z})_2$ by the multiplicative group $\mathbb{Z}^*$ we obtain:

(14.1). *Theorem. Let $R = R_S$ be an arithmetic Dedekind domain. Then*

$$Br(R)_2 = \{[B] \in Br(F)_2 \mid [B \otimes_F F_v] = 1 \text{ for all } v \in \Omega - (\Omega_{re} \cup S_{fin})\},$$

and the sequence

$$1 \to Br(R)_2 \longrightarrow \prod_{v \in \Omega_{re} \cup S_{fin}} Br(F_v)_2 \cong \prod_{\Omega_{re} \cup S_{fin}} \mathbb{Z}^* \xrightarrow{\text{product}} \mathbb{Z}^*$$

is exact.

Remark. It follows that

$$Br(R)_2 \cong (\mathbb{Z}^*)^{r+f-1},$$

where this (restricted) product is understood to be 1 if $r + f \leq 1$. Notice that $Br(R)_2$ is finite if and only if R is a Hasse domain.

The Λ-Construction: Let [B] in $Br(R)_2 \subseteq Br(F)_2$ be arbitrary. By the remarks that precede Theorem (11.13), either [B] = 1, or [B] = [D], for a quaternion division algebra D over F. Assume first that [B] = 1. Take M = H $\perp$ H with H a hyperbolic plane over R. By Exercises 1 and 2 of Chapter 5E, C(M) $\cong$ Mat$_4$(R). So [C(M)] = 1 = [B]. By Example 3 in Chapter 7C and the multiplicativity of the Arf invariant, Arf M = 1. Assume next that [B] = [D]. By Theorem 10.4 in Reiner, there is a maximal R-order Λ in D. Since [D] $\in$ Br(R), [D $\otimes_F$ F$_v$] = 1 for all v $\in$ Ω – S. Therefore by Theorem 25.7 in Reiner, D(Λ/R) = R. It follows by results of Auslander-Buchsbaum (for a detailed exposition see Fossum), that Λ is separable over R. Since Λ spans D over F, Cen $\Lambda \subseteq$ Cen D = F. By Theorem 8.6 in Reiner, every element of Λ is integral over R. Since R is integrally closed in F, Cen Λ = R, and therefore Λ is an Azumaya algebra over R. By (9.5), Λ is finitely generated projective, and since $\Lambda \otimes_R F \cong D$, Λ has rank 4. By Proposition (7.3.6) in Chapter 1 of Knus [1991], Λ with its norm form is a quadratic space over R. Denote this space by M. By Theorem (9.11), C(M) is Azumaya over R and by Example (1.2) of Knus [Advances Math. 1988], C(M) $\cong$ Mat$_2$(Λ). Since Mat$_2$(Λ) $\cong$ Mat$_2$(R) $\otimes_R \Lambda$, [C(M)] = [Λ]. So [C(M)] = [Λ] = [D] = [B]. Denote by V the quadratic space over F given by the quaternion algebra D and its norm form. By Exercise 5 of Chapter 11F, the Arf invariant of V is trivial. Since A(M) $\otimes_R F \cong$ A(M $\otimes_R F) \cong$ A(V), and Qu(R) $\to$ Qu(F) is injective (apply 12.10), it follows that the Arf invariant of M is trivial.

The Λ-Construction provides the following theorem.

(14.2). *Theorem. Every element in* Br(R)$_2$ *is equal to* [C(M)] *for a quadratic space M over R of rank 4 and trivial Arf invariant.*

Remark. Whereas D is unique for the given [B], M depends on the choice of the maximal order Λ. Indeed, every Λ provides an M, and nonisomorphic Λ

provide distinct M. This is a consequence of Theorem (3.10) of Knus-Paques, which says that $\Lambda \cong \Lambda'$ if and only if $M \cong {}^uM'$ for some $u \in R^*$.

Recall that if an Azumaya algebra B over R has an antiautomorphism, then $[B] \in Br(R)_2$. Therefore

(14.3). *Corollary. Any Azumaya algebra over R which has an anti-automorphism is Brauer equivalent to the Clifford algebra of a quadratic space over R of rank 4 and trivial Arf invariant.*

We continue our discussion by specializing the Λ-Construction to certain specific [B] in $Br(R)_2$. Set

$$\Omega_{re} = \{v_1, ..., v_r\} \quad \text{and} \quad S_{fin} = \{v_i \mid i \geq r+1\}.$$

Suppose that card $(\Omega_{re} \cup S_{fin}) \geq 2$. Fix any two distinct v_i and v_j in $\Omega_{re} \cup S_{fin}$. In view of Theorem (14.1), there is an element $[B_{i,j}] \in Br(R)_2$, such that $[B_{i,j} \otimes_F F_w] \neq 1$ for w equal to v_i and v_j and $[B_{i,j} \otimes_F F_w] = 1$ for all other $w \in \Omega$. The image of $[B_{i,j}]$ under the composite

$$Br(R)_2 \longrightarrow \prod_{v \in \Omega_{re} \cup S_{fin}} Br(F_v)_2 \cong \prod_{\Omega_{re} \cup S_{fin}} \mathbb{Z}^*$$

is $(1,...,1,-1,1,...,1,-1,1,...)$ with -1 in the i-th and j-th position and 1 everywhere else. Since $[B_{i,j}] \neq 1$, the Λ-construction provides a quaternion division algebra $D_{i,j}$ over F with $[D_{i,j}] = [B_{i,j}]$, such that $D_{i,j} \otimes_R F_w$ is a division algebra over F_w precisely for w equal to v_i and v_j and $D_{i,j} \otimes_R F_w \cong Mat_2(F_w)$ for all other $w \in \Omega$. It also provides a quadratic space

$$M_{i,j}$$

of rank 4 and trivial Arf invariant over R, such that $[C(M_{i,j})] = [D_{i,j}]$ and such that the quadratic spaces $M_{i,j} \otimes_R F$ and $V_{i,j}$, the latter being $D_{i,j}$ with its norm form, are isometric. By a basic property of quaternion algebras (see Chapter 11C), $V_{i,j} \otimes_R F_w$ is anisotropic for w equal to v_i and v_j and hyperbolic for all other $w \in \Omega$. In the special case where $j = i + 1$, we denote $[B_{i,j}]$ by $[B_i]$, $D_{i,j}$ by D_i, $V_{i,j}$ by V_i, and $M_{i,j}$ by

$$M_i.$$

The image of $[B_i] = [D_i]$ is $(1,...,1,-1,-1,1,...)$ with the first -1 in the i-th position. An easy induction shows that any element in the kernel of

$$\prod_{\Omega_{re} \cup S_{fin}} \mathbb{Z}^* \xrightarrow{\text{product}} \mathbb{Z}^*$$

is a finite product of tuples of the form $(1,...,1,-1,-1,1,...)$. Since no product of distinct such elements can be $(1,1,...)$, these elements constitute (in additive language), a $\mathbb{Z}_2$-basis for the kernel.

The Λ-Construction produces for certain quadratic spaces V over F (namely, those coming from quaternion algebras) a quadratic space M over R such that $V \cong M \otimes_R F$. In the special case where F is an algebraic number field and $R = R_\infty$, there is the following generalization due to Fröhlich [1971]:

(14.4). *Theorem. Let V be a quadratic space over F with $\dim V$ even. Let $A(V)$ and $C(V)$ be the Arf and Clifford algebras of V. If there exist a separable quadratic algebra A over R and an Azumaya algebra C over R such that $A(V) \cong A \otimes_R F$ and $C(V) \cong C \otimes_R F$, then there exists a quadratic space M over R such that $V \cong M \otimes_R F$.*

C. Analyzing Wq(R)

We continue to let $R = R_S$ be an arithmetic Dedekind with field of fractions F. As before, $r = \text{card } \Omega_{re}$ and $f = \text{card } S_{fin}$ (with the latter finite or infinite). The invariants defined in Chapter 13 in combination with the Λ-Construction provide considerable information about $Wq(R)$. This analysis is the goal of this section.

Consider the homomorphism

$$Wq(R) \rightarrow Wq(F)$$

defined by extension of scalars. Refer to Knebusch [1969/70] or Chapter I in Baeza [1978] for the fact that it is injective. So we regard $Wq(R) \subseteq Wq(F)$. It follows from the definitions that the Clifford and Arf invariants of $Wq(F)$ restrict to those of $Wq(R)$. We will denote the former by $Cliff_F$ and Arf_F and the latter by $Cliff$ and Arf. The total signature of $Wq(F)$ is denoted sig_F. Its restriction to $Wq(R)$ is

$$sig : Wq(R) \rightarrow \mathbb{Z}^r \text{ (mod 2)}.$$

By Theorem (13.13), the kernel of sig is $Wq(R)_{tor}$. So there is the exact sequence

$$0 \to Wq(R)_{tor} \to Wq(R) \xrightarrow{\ sig\ } sig\ Wq(R) \to 0.$$

Since $sig\ Wq(R))$ is a free $\mathbb{Z}$-module of finite rank, the sequence splits. Therefore,

$$Wq(R) \cong Wq(R)_{tor} \oplus sig\ Wq(R).$$

If $r = 0$, then sig_F is the trivial map and hence $Wq(R) = Wq(R)_{tor}$.

We will now see how the invariants sig, $Cliff$, and Arf together unfold the structure of $Wq(R)$.

(14.5). $Cliff: Wq(R) \to HW(R)$ *is injective on* $Wq(R)_{tor}$. *If* $r = 0$, *e.g., if* F *is an algebraic function field, then* $Wq(R) = Wq(R)_{tor} \cong HW(R)$. *If* $2 \notin R^*$, *then* $Wq(R)_{tor}$ *has exponent* 4.

Proof. For the first statement, apply Theorem (13.13). The second statement is clear, and the third follows from Remark 2 in Chapter 13C.   QED.

(14.6). *Theorem.* $Cliff: \operatorname{Ker} Arf \to Br(R)_2$ *is surjective, i.e.,* $Hs(R) = Br(R)_2$.

Proof. Apply Theorem (14.2).   QED.

As a consequence of (13.8), HW(R) fits into an exact sequence

$$1 \to Br(R)_2 \to HW(R) \to QU(R) \to 1.$$

Order the valuations in $\Omega_{re} \cup S_{fin}$ as in Section B. For $v = v_i$ in $\Omega_{re} \cup S_{fin}$, denote the quadratic space M_i supplied by the Λ-Construction by M_v. Observe that if v_i is in S_{fin}, then v_{i+1} is also. The theorem and proof that follow assert, roughly speaking, that the $[M_v]$ with $v \in S_{fin}$ determine $(\operatorname{Ker} Arf)_{tor}$ and the $[M_v]$ with $v \in \Omega_{re}$ determine the torsion-free part of $\operatorname{Ker} Arf$. Note that

$$(\operatorname{Ker} Arf)_{tor} = \operatorname{Ker} Arf \cap Wq(R)_{tor} = \operatorname{Ker} Arf \cap \operatorname{Ker} sig.$$

(14.7). *Theorem.* (1) *If* $f = 0$ *and* $r > 0$, *then there is an exact sequence*

$$0 \to \operatorname{Ker} Arf \xrightarrow{\ sig\ } (4\mathbb{Z})^r \to \mathbb{Z}^* \to 0,$$

where $(4\mathbb{Z})^r \to \mathbb{Z}^*$ *is given by* $(4k_1,..., 4k_r) \to (-1)^{k_1} ... (-1)^{k_r}$.

(2) *In all other cases, there is a split exact sequence*

$$0 \to (\operatorname{Ker} Arf)_{tor} \longrightarrow \operatorname{Ker} Arf \xrightarrow{\ sig\ } (4\mathbb{Z})^r \to 0,$$

and $(\operatorname{Ker} Arf)_{tor}$ *is an elementary Abelian 2-group with* $\mathbb{Z}_2$-*basis* $\{[M_v] \mid v \in S_{fin}\}$. *In particular if* $f \leq 1$, *then* $(\operatorname{Ker} Arf)_{tor}$ *is trivial.*

Proof. A) We prove first that $(\operatorname{Ker} Arf)_{tor}$ is trivial if $f \leq 1$ and that it has the required basis if $f \geq 2$. Assume that $[M]$ is in $(\operatorname{Ker} Arf)_{tor} \subseteq \operatorname{Ker} sig$. So M is a quadratic space of even rank and trivial Arf invariant. Since Arf $M = 1$, Arf $(M \otimes_R F) = 1$ also. Recall the invariants c_v from Chapter 13E. Since $[C(M)] \in \operatorname{Br}(R)_2$, $c_v(M \otimes_R F) = 1$ for all $v \in \Omega - S$, by Theorem (14.1). Therefore $c_v(M \otimes_R F) \neq 1$ is possible only for $v \in S_{fin}$. Again by Theorem (14.1), $\prod\limits_{v \in S_{fin}} c_v(M \otimes_R F) = 1$. If $f \leq 1$, then clearly, $c_v(M \otimes_R F) = 1$ for all $v \in \Omega_{fin}$ and therefore, $w(M \otimes_R F) = 1$. Observe that $[M \otimes_R F] \in \operatorname{Ker} sig_F$. Theorem (13.11) now implies that $[M \otimes_R F] = 0$. Therefore, $[M] = 0$ by the injectivity of $Wq(R) \to WQ(F)$. So if $f \leq 1$, then $(\operatorname{Ker} Arf)_{tor} = \{0\}$ as required.

Assume that $f \geq 2$. Since $(\operatorname{Ker} Arf)_{tor} \subseteq \operatorname{Ker} sig$, the image of $[C(M)]$ under the composite

$$\operatorname{Br}(R)_2 \longrightarrow \prod\limits_{v \in \Omega_{re} \cup S_{fin}} \operatorname{Br}(F_v)_2 \cong \prod\limits_{\Omega_{re} \cup S_{fin}} \mathbb{Z}^*$$

has the form $(1,...,1,*,*,...)$, where the first r co-ordinates (i.e., those corresponding to the valuations in Ω_{re}) are all 1, and the S_{fin} co-ordinates are 1 for all but a finite (even) number of them where -1 is possible. The image of $[M]$ under the composite of *Cliff*: $\operatorname{Ker} Arf \to \operatorname{Br}(R)_2$ and the map above has (of course) the same form. Consider the quadratic space M_i for $i \geq r + 1$. By the Λ-Construction, the image of $[M_i]$ under the composite is $(1,...,1,-1,-1,1,...)$ with the first -1 in the i-th coordinate. Notice that the image of [M] is a product of images of appropriate $[M_i]$ in a unique way. It follows that these tuples (for all $i \geq r + 1$) consitute a $\mathbb{Z}_2$-basis of the image of (Ker $Arf)_{tor}$ in $\prod\limits_{\Omega_{re} \cup S_{fin}} \mathbb{Z}^*$. By (14.5), $Cliff :$ Ker $Arf \to$ Br(R)$_2$ is injective on (Ker $Arf)_{tor}$. It follows that $\{[M_i] \mid i \geq r + 1\} = \{[M_v] \mid v \in S_{fin}\}$ is a $\mathbb{Z}_2$-basis of (Ker $Arf)_{tor}$.

B) Consider case (2) of the theorem. If $r = 0$, then sig is the trivial map, (Ker $Arf)_{tor} =$ Ker Arf, and the split exactness of the sequence is trivial. So assume that $r \geq 1$ and $f \geq 1$. Take v_i in Ω_{re}, and v_{r+1} in S_{fin}. The quadratic module $M_{i,r+1}$ provided by the Λ-Construction has the property that $sig [M_{i,r+1}] = (0,...,0,4,0,...,0)$, with the 4 in the i-th position. Since these tuples generate $(4\mathbb{Z})^r$ additively, $sig($Ker $Arf) = (4\mathbb{Z})^r$. In view of (A), the sequence is exact. Since these tuples are a basis of $(4\mathbb{Z})^r$, the sequence splits.

C) Now to case (1). Since $f = 0$, F is an algebraic number field and $R = R_\infty$ is its ring of integers. Let $[M] \in$ Ker Arf be arbitrary. By Theorem (14.1), $c_v(M \otimes_R F) = 1$ for all $v \in \Omega - S$, i.e., for all $v \in \Omega_{fin}$. So by Theorem (13.14), $sig [M]$ is an element of $X = \{(4k_1,..., 4k_r) \mid (-1)^{k_1}... (-1)^{k_r} = 1\}$. In view of (A), it remains to show that sig maps onto X. Consider the quadratic spaces M_i for $1 \leq i < r$ as well as $M_{1,r}$. If $r \geq 3$, check by an induction that the elements $sig [M_i] = (0,...,0,4,4,0,...,0)$ and $sig [M_{1,r}] = (4,0,...,0,4)$ generate X. Suppose $r = 2$. Here $sig [M_1] = (4,4) = sig [M_{1,r}]$ does not suffice. Since it is not hard to see that $(4,4)$ and $(8,0)$ together generate X, it suffices to construct a quadratic space M over R of rank 8 and trivial Arf invariant such that $sig [M] = (8,0)$. The strategy will be to specify the required invariants locally, observe that there is a quadratic space V over F with these invariants, and show that V comes via extension of scalars from an M. Let $V_{v_1} = <1> \perp ... \perp <1>$ be the orthogonal sum of eight copies of $<1>$ over F_{v_1}. For all other v let $V_v = H \perp ... \perp H$ be the orthogonal sum of four hyperbolic planes H over F_v. All $c(V_v)$ are trivial. In the case of v_1 this follows by (11.10) and more explicitly by its proof. For all other v it follows by (11.3). By an application of Theorem 72:1 in O'Meara [1971], there exists a quadratic space V over F such that $V \otimes_F F_v \cong V_v$ for all $v \in \Omega$. Its discriminant is trivial. So by Theorem (10.12) and (12.10), its Arf invariant is also trivial. Since $\beta :$ Br(F) $\longrightarrow \prod\limits_{v \in \Omega}$ Br(F$_v$) is injective, [C(V)] $\in$ Br(F) is trivial. By Theorem (14.4) there is a quadratic space M over R such that $M \otimes_R F \cong V$. By (13.9) and (12.10), the Arf invariant of M is trivial.

Therefore [M] is in Ker *Arf*. It is clear that $sig [M] = (8,0)$. If $r = 1$, then X $= 8\mathbb{Z}$. Construct M as before.   QED.

(14.8). *Corollary. sig* Wq(R) *is a free Abelian group of rank* r.
Proof. Since sig Wq(R) $\subseteq (\mathbb{Z})^r$ contains a free subgroup of rank r, namely, sig (Ker *Arf*), it has rank r.   QED.

(14.9). *Corollary. Arf* : Wq(R)$_{tor}$ $\to$ QU(R) *is injective* $\Leftrightarrow$ f ≤ 1.

This fact, together with (14.5) and (13.7), provides

(14.10). *Corollary. If* F *is an algebraic function field and* f = 1, *then* Wq(R) $\cong$ QU(R).

A combination of Theorems (13.7) and (14.7) and Exercise 4 provides the following structural configurations:

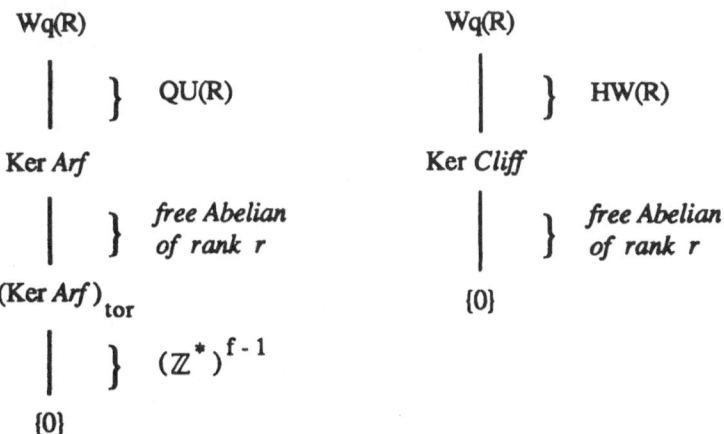

D. Computing Qu(R_∞) and Wq(R_∞)

For an algebraic function field F, the structure of Wq(R) for the smallest Dedekind domains of F is given by (14.10). In this section, F is an algebraic number field, $R = R_\infty$, and the focus will be on the structure of Wq(R) for this smallest Dedekind domain in F. Observe that $2 \notin R^*$, and hence that Wq(R) = Wq$_{ev}$(R). So *Arf* maps Wq(R) into Qu(R). By (14.9), the restriction of *Arf* : Wq(R) $\to$ Qu(R) to Wq(R)$_{tor}$ is injective. The goal is the analysis of Qu(R) and its subgroup *Arf* (Wq(R)$_{tor}$).

We begin by recalling how the concept unramified, defined in Chapter 12E for prime ideals of R, can be extended to the valuations of F. See the references already provided in Section A for the details. Let E be a finite Galois extension of F. Let $v \in \Omega$. Then v lifts to a valuation, say w, of E. There are (up to equivalence) only finitely many such w and we let T_v be a set of representatives. Let $w \in T_v$ and let E_w be a completion of E at w. The closure of F in E_w is a completion F_v of F at v. Denote the extension of w to E_w by w also, and consider the value groups $w(E_w^*) \supseteq w(F_v^*)$. Suppose $v \in \Omega_{fin}$. Then v is said to be *unramified in* E if $w(E_w^*) = w(F_v^*)$ for all $w \in T_v$. Let $\mathfrak{p}$ be the prime ideal of R that corresponds to the valuation v. It turns out that v is unramified in E if and only if $\mathfrak{p}$ is an unramified prime ideal in the integral closure of R in E. Of course, if $F_v = E_w$ for all $w \in T_v$, then v is unramified in E. This is how the concept is extended to an Archimedean v. Let v in Ω be Archimedean. Then all w in T_v are Archimedean, and we say that v is *unramified in* E if $F_v = E_w$ for all w in T_v. If v is complex, then v is certainly unramified in E; if v is real, then v is unramified if and only if all the valuations in T_v are also real. We say that the extension E of F is *unramified* if all the valuations in Ω are unramified in E.

By (12.10), the maps $Qu(R) \to Qu(F)$ and $Qu_f(F) \xrightarrow{\delta} F^*/(F^*)^2$ are both injective. Therefore the composite

$$Qu(R) \to F^*/(F^*)^2$$

is injective. What is the image? Let A be a separable quadratic algebra over R and consider $[A] \in Qu(R)$. If $[A] \neq [1, 0] = 1$, then by Chapter 12D, especially (12.9), $E = A \otimes_R F$ is a quadratic Galois extension of F and A is the integral closure of R in E. Also, $E \cong F[X]/(X^2 - \beta)$ for some β in F^*. By Theorem (12.12) and the remarks that follow it, all finite $v \in \Omega$ are unramified in $E \cong F(\sqrt{\beta})$. Let

$$F_0^* = \{\beta \in F^* \mid \text{all finite } v \text{ of } \Omega \text{ are unramified in } F(\sqrt{\beta})\}.$$

We have shown that the image of $Qu(R) \to F^*/(F^*)^2$ is contained in $F_0^*/(F^*)^2$. Another application of Theorem (12.12) shows that

$$Qu(R) \to F_0^*/(F^*)^2$$

is onto. It is therefore an isomorphism.

We continue by establishing a connection between Qu(R) and the ideal class group of R. Recall the group Inv(R) of invertible fractional ideals of R and the subgroup P(R) of principal fractional ideals. The cardinality of the ideal class group

$$Cl(R) = Inv(R)/Pr(R)$$

is the *class number* h of F. A related group is defined as follows. An element t in F^* is called *totally positive* if it is a square in F_v^* for all real $v \in \Omega$. Let $Pr^+(R)$ denote the subgroup of Inv(R) consisting of all principal ideals of the form tR with t totally positive. The quotient group

$$Cl^+(R) = Inv(R)/Pr^+(R)$$

is called the *narrow ideal class group* of F and its order is the *narrow class number* h^+ of F. The natural map $Cl^+(R) \to Cl(R)$ provides the exact sequence

$$1 \to Pr(R)/Pr^+(R) \to Cl^+(R) \to Cl(R) \to 1.$$

The quotient $Pr(R)/Pr^+(R)$ has order h^+/h. It is clear that it is an elementary Abelian 2-group. Therefore the "odd" parts of the class numbers h^+ and h are the same.

Some facts from class field theory are next. See Neukirch or Iyanaga, for instance. The number field F has a unique maximal Abelian unramified extension H. This is the *Hilbert class field* of F. The extension H/F is finite Galois and its Galois group is isomorphic to Cl(R). There is also a unique Abelian extension H^+ of F which is maximal with the property that all finite valuations in Ω are unramified in H^+. The extension H^+ over F is also finite Galois. Its Galois group is isomorphic to $Cl^+(R)$. Clearly, $H^+ \supseteq H$.

(14.11). *Theorem.* $Qu(R) \cong F_0^*/(F^*)^2 \cong Cl^+(R)_2$.

Proof. The first isomorphism is already established. We verify the second. It follows from the definition of the field H^+ that every quadratic Galois extension of F which is unramified at all finite valuations has a (unique) F-isomorphic copy in H^+. By the Galois correspondence these are in one-to-one correspondence with the subgroups of index 2 of $Cl^+(R)$. Denote $Cl^+(R)$ by G and form the quotient G/G^2. In additive language, G/G^2 is a vector space over $\mathbb{Z}_2$. The subgroups of index 2 in G are in one-to-one correspondence with the hyperplanes in this space, and therefore by duality with the lines of this space. Since each line contains a unique nonzero vector, and $g \to g^2$ induces an isomorphism $G/G_2 \cong G^2$, we have now shown that the unramified quadratic

Galois extensions of F are in one-to-one correspondence with the nontrivial elements of G_2. It follows that $F_0^*/(F^*)^2$ and G_2 have the same cardinality. Since they are both elementary Abelian 2-groups, they are isomorphic.   QED.

Remark. The proof above is a counting argument which gives little indication of the underlying realities. For some deeper connections and an explicit construction of the isomorphism between $F_0^*/(F^*)^2$ and the dual of the $\mathbb{Z}_2$-space $Cl^+(R)/Cl^+(R)^2$ refer to Exercises 11 – 13.

Let G be a finite Abelian group. Let $k = 2\text{-}rank$ G, i.e., k is the number of factors in the decomposition of the Sylow 2-subgroup of G as a product of cylic groups of 2-power order. Refer to Exercise 8 for the fact that card $G_2 = 2^k$. Theorem (14.11) asserts, in particular, that

$$\text{card } Qu(R) = 2^{k^+}, \text{ where } k^+ = 2\text{-rank } Cl^+(R).$$

Remark. The number k^+ was already investigated by Gauss. In the *Disquisitiones Arithmeticae*, he computes k^+ in the case $F = \mathbb{Q}(\sqrt{d})$ as follows: Let $\mathfrak{d}_{R/\mathbb{Z}} = D\mathbb{Z}$, with $D > 0$, be the discriminant of F. Let ρ be the number of primes that divide D (so this is the number of primes of $\mathbb{Z}$ that ramify in F). Then $k^+ = \rho - 1$. For some insight into the extensive investigations that have taken place in this area since, refer to Armitage-Fröhlich, Conner-Hurrelbrink, Gras, Hasse, and their bibliographies.

We now turn to the determination of the image of $Wq(R)_{tor}$ under

$$Arf : Wq(R) \to Qu(R).$$

Denote this image by $Qu_t(R)$. Let F_t^* denote the group of all totally positive elements. Clearly, $F_t^* \supseteq (F^*)^2$.

(14.12). *Theorem.* $Wq(R)_{tor} \cong Qu_t(R) \cong (F_0^* \cap F_t^*)/(F^*)^2 \cong Cl(R)_2$.

Proof. We show first that the composite $Wq(R) \xrightarrow{Arf} Qu(R) \to F_0^*/(F^*)^2$ maps $Wq(R)_{tor}$ into $(F_0^* \cap F_t^*)/(F^*)^2$. Let $[M] \in Wq(R)_{tor}$. Let $A = A(M)$ be the Arf algebra of M. We may assume that $[A] \neq 1$ in $Qu(R)$ and hence by (12.9), that $E = A \otimes_R F$ is a quadratic Galois extension of F. Put $E = F(\sqrt{\beta})$. From above, $\beta \in F_0^*$. Let $V = M \otimes_R F$. By properties of the Arf algebra, $A(V) \cong A \otimes_R F = F(\sqrt{\beta})$. Let $v \in \Omega$ be real. Since $[M] \in Wq(R)_{tor}$, the signature of $V \otimes_F F_v$ is 0. So $V \otimes_F F_v$ is an orthogonal sum of hyperbolic planes. By Example 3 in Chapter 7C, $[A(V \otimes_F F_v)] = 1$ in $Qu(F_v)$. So $A(V) \otimes_R F_v \cong F_v(\sqrt{\beta})$ is the trivial separable quadratic algebra over F_v. It follows by (3.3) that $\beta \in (F_v^*)^2$. We have proved that $\beta \in F_t^*$.

We show next that $Wq(R)_{tor} \xrightarrow{\ Arf\ } Qu_t(R) \to F_0^*/(F^*)^2$ maps onto $(F_0^* \cap F_t^*)/(F^*)^2$. Since we already know that the two maps are injective, this will establish the first two isomorphisms of the theorem. Let $\beta \in F_0^* \cap F_t^*$. We can assume that $\beta \notin (F^*)^2$ and consider the field $E = F(\sqrt{\beta})$. Since no finite v of Ω ramifies in E, the integral closure A of R in E is a separable quadratic algebra. Denote by M the quadratic module obtained by taking A with its norm form. By Exercise 2 of Chapter 10E, M is a quadratic space. Since $M \otimes_R F \cong F(\sqrt{\beta})$, $M \otimes_R F$ is isometric to $<1> \perp <-\beta>$ by the Example of Chapter 4D. The choice of β implies that the total signature of this quadratic space is 0. So $[M] \in Wq(R)_{tor}$. By the proof of Theorem (13.7), $Arf\,[M] = [A] \in Qu(R)$. It only remains to note that $Qu(R) \to F_0^*/(F^*)^2$ takes $[A]$ to $\beta(F^*)^2$.

To establish the third isomorphism follow the counting argument in the proof of Theorem (14.11). Refer to Exercise 14 for a more explicit approach to this isomorphism.   QED.

(14.13). *Theorem.* $Wq(R_\infty) \cong Cl(R_\infty)_2 \oplus G$, *where* G *is a free Abelian group of rank* r.

Proof. Combine Theorem (14.12) with Corollary (14.8). QED.

E. Connections between $W(R)$ and $Wq(R)$

We return to an arithmetic Dedekind domain $R = R_S$ in a global field F. As before, $r = $ card Ω_{ev}, $c = $ card Ω_{co}, and $f = $ card S_{fin} (possibly infinite). We will assume that char $R = $ char $F \neq 2$. Refer to Chapter 13D and consider the commutative diagram

$$
\begin{array}{ccc}
Wq(R) & \longrightarrow & Wq(F) \\
\downarrow & & \downarrow \\
W(R) & \longrightarrow & W(F)
\end{array}\ .
$$

The horizontal maps are given by change of scalars. The map on the right is an isomorphism, and the map on the bottom is injective. See the reference to Knebusch and Baeza in Chapter 14C for the last fact. So the other two maps of the diagram are also injective. We therefore consider $Wq(R) \subseteq W(R)$. The focus in this section will be on the quotient group $W(R)/Wq(R)$. If $2 \in R^*$, then $Wq(R) = W(R)$. We will therefore assume that $2 \notin R^*$. In particular, see Section A, F will be an algebraic number field.

By the rank homomorphism, $W(R)/W_{ev}(R) \cong \mathbb{Z}_2$. By (4.18), $Wq(R) = Wq_{ev}(R)$, and hence $W_{ev}(R) \supseteq Wq(R)$. We will analyze

$$
W_{ev}(R)/Wq(R).
$$

By (13.9), we obtain the commutative diagram

$$
\begin{array}{ccccccccc}
1 & \rightarrow & \operatorname{Ker} dis & \longrightarrow & W_{ev}(R) & \xrightarrow{\ dis\ } & \operatorname{Dis}(R) & \rightarrow & 1 \\
& & \uparrow & & \uparrow & & \uparrow & & \\
1 & \rightarrow & \operatorname{Ker} Arf & \longrightarrow & Wq(R) & \xrightarrow{\ Arf\ } & \operatorname{Qu}(R) & \rightarrow & 1
\end{array}
$$

By (12.10), $\operatorname{Qu}(R) \to \operatorname{Dis}(R)$ is injective, and it follows that there is an induced exact sequence

$$
1 \rightarrow (\operatorname{Ker} dis)/(\operatorname{Ker} Arf) \longrightarrow W_{ev}(R)/Wq(R) \longrightarrow \operatorname{Dis}(R)/\operatorname{Qu}(R) \rightarrow 1.
$$

We consider $(\operatorname{Ker} dis)/(\operatorname{Ker} Arf)$ first. Since

$$
\operatorname{Ker} dis \cong (\operatorname{Ker} dis)_{tor} \oplus sig\,(\operatorname{Ker} dis)
$$

and similarly for $\operatorname{Ker} Arf$, it follows that

$$
(\operatorname{Ker} dis)/(\operatorname{Ker} Arf) \cong (\operatorname{Ker} dis)_{tor}/(\operatorname{Ker} Arf)_{tor} \oplus sig\,(\operatorname{Ker} dis)/sig\,(\operatorname{Ker} Arf)\ .
$$

By facts collected toward the conclusions of Chapters 13D and 13E, Ker dis_F = Ker $Arf_F = I^2F$, and sig (Ker dis) is contained in $(4\mathbb{Z})^r$. If $f \geq 1$ or $r = 0$, then by Theorem (14.7),

$$sig \text{ (Ker } dis)/sig \text{ (Ker } Arf)$$

is trivial. Suppose $f = 0$ and $r \geq 1$. Observe that in this case $R = R_\infty$. By the discussion on page 97 of Milnor-Husemoller, sig (Ker dis) = $(4\mathbb{Z})^r$, so that, again by Theorem (14.7), sig (Ker dis)/sig (Ker Arf) has order 2. We concentrate next on

$$(\text{Ker } dis)_{tor} / (\text{Ker } Arf)_{tor}.$$

A careful analysis of the discussion in §3 and §4 of Chapter IV of Milnor-Husemoller shows that the composite

$$(\text{Ker } Arf_F) \xrightarrow{Cliff_F} Br(F)_2 \longrightarrow \prod_{v \in \Omega} Br(F_v)_2 \dashrightarrow \prod_\Omega \mathbb{Z}^*$$

takes (Ker dis)$_{tor}$ onto the kernel of $\prod_{S_{fin} \cup (\Omega - S)_{dy}} \mathbb{Z}^* \xrightarrow{\text{product}} \mathbb{Z}^*$, where $(\Omega - S)_{dy}$ denotes the (finite) set of dyadic valuations in $\Omega - S$. Since Ker $Cliff_F$ is torsion free, this provides an isomorphism

$$(\text{Ker } dis)_{tor} \longrightarrow \prod_{f+d-1} \mathbb{Z}^*,$$

where $d =$ card $(\Omega - S)_{dy}$. By part (A) of the proof of Theorem (14.7), this isomorphism restricts to an isomorphism

$$(\text{Ker } Arf)_{tor} \longrightarrow \prod_{f-1} \mathbb{Z}^*,$$

where $\prod_{f-1} \mathbb{Z}^*$ is understood to be 1 if $f = 0$. It follows that

$$(\text{Ker } dis)_{tor} / (\text{Ker } Arf)_{tor} \cong \begin{cases} \prod_{d-1} \mathbb{Z}^*, \text{ if } f = 0 \\ \prod_{d} \mathbb{Z}^*, \text{ if } f \geq 1 . \end{cases}$$

The following are easy consequence of the preceding discussion.

(14.14). W(R)/Wq(R) *is an Abelian group of exponent* 8 *and*

$$\text{card } (W(R)/Wq(R)) = \begin{cases} 2^d \cdot \text{card } (\text{Dis}(R)/\text{Qu}(R)), \text{ if } f = 0 \text{ and } r = 0 \\ 2^{d+1} \cdot \text{card } (\text{Dis}(R)/\text{Qu}(R)), \text{ if } f \geq 1 \text{ or } r \geq 1. \end{cases}$$

Remark. The fact that R/4R is a finite ring shows in combination with (12.7) and (12.10) that the group Dis(R)/Qu(R) is finite.

For the remainder of this section we specialize to the case $R = R_\infty$. So now $f = 0$. Here the quotient Dis(R)/Qu(R) can be calculated explicitly. Since it is an elementary Abelian 2-group, it suffices to specify the cardinality. By a combination of Theorem (14.11), (12.4), and consequences of the Dirichlet Unit Theorem already pointed out in Section A, we get

$$\text{card } (\text{Dis}(R)/\text{Qu}(R)) = \frac{2^s \text{card Cl}(R)_2}{\text{card Cl}^+(R)_2} = 2^{s - (k^+ - k)},$$

where $s = \text{card } S$, $k^+ = 2\text{-rank Cl}^+(R)$, and $k = 2\text{-rank Cl}(R)$. To analyze $k^+ - k$, we return to the exact sequence

$$1 \to \text{Pr}(R)/\text{Pr}^+(R) \to \text{Cl}^+(R) \to \text{Cl}(R) \to 1.$$

Let Syl_2^+ and Syl_2 be the respective Sylow 2-groups of $\text{Cl}^+(R)$ and $\text{Cl}(R)$. Let $C_1 \times ... \times C_{k^+}$ be a decomposition of Syl_2^+ into cyclic factors. The images of the C_i under $\text{Cl}^+(R) \to \text{Cl}(R)$ provide a cyclic decomposition of Syl_2. So $k^+ \geq k$. What can one say about $k^+ - k$? If C_i is a cyclic factor of order greater than 2, then C_i cannot vanish under $\text{Cl}^+(R) \to \text{Cl}(R)$ because $\text{Pr}(R)/\text{Pr}^+(R)$ is an elementary Abelian 2-group. So only cyclic factors of order 2 can vanish. This means that $k^+ - k$ is precisely the number of C_i (of order 2) contained in $\text{Pr}(R)/\text{Pr}^+(R)$. Therefore, $2^{(k^+ - k)} \leq h^+/h$, and hence

$$\text{card } (\text{Dis}(R)/\text{Qu}(R)) = 2^{s - (k^+ - k)} \geq \frac{2^s h}{h^+}.$$

If $r = \text{card } \Omega_{re} = 0$, then everything in F^* is trivially totally positive. So $k^+ = k$ and card (Dis(R)/Qu(R)) = 2^s. More generally, by an easy exercise,

(14.15). card $(\text{Dis}(R)/\text{Qu}(R)) = \dfrac{2^s h}{h^+}$ *if and only if the exact sequence*

$$1 \to \text{Pr}(R)/\text{Pr}^+(R) \to \text{Cl}^+(R) \to \text{Cl}(R) \to 1$$

splits.

Combining the order formula for card $(\text{Dis}(R)/\text{Qu}(R))$ with (14.14) — note that $s = c$ if $r = 0$ — we get

$$\text{card } (W(R)/Wq(R)) = \begin{cases} 2^{d+c}, & \text{if } r = 0. \\ 2^{d+s+1-(k^+-k)}, & \text{if } r \geq 1 . \end{cases}$$

Since $W(R) \cong W(R)_{\text{tor}} \oplus sig(W(R))$ and $Wq(R) \cong Wq(R)_{\text{tor}} \oplus sig(Wq(R))$, it is clear that

$$W(R)/Wq(R) \cong (W(R)_{\text{tor}}/Wq(R)_{\text{tor}}) \oplus (sig\ W(R)/sig\ Wq(R)).$$

How does the order of $W(R)/Wq(R)$ distribute over the two summands? Combining Theorem (4.1) in Chapter IV of Milnor-Husemoller with Theorem (14.12) shows that

$$\text{card } (W(R)_{\text{tor}}/Wq(R)_{\text{tor}}) = \begin{cases} 2^{d+c}, & \text{if } r = 0. \\ 2^{d+c-1+(k^+-k)}, & \text{if } r \geq 1. \end{cases}$$

Consequently,

$$\text{card } (sig\ W(R)/sig\ Wq(R)) = \begin{cases} 0, & \text{if } r = 0. \\ 2^{r-2(k^+-k-1)}, & \text{if } r \geq 1. \end{cases}$$

Remark. Implicit in this formula is the inequality $r \geq 2(k^+ - k - 1)$. Refer to Armitage-Fröhlich for the slightly sharper $[r/2] \geq k^+ - k$, where $[\ ]$ is the greatest integer function.

In view of (14.14), the group $W(R)/Wq(R)$ is a finite Abelian 2-group. The preceding order formulas provide (of course) no insight into the nature of its decomposition into cyclic factors. Indeed, the structures of the quotients $W(R)_{\text{tor}}/Wq(R)_{\text{tor}}$ and $sig\ W(R)/sig\ Wq(R)$ appear to be subtle. In the following table they are listed for $\mathbb{Z}$ (see the case $n = 1$), where they were known previously, and for the rings of integers of the quadratic extensions $\mathbb{Q}(\sqrt{n})$ with $|n| \leq 17$. The structure of $W(R)$ comes from Chapter IV in Milnor-Husemoller, primarily from the table on page 96. For the information about W(R)/Wq(R) use: the preceding results; the formula for the discriminant of $\mathbb{Q}(\sqrt{n})$; the result of Gauss from the Remark that precedes Theorem (14.12); the fact that 2 splits in $\mathbb{Q}(\sqrt{n})$ if and only if $n \equiv 1 \mod 8$ (see §5.4 in Chapter V of Samuel); the class number tables from pages 422–425 of Borevich-Shafarevich; Exercises 6 and 9 in the real cases; and Exercise 7 in combination with Lemma (20.1) from Conner-Hurrelbrink in the complex cases.

A cyclic summand of order k of either $W(R)_{tor}$ or $W(R)_{tor}/Wq(R)_{tor}$ is denoted by k and a cyclic summand of oder k of $sig\,(W(R)/sig\,Wq(R))$ is denoted by $\mathbb{Z}_k$.

n =	W(R)	W(R)/Wq(R)	n =	W(R)	W(R)/Wq(R)
1	$\mathbb{Z}$	$\mathbb{Z}_8$	−1	$2 \oplus 2$	$2 \oplus 2$
2	$\mathbb{Z} \oplus \mathbb{Z}$	$\mathbb{Z}_4 \oplus \mathbb{Z}_4$	−2	4	4
3	$2 \oplus \mathbb{Z} \oplus \mathbb{Z}$	$2 \oplus \mathbb{Z}_2 \oplus \mathbb{Z}_2$	−3	4	4
5	$\mathbb{Z} \oplus \mathbb{Z}$	$\mathbb{Z}_4 \oplus \mathbb{Z}_4$	−5	$2 \oplus 4$	$2 \oplus 2$
6	$2 \oplus \mathbb{Z} \oplus \mathbb{Z}$	$2 \oplus \mathbb{Z}_2 \oplus \mathbb{Z}_2$	−6	$2 \oplus 4$	4
7	$2 \oplus \mathbb{Z} \oplus \mathbb{Z}$	$2 \oplus \mathbb{Z}_2 \oplus \mathbb{Z}_2$	−7	8	8
10	$2 \oplus \mathbb{Z} \oplus \mathbb{Z}$	$2 \oplus \mathbb{Z}_4 \oplus \mathbb{Z}_4$	−10	$2 \oplus 4$	4
11	$2 \oplus \mathbb{Z} \oplus \mathbb{Z}$	$2 \oplus \mathbb{Z}_2 \oplus \mathbb{Z}_2$	−11	4	4
13	$\mathbb{Z} \oplus \mathbb{Z}$	$\mathbb{Z}_4 \oplus \mathbb{Z}_4$	−13	$2 \oplus 4$	$2 \oplus 2$
14	$2 \oplus \mathbb{Z} \oplus \mathbb{Z}$	$2 \oplus \mathbb{Z}_2 \oplus \mathbb{Z}_2$	−14	$2 \oplus 2 \oplus 4$	4
15	$2 \oplus 2 \oplus \mathbb{Z} \oplus \mathbb{Z}$	$2 \oplus \mathbb{Z}_2 \oplus \mathbb{Z}_2$	−15	$2 \oplus 8$	$2 \oplus 4$
17	$2 \oplus \mathbb{Z} \oplus \mathbb{Z}$	$2 \oplus \mathbb{Z}_4 \oplus \mathbb{Z}_4$	−17	$2 \oplus 4$	4

F. Exercises

In the following exercises, F is a global field and $R = R_S$ is a Dedekind domain in F. Also, card $S = s$, card $\Omega_{re} = r$, card $\Omega_{co} = c$, card $S_{fin} = f$, $k = 2$-rank $Cl(R)$, and $k^+ = 2$-rank $Cl^+(R)$.

1. Let R be a Hasse domain. Suppose char $R \neq 2$. Then HW(R) is a finite group. If $2 \in R^*$, then card $HW(R) = 2^{r+f+s+k}$, and if $2 \notin R^*$, then

card $HW(R) \le 2^{r+f+ \ s-1+k-u}$, where card $(R/4R)^*/((R/4R)^*)^2 = 2^u$. If char $R = 2$, then card $HW(R) = 2^{r+f-1} \cdot$ card $R/\wp(R)$. (I don't know whether this last group is finite, but it seems reasonable that this should be so.)

2. Suppose char $R \ne 2$. Then, R is a Hasse domain $\Leftrightarrow$ $Br(R)_2$ is finite $\Leftrightarrow$ $HW(R)$ is finite $\Leftrightarrow$ $Wq(R)_{tor}$ is finite $\Leftrightarrow$ $Wq(R)$ is a finitely generated group.

3. The following holds: $f \le 1 \Leftrightarrow Br(R)$ is finite $\Leftrightarrow$ Ker Arf is torsion free.

4. Show that $sig : Wq(R) \to (\mathbb{Z})^r$ (mod 2) restricts to an isomorphism from Ker $Cliff$ onto $(8\mathbb{Z})^r$.

5. Suppose that F is an algebraic number field. Let v be a real valuation of F. Let $\sigma_v : F \to \mathbb{Z}^*$ be defined as follows $\sigma_v(d) = 1$ if $d \in (F_v^*)^2$ and $\sigma_v(d) = -1$ otherwise. Show that this leads to an exact sequence

$$1 \to F_t^* \to F \to \prod_{v \in \Omega_{re}} \mathbb{Z}^* \xrightarrow{\text{product}} \mathbb{Z}^*.$$

6. Suppose that r is even or $f > 0$. Show that $2(sig\ W_{ev}(R)) \subseteq sig\ Wq(R)$. Conclude that $sig\ W(R)/sig\ Wq(R)$ has exponent 4.

7. Let char $F \ne 2$. Consider the element $[<2>] \in Wq(F)$. If $-1 \in (F^*)^2$, then $2[<2>] \in Wq(R)$. If $-1 \notin (F^*)^2$, then $2[<2>] \in Wq(R)$ if and only if all finite valuations of F are unramified in $F(\sqrt{-1}\ )$.

8. Prove that if G is a finite Abelian group, then card $G_2 = 2^k$, where $k = $ 2-rank G.

In Exercises 9 – 15, F is an algebraic number field.

9. Let $R = R_\infty$. Prove that if there is a $\beta \in F_0^*$ such that $\beta \notin (F_v^*)^2$ for all real v, then $sig\ W(R)/sig\ Wq(R)$ is an elementary Abelian 2-group. Show that if $r = 2$ and $k^+ > k$, then such a β exists.

10. Let $R = R_\infty$. Let $v \in \Omega$ be finite. Consider the completion F_v, its ring of integers R_v, and its maximal ideal $\mathfrak{p}_v$. Choose $\pi_v \in R_v$ such that $\mathfrak{p}_v$

$= \pi_v R_v$. Let $d \in F$. Put $d = \varepsilon \pi^j$ with $\varepsilon \in R_v^*$, and define $\mathrm{ord}_v\, d = j$. By the product formula (see page 66 in O'Meara [1971]), $\mathrm{ord}_v\, d = 0$ for all but finitely many v. Set

$$F_{ev}^* = \{d \in F^* \mid \mathrm{ord}_v\, d \text{ is even (or zero) for all finite } v \in \Omega\}.$$

Recall that $\mathrm{Dis}(R) \to F^*/(F^*)^2$ is injective. Show that the image is $F_{ev}^* /(F^*)^2$. Deduce that $\mathrm{Dis}(R)/\mathrm{Qu}(R) \cong F_{ev}^*/F_0^*$.

11. Let $R = R_\infty$. Let $\mathfrak{p}$ be a prime ideal of R and let $v \in S$ be the finite valuation corresponding to $\mathfrak{p}$. For $d \in F_0^*$, define the 2-*power residue symbol* $\left(\frac{d}{\mathfrak{p}}\right)$ to be equal to 1 if $d \in (F_v^*)^2$ and -1 otherwise. Define $\left(\frac{d}{\mathfrak{a}}\right)$ for any $\mathfrak{a} \in \mathrm{Inv}(R)$ by using the uniqueness of the factorization into primes. Let $F_0^*/(F^*)^2 \times \mathrm{Inv}(R) \to \mathbb{Z}^*$ be defined by $(d(F^*)^2, \mathfrak{a}) = \left(\frac{d}{\mathfrak{a}}\right)$. Show that this pairing is multiplicative in both variables.

12. Let $R = R_\infty$. Let $r \in F_t^*$. Consider the factorization of $rR = \mathfrak{p}_1^{m_1} \dots \mathfrak{p}_k^{m_k}$ into primes and let $T = \{v_1, ..., v_k\}$ be the valuations corresponding to $\mathfrak{p}_1, ..., \mathfrak{p}_k$. Let $d \in F_0^*$. Note that $\left(\frac{d}{rR}\right) = \prod_{1 \le i \le k} \left(\frac{d}{\mathfrak{p}_i}\right)^{m_i}$. By classical number theory (refer to Iyanaga, for example), $\left(\frac{d}{rR}\right) = \prod_{1 \le i \le k} \left(\frac{r, d}{v_i}\right)$, where $\left(\frac{r, d}{v_i}\right)$ is the Hilbert symbol (see Exercise 3 of Chapter 13F). Show that $\left(\frac{r, d}{v}\right) = 1$ for all $v \in \Omega - T$. Therefore, $\prod_{v \in T} \left(\frac{r, d}{v}\right) = 1$ and $\left(\frac{d}{rR}\right) = 1$.

13. Show that the pairing $F_0^*/(F^*)^2 \times \mathrm{Inv}(R) \to \mathbb{Z}^*$ of Exercise 11 is trivial on the elements $(d(F^*)^2, rR)$ with $r \in F_t^*$. So there are induced pairings $F_0^*/(F^*)^2 \times Cl^+(R) \to \mathbb{Z}^*$ and $F_0^*/(F^*)^2 \times Cl^+(R)/Cl^+(R)^2 \to \mathbb{Z}^*$. By facts from class field theory, the latter is a perfect duality, so that $F_0^*/(F^*)^2$ is isomorphic to the dual of $Cl^+(R)/Cl^+(R)^2$.

14. Consider the restriction $(F_0^* \cap F_t^*)/(F^*)^2 \times \text{Inv}(R) \to \mathbb{Z}^*$ of the pairing of Exercise 11. Show that this pairing is trivial on all elements $(d(F^*)^2, rR)$. Therefore, there are induced pairings $(F_0^* \cap F_t^*)/(F^*)^2 \times \text{Cl}(R) \to \mathbb{Z}^*$ and $(F_0^* \cap F_t^*)/(F^*)^2 \times \text{Cl}(R)/\text{Cl}(R)^2 \to \mathbb{Z}^*$. By facts from class field theory, the latter is a perfect duality. Therefore, $(F_0^* \cap F_t^*)/(F^*)^2$ is isomorphic to the dual of $\text{Cl}(R)/\text{Cl}(R)^2$.

15. Let R be a Hasse domain. Consider the field extension $F(\sqrt{R^*})$ of F obtained by adjoining to F representatives of the s square classes of $R^*/(R^*)^2$. Show that the degree of $F(\sqrt{R^*})$ over F is 2^s. The extension $H \cap F(\sqrt{R^*})$ is the unique maximal unramified extension of F in $F(\sqrt{R^*})$. The field degree of $H \cap F(\sqrt{R^*})$ over F is a 2-power, say 2^u. Show that card $\text{Qu}_f(R) = 2^u$.

Hints:

1. Use the exact sequence in (13.8), Theorem (14.6), information about $\text{Br}(R)_2$, and information about $\text{QU}(R)$ and $\text{Qu}(R)$.

2. Label the statements in order by (1), (2), (3), (4), and (5). By the Remark following Theorem (14.1), (1) $\Rightarrow$ (2); (2) $\Rightarrow$ (3) by arguing as in Exercise 1; (3) $\Rightarrow$ (4), by (14.5); (4) $\Rightarrow$ (Ker $Arf)_{\text{tor}}$ is finite $\Rightarrow$ (1) by Theorem (14.7). Finally, by (14.8), (4) $\Leftrightarrow$ (5).

3. Label the statements in order by (1), (2), and (3). By the exact sequence that precedes Theorem (14.1), (1) $\Leftrightarrow$ (2). By Theorem (14.7), (1) $\Leftrightarrow$ (3).

4. By facts from Chapter 13E, the restriction of sig_F to $\text{Ker}\, Cliff_F$ is injective and $sig_F (\text{Ker}\, Cliff_F)$ is contained in (in fact is equal to) $(8\mathbb{Z})^r$. Only the surjectivity of sig remains. We may assume that $r > 0$. Consider the classes $[M_i]$, $[M_{i,r+1}]$, $[M_{1,r}]$ from part (B) of the proof of Theorem (14.7). Since $Cliff$ maps $\text{Ker}\, Arf$ into $\text{Br}(R)_2$, the elements $2[M_i]$, $2[M_{i,r+1}]$, and $2[M_{1,r}]$ are in $\text{Ker}\, Cliff$. Their images under sig generate $(8\mathbb{Z})^r$, except when $f = 0$ and $r \le 2$. In the remaining cases, the class $[M]$ of the special module M (see part (C) of the same proof) is in $\text{Ker}\, Cliff$. It and $2[M_1]$ generate $(8\mathbb{Z})^r$.

5. Apply Theorem 11:8 in O'Meara [1971], for example.

6. If $r = 0$, sig is trivial and there is nothing to prove. Assume that $r > 0$. So F is an algebraic number field. Let $[M] \in W_{ev}(R)$. By Exercise 9 in Chapter 13F, $[M] = [(\mathfrak{a}, f_d) \perp (R, f_{-1})] \perp [N]$, where $[N] \in \text{Ker } dis$. By Chapter 13E, $sig [N] \in (4\mathbb{Z})^r$. By Theorem (14.7), $2sig [N] \in sig Wq_{ev}(R)$. For a real v, $sig_v ((\mathfrak{a}, f_d) \perp (R, f_{-1}) \otimes_R F) = 0$ if $d \in (F_v^*)^2$ and -2 if $d \notin (F_v^*)^2$. By Exercise 5, d must be a square at an even number of v. So if r is even, then by Theorem (14.7), $2sig ((\mathfrak{a}, f_d) \perp (R, f_{-1})) \in sig (\text{Ker } Arf)$. If $f > 0$, the same conclusion holds. This proves the first statement. Since $4sig [(R, f_1)] = (4, ..., 4) \in sig (\text{Ker } Arf)$ when r is even or $f > 0$, the second statement follows.

7. When is $2[<2>] = [<2> \perp <2>]$ in Wq(R) ? If $-1 \in (F^*)^2$, then $<2> \perp <2>$ is isotropic, and hence a hyperbolic plane over F. See Chapter 11B. So $[<2> \perp <2>] = 0$ is in Wq(R). So assume $-1 \notin (F^*)^2$. By (7.4), the Arf algebra (over F) of $<2> \perp <2>$ is $F[X]/(X^2 + 16) \cong F(\sqrt{-1})$. Suppose $[<2> \perp <2>] = [M]$ for some $[M]$ in Wq(R). The Arf algebra $A = A(M)$ is a separable quadratic algebra and $A \otimes_R F \cong F(\sqrt{-1})$. As a consequence of Theorem (12.12), all finite valuations of F are unramified in $F(\sqrt{-1})$. Conversely, if all finite valuations of F are unramified in $F(\sqrt{-1})$, then the integral closure of R in $F(\sqrt{-1})$ is a separable quadratic R-algebra. Let M be A equipped with its norm form and note that $M \otimes_R F \cong <2> \perp <2>$ over F.

8. Let $C_1 \times ... \times C_k$ be the decomposition of the Sylow 2-group of G into cyclic factors. Since C_k is cyclic, there is a unique $c_k \in C_k$ with $c_k^2 = 1$. It follows that G_2 has order 2^k.

9. By Exercise 5, the existence of β implies that r is even. In view of Exercise 6, it must be shown that $2sig [(R, f_1)] \in sig Wq(R)$. Let A be the integral closure of R in $E = F(\sqrt{\beta})$. By Theorem (12.12), A is separable quadratic. Now A with its norm is a quadratic space over R. By Chapter 12D, $A \otimes_R F \cong F[X]/(X^2 - \beta)$ as quadratic algebras. By the Example of Chapter 4D, $A \otimes_R F \cong <2> \perp <-2\beta>$ as quadratic spaces over F. Hence, $2sig [(R, f_1)] = (2, ..., 2) = sig [A] \in sig Wq(R)$. Now to

the second statement. If $k^+ > k$, then by Exercise 8 and the proof of Theorem (14.11), $Cl^+(R)$ has more subgroups of index 2 than $Cl(R)$. So by the properties of the fields H and H^+ of Section D, there is a quadratic extension, say $F(\sqrt{\beta})$, of F which is unramified at all finite valuations, but not at some real valuation, say v. So $\beta \notin (F_v^*)^2$, and by Exercise 5, $\beta \notin (F_{v'}^*)^2$ for the other real valuation v'.

10. Refer to Chapter 4C, and let $[\mathfrak{a}, d]$ be any element in $Dis(R)$. Let $\mathfrak{a} = \mathfrak{p}_1^{j_1} \dots \mathfrak{p}_k^{j_k}$ be the factorization of $\mathfrak{a}$ into primes. Observe that $d R = \mathfrak{a}^{-2} = \mathfrak{p}_1^{-2j_1} \dots \mathfrak{p}_k^{-2j_k}$. Let $v_1, \dots, v_k$ be the finite valuations corresponding to the primes $\mathfrak{p}_1, \dots, \mathfrak{p}_k$. Note that $ord_{v_i} d = -2j_i \neq 0$ for $1 \leq i \leq k$, and that $ord_v d = 0$ for all other finite v in Ω. Therefore, the image of $Dis(R) \to F^*/(F^*)^2$ lies in $F_{ev}^*/(F^*)^2$. Fix $d \in F_{ev}^*$ and let $v_1, \dots, v_k$ be the finite valuations in Ω such that $ord_{v_i} d = 2j_i \neq 0$. Let $\mathfrak{p}_1, \dots, \mathfrak{p}_k$ be the corresponding prime ideals of R. The unique factorization of the fractional ideal dR of F into primes is $dR = \mathfrak{p}_1^{2j_1} \dots \mathfrak{p}_k^{2j_k}$. Put $\mathfrak{a} = \mathfrak{p}_1^{-j_1} \dots \mathfrak{p}_k^{-j_k}$ and observe that $dR = \mathfrak{a}^{-2}$.

11. That it is multiplicative in the second variable is easy. That it is multiplicative in the first, follows from the following considerations: for a fixed $\mathfrak{p}$ and fixed d and d' in F_0^* both nonsquares in F_v^*, dd' is in $(F_v^*)^2$. To see this, refer to O'Meara [1971]. For $F_v(\sqrt{d}) = F_v(\sqrt{d'})$, see 32:11. Since $F_0^* \subseteq F_{ev}^*$, d and d' can be taken to be units of F_v; so dd' is a square by 63:3 and 63:4.

12. Suppose v is real. Since r is totally positive, $\left(\frac{r, d}{v}\right) = 1$ by (11.5). Suppose $v \in \Omega - T$ is finite. Then $v(r) = 1$, since the prime that corresponds to v does not appear in the factorization of rR. So r is a unit in the integers of F_v. Since $F(\sqrt{d})$ is an unramified extension of F, r is a norm from $F(\sqrt{d})$ by 63:16 of O'Meara [1971]. So by (11.5), $\left(\frac{r, d}{v}\right) = 1$. Therefore, by Hilbert reciprocity, $\prod_{v \in T} \left(\frac{r, d}{v}\right) = 1$.

13. Use Exercise 12 and page 49 in Conner-Hurrelbrink.

14. Similar to 13.

15. By page 183 in O'Meara [1971], $[F(\sqrt{R^*}) : F] = 2^s$. Let

$$\{r(R^*)^2, \ldots, r_u(R^*)^2\}$$

be the square classes in $R^*/(R^*)^2$ such that $F(\sqrt{r_i})/F$ is unramified at all finite primes. Then $\{r_1(R^*)^2, \ldots, r_u(R^*)^2\}$ is a subgroup of $R^*/(R^*)^2$. This follows by looking inside the extension field H^+ of F. If $F(\sqrt{u})$ and $F(\sqrt{w})$ are both unramified at all finite valuations in Ω, they are both subfields of H^+. So $F(\sqrt{uw}) \subseteq H^+$. For the equality card $Qu_f(R) = 2^u$, see page 183 in O'Meara again.

15
Applications of Clifford Modules

Overview

Clifford modules, i.e., representations of Clifford algebras, were already used in the structure theory of the Clifford algebra; see Chapter 8D. It is an amazing fact that Clifford algebras and Clifford modules over the real and complex numbers lie at the core of an astonishing variety of problems in differential geometry and topology. In particular, they play an important role in the analysis of vector fields on spheres, Lie groups and algebras, Bott periodicity, partial differential equations, immersions of manifolds into spheres, curvature properties in Riemannian geometry, and the structure of isoparametric submanifolds. This chapter will be nothing more than a sequence of glimpses at some of these connections. For details, refer to the impressive recent volumes by Berline-Getzler-Vergne, Gilbert-Murray, and Lawson-Michelsohn; to Kazdan's survey article; and to Stolz [1990] and Thorbergsson [1991]. The main references for the discussion that follows are Lawson-Michelsohn and Husemoller. In the analytical context the Clifford algebra is defined slightly differently, i.e., the requirement $x^2 = q(x)$ is replaced by $x^2 = -q(x)$. This leads to some notational differences between the references and the following discussion.

A. Clifford Modules

Let F be a field and let D be a finite dimensional division algebra over F. Let V be a quadratic space over F and $C(V)$ be its Clifford algebra. Let W be a finite-dimensional right vector space over D and

$$\rho : C(V) \to \text{End}_D W$$

be a representation of $C(V)$. Such a representation is called a D-*representation of* $C(V)$. Defining $cw = (\rho c)w$ for $c \in C(V)$ and $w \in W$, makes W into a $(C(V)$-$D)$-bimodule. Any such bimodule is a *Clifford module*. The product cw is a *Clifford multiplication*. Observe, conversely, that a $(C(V)$-$D)$-bimodule defines a D-representation of $C(V)$. If a basis of V is fixed, then as already observed in Chapter 8D, the concept of a representation is equivalent to that of a Clifford system.

Let $\rho : C(V) \to \text{End}_D W$ be a D-representation. We call ρ *irreducible*, if the only ρ-invariant D-subspaces U of W, i.e., those with the property that $(\rho c)U \subseteq U$ for all c, are $U = \{0\}$ and $U = W$. If there are additional ρ-invariant subspaces, then ρ is *reducible*. Certain properties of $C(V)$, or more accurately of the group of units $C(V)^*$, allow the conclusion that any ρ-invariant D-subspace has a complementary ρ-invariant subspace. In other words, ρ is reducible if and only if there are nonzero ρ-invariant D-subspaces U_1 and U_2 of W such that $W = U_1 \oplus U_2$. Note in particular that every representation is a direct sum of irreducible representations.

Suppose char $F \neq 2$ and $D = F$. Given the periodicity phenomena for the Clifford algebras $C^{n,0}$ already observed in Chapter 11D, it is not very difficult to answer the following question: for a given integer N what is the largest integer $n = n_N$ such that $C^{n,0}$ has a representation

$$\rho : C^{n,0} \to \text{End}_F W$$

with dim $W = N + 1$. It turns out that n_N depends only on the answers to the following questions:

(i) is $-1 \in (F^*)^2$?   (ii) is $\left(\dfrac{-1,-1}{F} \right)$ a division algebra ?

For example, if the answer to (i) is no and the answer to (ii) is yes, then n_N is determined as follows: Set $N + 1 = 2^{4a+b} n_0$, with $0 \le b \le 3$ and n_0 odd. Then

$$n_N = 8a + 2^b - 1.$$

This is the *Radon-Hurwitz number* of N. See §4 in Chapter 5 of Lam for a complete discussion of this matter.

Following Lawson-Michelsohn, we denote the Clifford algebras $C^{r,s}$ and $C^{n,0}$ by $Cl_{r,s}$ and Cl_n, respectively, if $F = \mathbb{R}$, and $C^{n,0}$ by $\mathbb{C}l_n$ if $F = \mathbb{C}$ (note that only these $C^{r,s}$ arise).

The following theorem is a direct consequence of our discussion.

(15.1). *Theorem. For a given integer* N, *the Radon-Hurwitz number* n_N *is the largest integer* n *such that* $\mathbb{R}^{N+1}$ *is a* Cl_n-*module.*

Two representations $\rho : C(V) \to \mathrm{End}_D W$ and $\rho' : C(V) \to \mathrm{End}_D W'$ are *equivalent* if there is a D-linear isomorphism $T : W \to W'$ such that the composites $T \circ \rho(c)$ and $\rho'(c) \circ T$ are equal for all $c \in C(V)$.

(15.2). *Theorem.* Let $v_{r,s}$ *be the number of inequivalent irreducible* $\mathbb{R}$-*representations of* $Cl_{r,s}$ *and let* $v_n^{\mathbb{C}}$ *be the number of inequivalent irreducible* $\mathbb{C}$-*representaions of* Cl_n. *Then*

$$v_{r,s} = \begin{cases} 2, & \text{if } r + 1 - s \equiv 0 \mod 2 \\ 1, & \text{otherwise} \end{cases}$$

and

$$v_n^{\mathbb{C}} = \begin{cases} 2, & \text{if } n \text{ is odd} \\ 1, & \text{if } n \text{ is even}. \end{cases}$$

Consider $N = 3$, for example. Then $n_N = 8 \cdot 0 + 2^2 - 1 = 3$. So the largest n such that $\mathbb{R}^4$ is a Cl_n-module is $n = 3$. By Theorem (15.2), there are $v_{3,0} = 2$ inequivalent irreducible $\mathbb{R}$-representations of Cl_3. The quadratic module underlying Cl_3 is $<-1> \perp <-1> \perp <-1>$. Now turn to Hamilton's quaternions $H = \left(\frac{-1, -1}{R}\right)$ and denote its standard basis by $\{1, i, j, k\}$. For any h in H, let λ_h be the linear map on H defined by $\lambda_h(z) = hz$. Verify, using matrices, that λ_i, λ_j, and λ_k constitute a Clifford system for Cl_3. This provides an irreducible representation $\rho : Cl_3 \to \mathrm{End}_\mathbb{R} H$. The other irreducible representation is constructed via the Clifford system obtained by letting $i, j,$ and k act on H on the right.

B. Vector Fields on Spheres

The base field here is $\mathbb{R}$. Consider $\mathbb{R}^{N+1}$ and let

$$S^N = \{x \in \mathbb{R}^{N+1} \mid \|x\|^2 = 1\}$$

be the N-sphere. A *tangent vector field on* S^N is a continuous function $v : S^N \to \mathbb{R}^{N+1}$ such that $v(x)$ is tangent to S^N at x for all x.

Example. Let $N = 2k + 1$ be odd. Then $v : S^N \to \mathbb{R}^{N+1}$ defined by

$$v(x) = (x_1, -x_0, x_3, -x_2, \ldots, x_{2k-1}, -x_{2k})$$

is a tangent vector field on S^N which is nowhere 0. For N even, on the other hand, it is not very difficult to show that every tangent vector field on S^N vanishes somewhere.

Let $v_1, \ldots, v_n$ be tangent vector fields on S^N. They are said to be *pointwise linearly independent* if $\{v_1(x), \ldots, v_n(x)\}$ is an independent set of vectors in $\mathbb{R}^{N+1}$ for every x in S^N.

The following result relates the real representations of Cl_n to the tangent vector fields on S^N.

(15.3). *Suppose* $\mathbb{R}^{N+1}$ *is a* Cl_n-*module. Then there exist* n *pointwise linearly independent tangent vector fields on* S^N.

Sketch of a proof. Denote the Clifford multiplication by $c \cdot x$ for all $c \in Cl_n$ and $x \in \mathbb{R}^{N+1}$. One can choose the inner product $< , >$ on $\mathbb{R}^{N+1}$ such that $<e \cdot x, e \cdot y> = <x, y>$ for all $e \in \mathbb{R}^n$ with $\|e\| = 1$ and all x and y in $\mathbb{R}^{N+1}$ (note that $\mathbb{R}^n \subseteq Cl_n$). Let $\{y_1, \ldots, y_n\}$ be a basis for $\mathbb{R}^n$ and assign to each y_j the vector field v_j on $\mathbb{R}^{N+1}$ given by $v_j(x) = y_j \cdot x$. The choice of the inner product $< , >$ implies that the vector fields $v_1, \ldots, v_n$ are tangent to S^N. It is not hard to show that they are pointwise linearly independent.

$\qquad\qquad\qquad\qquad\qquad\qquad\qquad\qquad\qquad\qquad$ QED.

Combining this with (15.1) proves:

(15.4). *On the sphere* S^N *there exist* n_N *pointwise linearly independent tangent vector fields.*

This turns out to be the best result, for a deep theorem of Adams asserts:

(15.5). *Theorem. The number* n_N *is the largest possible number of pointwise linearly independent vector fields that can exist on* S^N.

We will sketch a proof of Adams' theorem in Section D. If N is even, then $N + 1$ is odd, so that $a = b = 0$. Therefore, $n_N = 0$ and the theorem is a consequence of the example.

C. Connections with Topological K-Theory

Fix a field F and a finite-dimensional division algebra D over F. Let W be a finite-dimensional vector space over D. Let D have the trivial grading and let $W = W_0 \oplus W_1$ be a grading of W. Consider a graded representation

$$\rho : C(V) \to \text{END}_D (W_0 \oplus W_1)$$

of $C(V)$ as defined in Chapter 8C. The concepts of irreducible and equivalent can be formulated for graded representations also.

Now let $F = \mathbb{R}$ and $D = \mathbb{C}$. Denote by

$$\hat{\mathfrak{m}}_n^{\mathbb{C}}$$

the free Abelian group generated by the equivalence classes of the irreducible graded representations of Cl_n. The fact that Cl_n embeds into Cl_{n+1} provides a homomorphism

$$i^* : \hat{\mathfrak{m}}_{n+1}^{\mathbb{C}} \to \hat{\mathfrak{m}}_n^{\mathbb{C}}.$$

The graded tensor product defines a multiplication

$$\hat{\mathfrak{m}}_n^{\mathbb{C}} \otimes \hat{\mathfrak{m}}_m^{\mathbb{C}} \to \hat{\mathfrak{m}}_{n+m}^{\mathbb{C}},$$

and this makes $\bigoplus_{n \geq 0} (\hat{\mathfrak{m}}_n^{\mathbb{C}}/i^*\hat{\mathfrak{m}}_{n+1}^{\mathbb{C}})$ into a graded ring (this is, of course, not a $\mathbb{Z}_2$-grading), which is denoted

$$(\hat{\mathfrak{m}}_*^{\mathbb{C}}/i^*\hat{\mathfrak{m}}_{*+1}^{\mathbb{C}}).$$

(15.6). *Theorem. The ring* $(\hat{\mathfrak{m}}_*^{\mathbb{C}}/i^*\hat{\mathfrak{m}}_{*+1}^{\mathbb{C}})$ *is generated by an element* x *from* $(\hat{\mathfrak{m}}_2^{\mathbb{C}}/i^*\hat{\mathfrak{m}}_3^{\mathbb{C}})$. *In fact, there is a graded* $\mathbb{Z}$-*algebra isomorphism*

$$(\hat{\mathfrak{m}}_*^{\mathbb{C}}/i^*\hat{\mathfrak{m}}_{*+1}^{\mathbb{C}}) \cong \mathbb{Z}[x].$$

This construction can be repeated in the case $F = D = \mathbb{R}$. Again consider the free Abelian group generated by the equivalence classes of the irreducible graded representations of Cl_n. This gives rise to the free Abelian group

$$\hat{\mathfrak{M}}_n,$$

the homomorphisms $i^* : \hat{\mathfrak{M}}_{n+1} \to \hat{\mathfrak{M}}_n$, and the graded ring

$$(\hat{\mathfrak{M}}_*/i^*\hat{\mathfrak{M}}_{*+1}) = \bigoplus_{n \geq 0} (\hat{\mathfrak{M}}_n/i^*\hat{\mathfrak{M}}_{n+1}).$$

This case is similar, but more complicated than the complex case. For example,

(15.7). *Theorem.* $(\hat{\mathfrak{M}}_*/i^*\hat{\mathfrak{M}}_{*+1}) \cong \mathbb{Z}[x,y,z]/(2x, x^3, xy, y^2 - 4z)$ *as graded $\mathbb{Z}$-algebras.*

The proofs of Theorems (15.6) and (15.7) are reasonably elementary (see Section 7 in Chapter 11 of Husemoller.)

There are the following important connections between the graded rings $(\hat{\mathfrak{M}}^{\mathbb{C}}_*/i^*\hat{\mathfrak{M}}^{\mathbb{C}}_{*+1})$ and $(\hat{\mathfrak{M}}_*/i^*\hat{\mathfrak{M}}_{*+1})$ and the theory of vector bundles. Let X be a compact pointed topological space. Consider the free Abelian group generated by the isomorphism classes of complex vector bundles over X. Let $K(X)$ be the quotient of this group by the subgroup generated by the elements $[E] + [E'] - ([E \oplus E'])$, where $\oplus$ is Whitney sum. Via the tensor product $K(X)$ has the structure of a ring. The dimension gives rise to a homomorphism $K(X) \to \mathbb{Z}$ whose kernel is written $\tilde{K}(X)$. Using the reduced suspension Σ, define $K^{-P}(X) = \tilde{K}(\Sigma^P X)$. Summing all the $K^{-P}(X)$ gives the graded ring $K^{-*}(X)$. The same constructions, with the isomorphism classes of real vector bundles as starting point, provide $KO(X)$, $\tilde{KO}(X)$, $KO^{-P}(X)$, and, finally, the graded ring $KO^{-*}(X)$. The structures of $K^{-*}(\text{pt})$ and $KO^{-*}(\text{pt})$ are given by the Atiyah-Bott-Shapiro isomorphisms

$$(\hat{\mathfrak{M}}^{\mathbb{C}}_*/i^*\hat{\mathfrak{M}}^{\mathbb{C}}_{*+1}) \cong K^{-*}(\text{pt}) \quad \text{and} \quad (\hat{\mathfrak{M}}_*/i^*\hat{\mathfrak{M}}_{*+1}) \cong KO^{-*}(\text{pt}).$$

They are established by the use the Bott periodicity theorems and the graded structures supplied by Theorems (15.6) and (15.7).

The preceding "topological K-Theory," which was introduced by Atiyah and Hirzebruch in the early 1960's, has some spectacular applications. These include the Riemann-Roch theorem for differentiable manifolds, the solution of the Hopf invariant one problem, and Adams' solution of the vector field problem on spheres.

The connection between K-Theory and vector fields on spheres is provided by a sequence of reformulations which ultimately asserts that the existence of k linearly independent tangent vector fields on S^N implies an estimate on the order of the group $\widetilde{KO}(\mathbb{RP}^k)$ of the real projective k-space $\mathbb{RP}^k$.

The first step is the observation that the existence of k linear independent tangent vector fields on S^N is equivalent to the existence of a section of the fiber bundle

$$(*) \qquad\qquad V_{k+1}(\mathbb{R}^{N+1}) \xrightarrow{\ \pi\ } S^N .$$

Here, $V_{k+1}(\mathbb{R}^{N+1})$ is the Stiefel manifold of $(k+1)$-frames in $\mathbb{R}^{N+1}$; i.e., a point in $V_{k+1}(\mathbb{R}^{N+1})$ is a tuple $(v_1,..., v_{k+1})$ of pairwise orthogonal unit vectors in $\mathbb{R}^{N+1}$, and π is the projection that sends $(v_1,..., v_{k+1})$ to v_{k+1}. Since a section $s: S^N \to V_{k+1}(\mathbb{R}^{N+1})$ maps a unit vector x in S^N to a tuple $(v_1(x),..., v_k(x), x)$ in $V_{k+1}(\mathbb{R}^{N+1})$, the $v_i(x)$ are pairwise orthogonal tangent vector fields on S^N. Conversely, k linearly independent vector fields on S^N determine (by a standard procedure) unit vector fields which are pairwise orthogonal, and, therefore, a section for the bundle $V_{k+1}(\mathbb{R}^{N+1}) \xrightarrow{\ \pi\ } S^N$.

In a range of dimensions the Stiefel manifold $V_{k+1}(\mathbb{R}^{N+1})$ turns out to be homotopy equivalent to the quotient space $\mathbb{RP}^N/\mathbb{RP}^{N-k-1}$, obtained by collapsing the subspace $\mathbb{RP}^{N-k-1}$ of $\mathbb{RP}^N$ to a point. More precisely, there is a map

$$R: \mathbb{RP}^N/\mathbb{RP}^{N-k-1} \longrightarrow V_{k+1}(\mathbb{R}^{N+1})$$

defined by sending the point represented by a unit vector x in $\mathbb{R}^{N+1}$ to $(R_x(e_1), \ldots, R_x(e_{k+1}))$, where R_x is the reflection in the hyperplane orthogonal to x, and $\{e_1,..., e_N\}$ is the standard basis of $\mathbb{R}^N$. This map induces an isomorphism on homotopy groups in a certain range of dimensions. In particular, if k is larger than the Radon-Hurwitz number n_N – which will be assumed from now on – then this range includes N. This implies that the fiber bundle $(*)$ has a section if and only if the map

$$R: \mathbb{RP}^N/\mathbb{RP}^{N-k-1} \xrightarrow{\ \pi\, \circ\, R\ } V_{k+1}(\mathbb{R}^{N+1})$$

has a homotopy right inverse $s': S^N \dashrightarrow \mathbb{RP}^N/\mathbb{RP}^{N-k-1}$; i.e., the composite $(\pi \circ R) \circ s'$ is homotopic to the identity.

The next step involves some machinery from algebraic topology; see Spanier, for example. Recall that the sphere bundle $S(\alpha)$ of a real vector bundle $\alpha : E \to X$ is the bundle obtained by restricting α to all unit vectors of E, relative to some continuous inner product on the fibres of α. The Thom space $T(\alpha)$ of α is constructed from the bundle $S(\alpha \oplus \theta)$, where θ is the one-dimensional product bundle $X \times \mathbb{R}$, by identifying the points with α-component zero and θ-component one. The Spanier-Whitehead dual $D(\mathbb{R}\mathbb{P}^N/\mathbb{R}\mathbb{P}^{N-k-1})$ is the Thom space of a bundle α over $\mathbb{R}\mathbb{P}^k$, whose Whitney sum with $N + 1$ copies of the canonical line bundle ξ_k is isomorphic to the product bundle. In other words, α represents the element $-(N + 1)[\xi_k]$ in $\widetilde{KO}(\mathbb{R}\mathbb{P}^k)$. The Spanier-Whitehead dual of the map s' is

$$D(s') : D(\mathbb{R}\mathbb{P}^N/\mathbb{R}\mathbb{P}^{N-k-1}) = T(\alpha) \longrightarrow D(S^N) = S^q,$$

where q is the fiber dimension of α. Composition with the projection $S(\alpha \oplus \theta) \to T(\alpha)$ produces a map $S(\alpha \oplus \theta) \to S^q$, whose restriction to each fiber (these are q-dimensional spheres) is a homotopy equivalence. It follows that $S(\alpha \oplus \theta)$ is fiber homotopy equivalent to the product bundle $\mathbb{R}\mathbb{P}^k \times \mathbb{R}^{q+1}$. This in turn implies that $\alpha \oplus \theta$ is isomorphic to this product bundle (this is true for bundles over $\mathbb{R}\mathbb{P}^k$, but not in general). We can conclude that α represents the trivial element in $\widetilde{KO}(\mathbb{R}\mathbb{P}^k)$. Therefore, $(N + 1)[\xi_k] = 0$ in $\widetilde{KO}(\mathbb{R}\mathbb{P}^k)$, i.e., the order of $[\xi_k]$ in $\widetilde{KO}(\mathbb{R}\mathbb{P}^k)$ divides $N + 1$.

Using a spectral sequence and the structure of $KO^{-*}(pt)$, one can show that $\widetilde{KO}(\mathbb{R}\mathbb{P}^k)$ is the cyclic group generated by $[\xi_k]$ and that its order c_k is given by the table

k	0	1	2	3	4	5	6	7	8
c_k	1	2	4	4	8	8	8	8	16

and the recursion $c_{k+8} = 16c_k$.

Putting all the pieces together, we now see that the existence of k linearly independent tangent vectorfields on S^N fo $k > n_N$ implies that c_k divides $N + 1$. Since c_k turns out to be equal to the dimension (as vector space over $\mathbb{R}$) of an irreducible module over Cl_k, this contradicts Theorem (15.1).

D. Lie Groups and Lie Algebras

Consider the real Clifford algebra Cl_n and its underlying quadratic space

$$V \cong <-1> \perp ... \perp <-1> \qquad n \text{ copies.}$$

Denote the quadratic form by q and, as on prior occasions, regard $V \subseteq Cl_n$. Providing the additive group of Cl_n with the "bracket" product

$$[c, d] = cd - dc,$$

for all c and d gives Cl_n the structure of a Lie algebra, which we denote $\mathfrak{c}_n$. (Any associative algebra can be made into a Lie algebra in this way.) The group of invertible elements $(Cl_n)^*$ of Cl_n is a real Lie group and $\mathfrak{c}_n$ is the associated Lie algebra. We single out a certain subgroup of $(Cl_n)^*$ which occupies a central role in much of the development of this chapter (mostly behind the scenes). Let $v \in V$ be any vector such that $q(v) = -1$. Since $v^2 = -1$, $v \in (Cl_n)^*$. Let

$$Spin_n$$

consist of 1 and all products of the form $v_1 \cdots v_k$ where k is even and $q(v_i) = -1$ for all i. Clearly, $Spin_n$ is a subgroup of $(Cl_n)^*$, indeed of the invertible elements of the even subalgebra of Cl_n. That it is compact, simply connected, and that it is the universal covering of the special orthogonal group of V, are some of the important properties of $Spin_n$.

It follows from Theorem (8.12) for example, that the even subalgebra of the Clifford algebra Cl_3 is isomorphic to the Clifford algebra Cl_2 and hence to the quaternion algebra $H = \left(\frac{-1, -1}{R}\right)$. It can be shown that the restriction of this isomorphism defines an isomorphism of Lie groups

$$Spin_3 \rightarrow Sp_1 \cong S^3,$$

where $Sp_1 \cong S^3$ is the sphere in H. There is an analogue of this isomorphism for $Spin_4$. The even subalgebra of the Clifford algebra Cl_4 is isomorphic to Cl_3, and therefore, by the results in Chapter 11D, to $H \oplus H$. By restricting to $Spin_4$ and applying the preceding isomorphism, we get the Lie group isomorphism

$$Spin_4 \rightarrow Spin_3 \times Spin_3.$$

There are other such "exceptional" isomorphisms between low dimensional Lie groups. We point out that they can be defined over any commutative ring. See Hahn-O'Meara or Knus [1991].

E. Dirac Operators

We restrict our attention to the field $\mathbb{R}$. Consider the real Clifford algebra $Cl_{r,s}$ and its underlying quadratic space

$$V \cong <-1> \perp \ldots \perp <-1> \perp <1> \perp \ldots \perp <1>.$$

Denote the quadratic form by q. Again, we consider $V \subseteq Cl_{r,s}$. The space $C^\infty(\Omega, Cl_{r,s})$ of smooth $Cl_{r,s}$-valued functions on an open set Ω in V is a $Cl_{r,s}$-module under pointwise multiplication. To each v in V there corresponds the directional derivative ∂_v in the direction v given by

$$\partial_v f(x) = \frac{d}{dt} f(x + tv)\Big|_{t=0}, \quad \text{for } x \in \Omega.$$

It acts on smooth scalar or vector-valued functions on Ω. It can be shown that ∂_v is linear, i.e., $\partial_v(\alpha f + \beta g) = \alpha \partial_v f + \beta \partial_v g$ for all α and β in $\mathbb{R}$, and that $v \rightarrow \partial_v$ is linear, i.e., $\partial_{\alpha v + \beta w} = \alpha \partial_v + \beta \partial_w$ for all α and β in $\mathbb{R}$ and v and w in V. Now let $\{x_1, \ldots, x_n\}$ be the basis of V that corresponds to the preceding $<\pm 1>$ decomposition and let $\{\partial_1, \ldots, \partial_n\}$ be the corresponding directional derivatives.

The *Dirac operator* associated with (V, q) is the first-order differential operator

$$D = \sum_{1 \leq i \leq n} q(x_i) x_i \partial_i$$

on $C^\infty(\Omega, Cl_{r,s})$ (with coefficients in V). The *Laplacian* Δ_q is the second-order constant-coefficient operator

$$\Delta_q = \sum_{1 \leq i \leq n} q(x_i)(\partial_i)^2.$$

By a routine expansion, $D^2 = \Delta_q$. Let $\{x_\alpha\}$ be the basis of $Cl_{r,s}$ provided by the basis $\{x_1,..., x_n\}$ and Proposition (5.3). It is not hard to see that if $f = \sum_\alpha f_\alpha(x)x_\alpha$ is a solution of $Df = 0$ in $C^\infty(\Omega, Cl_{r,s})$, where each f_α is real valued, then $\Delta_q f_\alpha = 0$ for all α.

Consider the special case $r = 2$ and $s = 0$. Now $Cl_2 \equiv H = \left(\dfrac{-1, -1}{\mathbb{R}}\right)$. Check that $\begin{bmatrix} 0 & 1 \\ -1 & 0 \end{bmatrix}$ and $\begin{bmatrix} 0 & i \\ i & 0 \end{bmatrix}$ form a Clifford system for Cl_2. The resulting representation $Cl_2 \to \mathrm{Mat}_2(\mathbb{C})$ is an isomorphism which satisfies $x \to \begin{bmatrix} 0 & 1 \\ -1 & 0 \end{bmatrix}$ and $y \to \begin{bmatrix} 0 & i \\ i & 0 \end{bmatrix}$, where x denotes x_1 and y denotes x_2. Substituting into $D = \sum\limits_{1 \le i \le n} q(x_i)x_i \partial_i$, we get

$$D = \begin{bmatrix} 0 & -\bar\partial \\ \partial & 0 \end{bmatrix}, \text{ where } \partial = \frac{\partial}{\partial x} - i\frac{\partial}{\partial y} \text{ and } \bar\partial = \frac{\partial}{\partial x} + i\frac{\partial}{\partial y}.$$

Similarly, $C^\infty(\Omega, Cl_2)$ can be split into two components of the form $C^\infty(\Omega, \mathbb{C})$ and it follows that the classical operators ∂ and $\bar\partial$ arise by restriction of the Dirac operator to its components.

Dirac introduced the operator D in order to study the wave operator

$$\frac{\partial^2}{\partial x^2} - \left(\frac{\partial^2}{\partial y^2} + \frac{\partial^2}{\partial z^2} + \frac{\partial^2}{\partial w^2}\right).$$

He considered the matrices

$$\sigma_0 = \begin{bmatrix} 1 & 0 \\ 0 & 1 \end{bmatrix} \quad \sigma_1 = \begin{bmatrix} 1 & 0 \\ 0 & -1 \end{bmatrix} \quad \sigma_2 = \begin{bmatrix} 0 & -i \\ i & 0 \end{bmatrix} \quad \sigma_3 = \begin{bmatrix} 0 & 1 \\ 1 & 0 \end{bmatrix}$$

in $\mathrm{Mat}_2(\mathbb{C})$. The matrices $\sigma_0, \sigma_1, \sigma_2,$ and σ_3 are known as the Pauli spin matrices. (Refer to Naber for the role which they play in the mathematics of special relativity.) They satisfy $\sigma_0^2 = \sigma_1^2 = \sigma_2^2 = \sigma_3^2 = I$ and $\sigma_k \sigma_n = -i\sigma_m$ for appropriate cyclic permutations. Using them, Dirac constructed the matrices

$$\gamma_0 = \begin{bmatrix} 0 & \sigma_0 \\ \sigma_0 & 0 \end{bmatrix} \text{ and } \gamma_m = \begin{bmatrix} 0 & \sigma_m \\ -\sigma_m & 0 \end{bmatrix}$$

in $Mat_4(\mathbb{C})$ for $m = 1, 2$, and 3. He now obtains the first-order linear operator

$$\gamma_0 \frac{\partial}{\partial x} + \gamma_1 \frac{\partial}{\partial y} + \gamma_2 \frac{\partial}{\partial z} + \gamma_3 \frac{\partial}{\partial w},$$

which has the property that its square is the wave operator.

This can be recast into the setting of Clifford algebras as follows. Consider $Cl_{3,1}$. Write the underlying space as $<1> \perp <-1> \perp <-1> \perp <-1>$ in the basis $\{x_0, x_1, x_2, x_3\}$. Check that the subset $\{\gamma_0, -\gamma_1, -\gamma_2, -\gamma_3\}$ of $Mat_4(\mathbb{C})$ is a Clifford system for $Cl_{3,1}$. The representation $Cl_{3,1} \to Mat_4(\mathbb{C})$ that it defines takes x_0 to γ_0 and x_i to $-\gamma_i$ for $i = 1, 2$, and 3. It follows that Dirac's operator is a special case of the operator $D = \sum_{1 \leq k \leq n} q(x_i) x_i \partial_i$.

Recall that a complex-valued function f of a complex variable $z = x + iy$ is analytic if and only if it satisfies the Cauchy-Riemann equation $\bar\partial f = 0$, where $\bar\partial = \frac{\partial}{\partial x} + i \frac{\partial}{\partial y}$. A higher-dimensional analogue of this is the equation $Df = 0$ for $f \in C^\infty(\Omega, Cl_n)$. This is the starting point of the theory of Clifford analytic functions. See Gilbert-Murray for details.

The article of Kazdan presents a nice survey about partial differential equations, Dirac operators and related concerns that also touch on some of the other topics considered in this chapter.

F. Spin Manifolds

Let X be a smooth, compact n-dimensional manifold without boundary and let g be a Riemannian metric on X. In each tangent space $T_x X$, the quadratic form $-g(v, v)$ gives rise to the Clifford algebra $C(T_x X)$. This results in a bundle $C(TX) \to X$ of algebras over X called the *Clifford bundle*. It carries the structural properties of the Clifford algebra such as the $\mathbb{Z}_2$-grading and the involutions, and is fundamental in the study of X.

To any bundle S of $\mathbb{Z}_2$-graded modules over $C(TX)$, one can associate a (generalized) Dirac operator

$$D : \Gamma(S) \to \Gamma(S),$$

where $\Gamma(S)$ is the space of smooth cross sections of S. The operator D is self-adjoint, graded, and splits into components D^+ and D^-, which are formal adjoints of one another. It is a fundamental result in the theory of elliptic operators that $\ker D = \ker D^+ \oplus \ker D^-$ is a finite-dimensional vector space over the real numbers. The *index* of D^+ is defined as

$$\text{ind } D^+ = \dim (\ker D^+) - \dim (\ker D^-).$$

It turns out that "classical" invariants of the manifold X, such as the Euler characteristic, the signature (if X is oriented), and the $\hat{A}$-genus of X (if X is a spin manifold) can be described as the index of the Dirac operator for suitable S. Later we will describe the construction of such a bundle S for spin manifolds.

Let $O(X)$ be the bundle over X whose fiber over a point x consists of the orthonormal bases of the tangent space $T_x X$. We note that each fiber $O(X)_x$ is homeomorphic to the orthogonal group O_n. In particular, $O(X)_x$ has two connected components corresponding to the two possible orientations of $T_x X$. An *orientation on* X is a choice of a sub-bundle $SO(X)$ of $O(X)$, whose fiber over a point x consists of one of the two connected components of $O(X)_x$. An *oriented manifold* is a manifold together with an orientation. A spin structure on an oriented manifold X is a choice of a double covering $Spin(X) \to SO(X)$, which when restricted to a fiber $SO(X)_x$ is the nontrivial double covering of $SO(X)_x$. Note that $SO(X)_x$ is homeomorphic to the special orthogonal group SO_n and that there is a unique nontrivial double covering of SO_n, namely, $Spin_n \to SO_n$. A *spin manifold* is an oriented manifold together with a spin structure $Spin(X) \to SO(X)$.

For example, the oriented unit circle in $\mathbb{R}^2$ has two inequivalent spin structures. The unit sphere in $\mathbb{R}^n$ with $n \geq 3$ has a unique spin structure. The real projective space $\mathbb{R}P^n$ has two spin structures if n is congruent to 3 mod 4 and none otherwise.

If X is a spin manifold, the projection map $Spin(X) \to X$ is a bundle with fibers homeomorphic to the spinor group $Spin_n$. Moreover, there is a canonical fiber-preserving action of $Spin_n$ on $Spin(X)$ which is transitive on each fiber. In addition, the tangent bundle TX can be recovered as the *associated vector bundle*

$$\text{pr}_1 : (Spin(X) \times \mathbb{R}^n)/Spin_n \to Spin(X)/Spin_n = X,$$

where $Spin_n$ acts on $\mathbb{R}^n$ via the projection map $Spin_n \to SO_n$, and pr_1 is the projection on the first factor.

Let Δ be a $\mathbb{Z}_2$-graded (left) module over the Clifford algebra Cl_n. Via the inclusion $Spin_n \subseteq Cl_n$ we can consider Δ as a representation of $Spin_n$ and form the associated vector bundle $S = (Spin(X) \times \Delta)/Spin_n$. The Clifford multiplication

$$\mathbb{R}^n \otimes \Delta \subseteq Cl_n \otimes \Delta \to \Delta$$

turns out to be $Spin_n$-equivariant, and hence induces a bundle map of associated vector bundles

$$(*) \qquad\qquad TX \otimes S \to S,$$

which makes S a bundle of $\mathbb{Z}_2$-graded modules over C(TX).

For $n \equiv 0 \mod 4$ there are two inequivalent irreducible $\mathbb{Z}_2$-graded modules Cl_n-modules Δ. The index of the associated Dirac operators can be calculated by the celebrated "index" theorem of Atiyah-Singer, and turns out to be equal to the characteristic number $\pm\hat{A}(X)$.

The notion of "index" can be generalized to give not only integral, but also torsion information about X. The kernel of D is a $\mathbb{Z}_2$-graded vector space and hence represents an element in $\hat{\mathfrak{M}}_0$. We note that $\hat{\mathfrak{M}}_0/i^*\hat{\mathfrak{M}}_1 \cong \mathbb{Z}$, where the isomorphism is defined by mapping a graded module $M = N \oplus N'$ to $\dim N - \dim N'$. In particular, the class $[\ker D] \in \hat{\mathfrak{M}}_0/i^*\hat{\mathfrak{M}}_1 \cong \mathbb{Z}$ maps to $\text{ind } D^+$.

This suggests the following generalization of definition of the index: Suppose D is a self-adjoint operator with finite-dimensional kernel which commutes with a (right) action of the Clifford algebra Cl_n. Then its Clifford index is defined by

$$\text{ind}_n D = [\ker D] \in \hat{\mathfrak{M}}_n/i^*\hat{\mathfrak{M}}_{n+1} \cong KO^{-n}(pt).$$

This Clifford index shares with the usual index the crucial property that it doesn't change when the operator D is deformed continuously (this is not so for ker D).

Geometrically we get an operator commuting with a (right) Cl_n-action, if we choose Δ to be Cl_n, considered as a left module over itself. Then the fibers of the associated vector bundle S are (right) modules over Cl_n and this action is compatible with the Clifford multiplication $(*)$ and hence with the associated Dirac operator $D : \Gamma(S) \to \Gamma(S)$. We denote the Clifford index of this operator by $\alpha(X) \in KO^{-n}(pt)$. By the deformation invariance of the Clifford index, $\alpha(X)$ depends only on the manifold X and not on the Riemannian metric used in the definition of the Dirac operator.

The Dirac operator of X is a useful tool when studying the scalar curvature of the Riemannian metric on X. More precisely, it follows from the

"Weizenböck formula" that the kernel of D is trivial if the scalar curvature is positive. This implies the following result of Hitchin:

(15.8). *Theorem. Let X be a compact spin manifold. If X admits a metric of positive scalar curvature, then $\alpha(X) = 0$.*

Recently, Stolz [1990] proved the following converse:

(15.9). *Theorem. If X is a simply connected compact spin manifold of dimension at least 5 with $\alpha(X) = 0$, then X admits a metric of positive scalar curvature.*

Additional interesting connections include:

(15.10). *Theorem. In every dimension $n \equiv 1$ or 2 (mod 8) with $n > 8$, there exist compact differentiable manifolds which are homeomorphic to S^n but which do not admit any Riemannian metric with positive scalar curvature.*

(15.11). *Theorem. Let X be a compact spin manifold such that $\alpha(X) = 0$. Then the only effective, compact, connected Lie transformation groups of X are tori.*

G. Isoparametric Hypersurfaces

One of the basic questions about curvature invariants in differential geometry concerns the determination of those spaces for which these invariants are constant. Examples of homogeneous spaces with this property are frequent. In this section we will see how Clifford modules can be used to construct inhomogeneous examples.

A *hypersurface*, i.e., a submanifold of codimension 1 in the Euclidean space $\mathbb{R}^n$, the unit sphere S^n, or the hyperbolic space H^n (not in the sense of quadratic forms, but in the sense of differential geometry) is said to be *isoparametric* if its principal curvatures are constant. These hypersurfaces are easy to classify in Euclidean and hyperbolic spaces. In Euclidean space they are open subsets of hyperplanes, hyperspheres, or spherical cylinders; and a similar classification holds in hyperbolic spaces. No classification is known for isoparametric hypersurfaces in spheres. Cartan started to investigate them in the late 30s, asking in particular whether − as in the Euclidean and hyperbolic situations − they are all homogeneous. It was only in 1976 that the first inhomogeneous examples were found by Ozeki and Takeuchi. In 1981, Ferus, Karcher and Münzer used Clifford modules to give a much larger class of examples (which include the examples of Ozeki and Takeuchi as special cases).

It is instructive to introduce the Clifford module approach to isoparametric hypersurfaces with an example. Let

$$V_{2,l}(\mathbb{R}) = \{(u, v) \mid u, v \in \mathbb{R}^l, \ <u, v> = 0, \ \|u\| = \|v\| = 1/\sqrt{2}\}$$

be the Stiefel manifold of orthogonal 2-frames in $\mathbb{R}^l$. Notice that there is a natural embedding of $V_{2,l}(\mathbb{R})$ into $\mathbb{R}^l \oplus \mathbb{R}^l = \mathbb{R}^{2l}$ obtained by mapping $(u, v) \in V_{2,l}(\mathbb{R})$ to $(u_1,..., u_l, v_1,..., v_l) \in \mathbb{R}^{2l}$. Obviously, the image of $V_{2,l}(\mathbb{R})$ lies in the unit sphere $S^{2l-1} \subseteq \mathbb{R}^{2l}$. The boundary of an ϵ-neighborhood, i.e., a *tube*, around this image is an isoparametric hypersurface in S^{2l-1}. This hypersurface is homogeneous. In fact, consider $\mathbb{R}^l \oplus \mathbb{R}^l = \mathbb{R}^{2l}$ as $\text{Mat}_{2,l}(\mathbb{R})$, the space of real $l \times 2$ matrices, and observe that SO_2 and SO_l act on $\text{Mat}_{2,l}(\mathbb{R})$ by multiplication on the right and left, respectively. The manifold $V_{2,l}(\mathbb{R})$ and the tubes around $V_{2,l}(\mathbb{R})$ are orbits of this action of $SO_2 \times SO_l$.

Clifford modules can be used to generalize Stiefel manifolds of orthonormal 2-frames. This is done as follows: Let $C = \{E_1,..., E_{m-1}\}$ be a Clifford system of skew symmetric $l \times l$ real matrices for the Clifford algebra Cl_{m-1} with its standard basis. Vectors u and v in $\mathbb{R}^l$ are called *Clifford orthogonal* if

$$<u, v> \ = \ <E_1 u, v> \ = \ ... \ = \ <E_{m-1} u, v> \ = \ 0.$$

The pairs (u, v) of Clifford orthogonal vectors satisfying $<u, u> = <v, v> = \frac{1}{2}$ form a submanifold $V_2(C)$ in S^{2l-1} called the *Clifford-Stiefel manifold* of C-orthonormal 2-frames in $\mathbb{R}^l$. The tubes around $V_2(C)$ in S^{2l-1} turn out to be isoparametric hypersurfaces. If m is 1 or 2, they are homogeneous (if m = 1 we get the earlier example of tubes around $V_{2,l}(\mathbb{R})$, and if m = 2 we get tubes around $V_{2,l}(\mathbb{C})$). If m = 4, both homogeneous (e.g., tubes around $V_{2,l}(H)$) and inhomogeneous examples arise. Finally, if m ≠ 1, 2, or 4, these hypersurfaces are always inhomogeneous !

It might seem profitable to define and make use of Stiefel manifolds $V_k(C)$ of orthonormal k-frames for k ≥ 3. This fails, however. Indeed, it is proved in Thorbergsson [1991] that all isoparametric submanifolds in spheres with codimension at least two are homogeneous, except for those that can be expressed as products of lower-dimensional ones. In the proof of this fact an incidence geometry − a Tits geometry − is associated to an isoparametric submanifold in a sphere of codimension at least two. The dimension of this incidence geometry is one larger than the codimension of the submanifold. In particular, if the codimension is at least two, this dimension is at least three. It is a classical fact that a projective space of dimension at least three is projectively equivalent to a projective plane over a field and is thus homogeneous. This fact has been generalized to Tits geometries and it is this generalization that is the key point in the proof of the fact just mentioned.

Incidence geometries can also be associated to the isoparametric hypersurfaces obtained by the Clifford module approach. This is done in Thorbergsson [1992]. A Clifford system of skew symmetric $l \times l$ real matrices $E_1,..., E_{m-1}$ gives rise to a Clifford system $P_0,..., P_m$ of symmetric $2l \times 2l$ real matrices as follows:

$$P_0(u, v) = (u, -v),\ P_1(u, v) = (v, u)\ ,\ ...\ ,\ P_{i+1}(u, v) = (E_i v, -E_i u)\ ...\ .$$

Let Σ be the unit sphere in the space of symmetric matrices spanned by $P_0,..., P_m$ with respect to the inner product $<A, B> = \frac{1}{2l}\ trace(AB)$. Define the sets P and L as follows:

$$P = V_2(C) = \{x \in S^{2l-1}\ |\ <Px, x> = 0\ \text{for all}\ P \in \Sigma\}\ \text{and}$$

$$L = \{L(x, P)\ |\ x \in P\ \text{and}\ P \in \Sigma\}$$

where $L(x, P) = \{y \in P\ |\ P(x - y) = -(x - y)\}$. We call the elements of P points and those of L lines. Let $p \in P$ and $L \in L$. In the usual terminology of incidence geometry, we say that L *contains* p if $p \in L$, that two lines *meet* if they have a common point, etc. The pair (P, L) satisfies the following properties:

(1) Let p and p' be distinct points. Then there is at most one line that contains both p and p'.

(2) Let p be a point and L a line not containing p. Then there is a unique line L' that meets L and contains p.

(3) There exist two disjoint lines.

(4) Every point is contained in at least three lines and every line contains at least three points.

The incidence geomertry (P, L) is two-dimensional and is, in view of (1) – (4), an example of a *polar plane*. If $m \neq 1, 2$, and 4, these geometries are new examples of inhomogeneous polar planes.

It is a fact that an incidence geometry can be associated to any of the *known* isoparametric hypersurfaces in a sphere. If the hypersurface is inhomogeneous, we have just described the construction, and if it is homogeneous, one can associate a Tits geometry. This raises two interesting questions. Is it possible to construct a meaningful incidence geometry for a general isoparametric hypersurface in a sphere, and if the answer is yes, what would the impact of the construction be on the problem of classifying such hypersurfaces ?

Bibliography

A. Albert,
- Structure of Algebras, Colloquium Publications Volume 24, Amer. Math. Soc., Providence, 1939.

S.A. Amitsur, L.H. Rowen, and J.-P. Tignol,
- Division algebras of degree 4 and 8 with involution, Israel J. Math. 33 (1979), 133–148.

J. Kr. Arason,
- A proof of Merkurjev's theorem, in Quadratic and Hermitian Forms (C. Riehm and I. Hambleton, editors), Canadian Math. Soc. Conference Proceedings Vol. 4, Providence 1984, pp. 121–130.

J. Kr. Arason, R. Elman, and B. Jacob,
- The graded Witt ring and Galois cohomology II, Trans. Amer. Math. Soc. 314 (1989), 745–780.

J. V. Armitage and A. Fröhlich,
- Class numbers and unit signatures, Mathematika 14 (1967), 94–98.

M. Atiyah, R. Bott, and A. Shapiro,
- Clifford modules, Topology, Vol. 3, Suppl. 1 (1964), 3–38.

M. Auslander and D. A. Buchsbaum,
- On ramification theory in Noetherian rings, Amer J. Math. 81 (1959), 749–765.

M. Auslander and O. Goldman,
- The Brauer group of a commutative ring, Trans. Amer. Math. Soc. 97 (1960), 367–409.

R. Baeza,
- Quadratic Forms over Semilocal Rings, Lecture Notes in Mathematics 655, Springer-Verlag, Berlin and New York, 1978.
- Discriminants of polynomials and of quadratic forms, J. Algebra 72 (1981), 17–28.
- The norm theorem for quadratic forms over a field of characteristic 2, Comm. Algebra 18 (1990), 1337–1348.

R. Baeza and R. Moresi,
- On the Witt equivalence of fields of characteristic 2, J. Algebra 92 (1985), 446–453.

268   Bibliography

A. Bak,
- K-Theory of Forms, Annals of Mathematical Studies 98, Princeton University Press, 1981.

H. Bass,
- Lectures on Topics in Algebraic K-Theory, Tata Institute of Fundamental Research, Bombay, 1967.
- Algebraic K-Theory, Benjamin, New York, 1968.
- Modules which support a non-singular form, J. Algebra 13 (1969), 246–252.
- Clifford algebras and spinor norms over a commutative ring, Amer. J. Math. 96 (1974), 156–206.

E. Bayer-Fluckiger,
- Principe de Hasse faible pour les systèmes de formes quadratiques, J. Reine Angew. Math. 378 (1987), 53–59.

M. Beattie,
- Computing the Brauer group of graded Azumaya algebras from its subgroups, J. Algebra 101 (1986), 339–349.

N. Berline, E. Getzler, and M. Vergne,
- Heat Kernels and Dirac Operators, Grundlehren der Mathematischen Wissenschaften, Vol. 298, Springer-Verlag, Berlin, Heidelberg, New York, 1991.

I. Bertuccioni,
- A short proof of a theorem of Suslin-Kopeiko, Arch. Math. 39 (1982), 9–10.

W. Bichsel and M.-A. Knus,
- Quadratic forms with values in line bundles, Recent Advances in Real Algebraic Geometry and Quadratic Forms, (W. Jacob and T.Y. Lam, editors) Berkeley, 1990-91, Contemp. Math., to appear.

Z. I. Borevich and I. R. Shafarevich,
- Number Theory, Academic Press, New York, London, 1966.

N. Bourbaki,
- Algèbre: Modules et Anneaux Semi-simples (Chap. 8), Hermann, Paris, 1958.
- Algèbre: Formes Sesqulinéaires et Formes Quadratiques (Chap. 9), Hermann, Paris, 1959.
- Commutative Algebra. Addison-Wesley, Reading, MA, 1972.
- Algebra, Part I. Addison-Wesley, Reading, MA, 1972.

C.J. Bushnell,
- Modular quadratic and hermitian forms over Dedekind domains I, J. Reine Angew. Math. 287 (1976), 169–186.
- Modular quadratic and hermitian forms over Dedekind domains II, J. Reine Angew. Math. 288 (1976), 24–36.

S. Caenepeel,
- A cohomological interpretation of the graded Brauer group I, Comm. Algebra 11 (1983), 2129–2149.

- A cohomological interpretation of the Brauer-Wall group, in Proceedings of the Second Belgian-Spanish week on Algebra and Geometry, Alxebra Santiago de Compostela 54 (1990), 31–46.
- Brauer-Long groups and $\mathbb{Z}$-gradings, Bull. Soc. Math. Belg. - Tijdschr. Belg. Wisk. Gen. 42B (1990), 123–136.
- Brauer groups, Hopf algebras and Galois theory, Vrije Universiteit Brussel, Fakulteit Wetenschappen, 1990-1.

S. Caenepeel and M. Beattie,
- A cohomological approach to the Brauer-Long group and the groups of Galois extensions and strongly graded rings, Trans. Amer. Math. Soc. 324 (1991), 747–775.

S. Caenepeel and F. Van Oystaeyen,
- Brauer groups and the cohomology of graded rings, Monographs and Textbooks in Pure and Applied Math. 121, Marcel Dekker, New York, 1988.
- A note on generalized Clifford algebras, Comm. Algebra, to appear.
- Quadratic forms with values in invertible modules, preprint.

J.W.C. Cassels,
- Rational Quadratic Forms, London, New York, San Francisco, Academic Press, 1978.

J.W.C. Cassels and A. Fröhlich,
- Algebraic Number Theory, Academic Press, London 1967.

A. Chalatsis and Th. Theohari-Apostolidi,
- Maximal orders containing local crossed products, J. Pure Appl. Alg. 50 (1988), 211–222.
- Integral representations of crossed-product orders, Comm. Algebra 16 (10) (1988), 2013–2022.

K. S. Chang,
- Discriminanten und Signaturen gerader quadratischer Formen, Arch. Math. 21 (1970), 59–65.

S. Chase, D. Harrison, and A. Rosenberg,
- Galois theory and Galois cohomology of commutative rings, Amer. Math. Soc. Memoirs 52 (1965), 15–33.

L. N. Childs,
- The Brauer group of graded Azumaya algebras II: Graded Galois extensions, Trans. Amer. Math. Soc. 204 (1975), 137–160.
- Representing classes in the Brauer group of quadratic number rings as smash products, Pacific J. Math. 125 (1986), 223–240.

J.-L. Colliot-Thélène and R. Parimala,
- Formes quadratiques multiplicatives et variétés algébriques: deux compléments, Bull. Soc. Math. France 108 (1980), 213–227.
- An appendix to Kato: A Hasse principle for two-dimensional global fields, J. Reine Angew. Math. 366 (1986), 181–183.

P. E. Conner and J. Hurrelbrink,
- Class Number Parity, Series in Pure Mathematics 8, World Scientific, Singapore, 1988.

G. Cornell and M. Rosen,
- Group-theoretic constraints on the structure of the class group, J. Number Theory, 13 (1981), 1–11.

T. Craven, A. Rosenberg, and R. Ware,
- The map of the Witt ring of a domain into the Witt ring of its field of fractions, Proc. Amer. Math. Soc. 51(1) (1975), 25–30.

F. DeMeyer and T. Ford,
- On the Brauer group of surfaces, J. Algebra 86 (1984), 259–271.
- Computing the Brauer-Long group of $\mathbb{Z}$-dimodule algebras, J. Pure Appl. Algebra 54 (1988), 197–208.

F. DeMeyer and E. Ingraham,
- Separable Algebras over Commutative Rings, Lecture Notes in Mathematics 181, Springer-Verlag, Berlin, Heidelberg, New York, 1971.

M. Deuring,
- Algebren, Springer-Verlag, Berlin, 1939; revised, 1968.

A. Earnest,
- Binary quadratic forms over rings of algebraic integers: A survey of recent results, in Number Theory (J.-M. De Koninck and C. Levesques, editors), Walter de Gruyter, Berlin, New York, 1989, 133–159.
- Ideal class groups of exponent two and one-class genera of binary quadratic lattices, Rocky Mountain J. Math. 19 (3) (1989), 669–673.

A. Earnest and J.S. Hsia,
- One-class spinor genera of positive quadratic forms, Acta Arith. 58 (1991), 133–139.

R. Elman and T.Y. Lam,
- Quadratic forms over formally real fields and Pythagorean fields, Amer. J. Math. 94 (1972), 1155–1194.
- Classification theorems for quadratic forms over fields, Comm. Math. Helv. 49 (1974), 373–341.

D. Estes,
- On the parity of the class number of the field of q-th roots of unity, Rocky Mountain J. Math. 19 (3) (1989), 675-682.

D. Estes and J.S. Hsia,
- Spinor genera under field extensions IV : Spinor class fields, Japan J. Math. 16(1990), 341–350

D. Ferus, H. Karcher, und H.-F. Münzer,
- Cliffordalgebren und neue isoparametrische Hyperflächen, Math. Z. (1981), 479–502.

T. Ford,
- On the Brauer group of a Laurent polynomial ring, J. Pure Appl. Algebra 51 (1988), 111–117.

R. Fossum,
- The Noetherian different of projective modules, thesis, University of Michigan, 1965.

A. Fröhlich,
- Discriminants of algebraic number fields, Math. Z. 74 (1960), 18–28.
- On the K-Theory of unimodular forms over rings of algebraic integers, Quart. J. Math. Oxford, 22 (1971), 401–23.

A. Fröhlich and M. J. Taylor,
- Algebraic Number Theory, Cambridge Studies in Advanced Mathematics 27, Cambridge University Press, Cambridge, New York, 1991.

A. Fröhlich and C.T.C Wall,
- Equivariant Brauer groups, Bull. Soc. Math. France, Mémoire 25 (1970), 91–96.
- Generalizations of the Brauer group I, preprint.

O. Gabber,
- Some theorems on Azumaya algebras, in Groupe de Brauer, Lecture Notes in Math. 844, Springer-Verlag, Berlin, Heidelberg, New York, 1981, 129–209.

D. Garbanati,
- Unit signatures, and even class numbers and relative class numbers, J. Reine Angew. Math. 274/275 (1975), 376–384.
- Units with norm −1 and signatures of units, J. Reine Angew. Math. 283/284 (1976), 164–175.

L.J. Gerstein,
- Stretching and welding indecomposable quadratic forms, J. Number Theory 37 (1991), 146–151.
- A note on splitting quadratic forms, J. Number Theory 29 (1988), 231–233.

W.-D. Geyer, G. Harder, M. Knebusch, and W. Scharlau,
- Ein Residuensatz für symmetrische Bilinearformen, Inventiones Math. 11 (1970), 319–328.

J. Gilbert and M. Murray,
- Clifford Algebras and Dirac Operators in Harmonic Analysis, Cambridge Studies in Advanced Mathematics 26, Cambridge University Press, Cambridge, New York, 1991.

G. Gras,
- Critère de parité du nombre de classes des extensions Abeliénnes réelles de Q de degré impair, Bull. Soc. Math. France, 103 (1975), 177–190.

M. Gromov and H. B. Lawson,
- Spin and scalar curvature in the presence of a fundamental group, Annals of Math., 111 (1980), 209–230.

A. Grothendieck,
- Le groupe de Brauer, I, II, III, pp. 46–188, Dix exposés sur la théorie des schémas, North Holland, Amsterdam, 1968.

U. Haag,
- Diskriminantenalgebren quadratischer Formen, Archiv Math. 57 (1991), 546–554.

A. Hahn and O.T. O'Meara,
- Classical Groups and K-Theory, Grundlehren der Mathematischen Wissenschaften, Vol. 291, Springer-Verlag, Berlin, Heidelberg, New York, 1989.

D. Harrison,
- Abelian extensions of commutative rings, Amer. Math. Soc. Memoirs, 52 (1965), 1–14.

H. Hasse,
- An algorithm for determining the structure of the 2-Sylow subgroup of the divisor class group of a quadratic number field, Symposia Mathematica 15 (1975), 341–352.

D. Haile,
- The Brauer Monoid over a field, J. Algebra 81 (1983), 521–539.
- On Azumaya algebras arising from Clifford algebras, J. Algebra 116 (1988), 372–384.

J. S. Hsia,
- Grothendieck groups of unimodular quadratic forms over local fields, J. Algebra 15 (1970), 328–334.
- On the classification of unimodular quadratic forms, J. Number Theory 12 (1980), 327–333.
- Arithmetic theory of integral quadratic forms, in Proceedings of the Queen's Number Theory Conference, Queen's Papers in Pure and Applied Math., Vol. 54, 1980, 173–204.

R. Hoobler,
- When is Br(X) = Br'(X) ?, in Brauer groups in ring theory and algebraic geometry, Lecture Notes in Math. 917, Springer-Verlag, Berlin, 1982.

I. Hughes and R. Mollin,
- Totally positive units and squares, Proc. Amer. Math. Soc 87(4) (1983), 613–616.

T. Hungerford,
- Algebra, Graduate Texts in Mathematics 73, Springer-Verlag, Berlin, Heidelberg, New York, 1989.

D. Husemoller,
- Fibre Bundles, McGraw-Hill, New York, London, 1966.

S. Iyanaga,
- The Theory of Numbers, North Holland, Amsterdam, Oxford, 1975.

N. Jacobson,
- Basic Algebra I, second edition, Freeman, San Francisco, 1985.
- Basic Algebra II, Freeman, San Francisco, 1985.

D. James,
- Diagonalizable indefinite integral quadratic forms, Acta Arith. 50 (1988), 309–314.

- Orthogonal decompositions of indefinite quadratic forms, Rocky Mountain J. Math. 19 (1989), 735–740.
- Quadratic forms over polynomial rings, Comm. Algebra 18 (1990), 247–251.
- Quadratic forms with cube-free discriminant, Proc. Amer. Math. Soc. 110 (1990), 45–52.
- Primitive representations by unimodular quadratic forms, J. Number Theory 44 (1993), 356-366.
- Representations by unimodular Z-lattices, Math. Z., to appear.

T. Kanzaki,
- On bilinear modules and Witt ring over a commutative ring, Osaka J. Math. 8 (1971), 485–496.
- On the quadratic extensions and the extended Witt ring of a commutative ring, Nagoya Math. J. 49 (1973), 127–141.

K. Kato,
- Symmetric bilinear forms, quadratic forms and Milnor K-theory in characteristic two, Inventiones Math. 66 (1982), 493–510.
- A Hasse principle for two-dimensional global fields, J. Reine Angew. Math. 366 (1986), 142–181.

J. Kazdan,
- Partial differential equations in differential geometry, in Differential Geometry, Springer Lecture Notes in Math. 1263 (1987), 134–170.

K. Kitamura,
- On the free quadratic extensions of a commutative ring, Osaka J. Math. 10 (1973), 15–20.

M. Knebusch,
- Grothendieck und Wittringe von nicht ausgearteten symmetrischen Bilinearformen. Sitzungsber. Heidelberg Akad. Wiss. Math. Naturwiss. Kl., 3. Abh. (1969/70), 89–157.
- Symmetric bilinear forms over algebraic varieties, in Conference on Quadratic Forms – 1976, Queen's Papers on Pure and Applied Mathematics 46, 1977, 103–283.
- Signaturen, reelle Stellen und reduzierte quadratische Formen, Jahresber. Deutsch. Math. Verein. 82 (1980) 3, 109–127.

M. Knebusch and M. Kolster,
- Witt Rings, Viehweg und Sohn, Braunschweig, Wiesbaden, 1982.

M. Knebusch, A. Rosenberg, and R. Ware,
- Grothendieck- and Witt rings of hermitian forms over Dedekind rings, Pacific J. Math. 43 (1972), 657–673.

M. Knebusch and W. Scharlau,
- Quadratische Formen und quadratische Reziprozitätsgesetze über algebraischen Zahlkörpern, Math. Z. 121 (1971), 346–368.

M. Kneser,
- Quadratische Formen. Ausarbeitung einer Vorlesung. Mathematisches Institut der Universität Göttingen, 1973/74.

274   Bibliography

- Composition of binary quadratic forms, J. Number Theory 15 (1982), 406–413.

M.-A. Knus,
- Quadratic forms, Clifford algebras and spinors, Seminarios de Mathematica, IMEEC, Unicamp, Campinas SP, Brazil, 1988.
- Pfaffians and quadratic forms, Advances in Math. 71 (1988), 1–20.
- Quadratic and Hermitian Forms over Rings, Grundlehren der Mathematischen Wissenschaften, Vol. 294, Springer-Verlag, Berlin, Heidelberg, New York, 1991.

M.-A. Knus and M. Ojanguren,
- Théorie de la Descente et Algèbres d'Azumaya, Lecture Notes in Mathematics 389, Springer-Verlag, Berlin, Heidelberg, New York, 1974.
- A Mayer-Vietoris sequence for the Brauer group, J. Pure Appl. Algebra 5 (1974), 345-360.
- Cohomologie étale et groupe de Brauer, in Groupe de Brauer, Lecture Notes in Math. 844, Springer-Verlag, Berlin, 198, 210-228.
- The Clifford algebra of a metabolic space, Archiv Math. 56 (1991), 440–445.

M.-A. Knus, M. Ojanguren, and R. Sridharan,
- Quadratic forms and Azumaya algebras, J. Reine Angew. Math. 303/304 (1978), 231–248.

M.-A. Knus and A. Paques,
- Quadratic spaces with trivial Arf invariant, J. Algebra 93 (1985), 267–291.

M.-A. Knus and R. Parimala,
- Quadratic forms associated with projective modules over quaternion algebras. J. Reine Angew. Math. 318 (1980), 20–31.

M.-A. Knus, R. Parimala, and R. Sridharan,
- On rank 4 quadratic spaces with given Arf and Witt invariants, Math. Ann. 274 (1986), 181–198.
- A classification of rank 6 quadratic spaces via Pfaffians, J. Reine Angew. Math. 398 (1989), 187–218.
- Pfaffians, central simple algebras and similitudes, Math. Z. 206 (1991), 589–606.

M. Kolster,
- Quadratic forms and Artin's reciprocity law, Math. Z. 180 (1982), 81–89.

V.I. Kopeiko,
- Quadratic spaces and quaternion algebras, Zap. Naucn. Sem. Leningrad Otdel. Mat. Steklov (LOMI) 75 1978, 110–120.

D. Kubert,
- The 2-divisibility of the class number of cyclotomic fields and the Stickelberger ideal, J. Reine Angew. Math. 369 (1986),192–218.

O. Laborde,
- Formes quadratiques et algèbres de Clifford, Bull. Sc. math., 2e série, 96 (1972), 199–208.

- Formes quadratiques, algèbres de Clifford et signatures, C. R. Acad. Sc. Paris, t. 278 (1974), 1599–1602.

T.Y. Lam,
- The Algebraic Theory of Quadratic Forms, Benjamin, Reading MA, 1973.

S. Lang,
- Algebraic Number Theory, Addison-Wesley, Reading, MA, 1970.

J. Lannes,
- Formes quadratiques d'enlacement sur l'anneau des entiers d'un corps de nombres, Ann. Sci. Éc. Norm. Sup. 8 (1973), 535–579.

H.-B. Lawson and M.-L. Michelsohn,
- Spin Geometry, Princeton University Press, Princeton, NJ, 1989.

D. W. Lewis,
- New improved exact sequences of Witt groups, J. Algebra 74 (1982), 206–210.
- A note on Clifford algebras and central division algebras with involution, Glasgow Math. J. 26 (1985), 171–176.
- New proofs of the structure theorems for Witt rings, Expo. Math. 7 (1989), 83–88.

F. Long,
- A generalization of the Brauer group of graded algebras, Proc. London Math. Soc. 29 (1974), 237–256.

O. Loos,
- Bimodule-valued hermitian and quadratic forms, Arch. Math., to appear.

Y.I. Manin,
- Cubic Forms, second edition, North Holland Mathematical Library, North Holland and Elsevier Science Publishers, Amsterdam, 1986.

B. McDonald,
- Linear Algebra over Commutative Rings, Pure Appl. Math. Dekker, New York, 1984.

A. S. Merkurjev,
- On the norm residue symbol of degree two, Sov. Math. Dokl. 24 (1981), 546–551.

A. Micali and P. Revoy,
- Modules Quadratiques, Bull. Soc. Math. France, Memoire 63, 1979.

A. Micali and J.-D. Thérond,
- Sur les groupes A(n), Col. sur les Formes Quadratiques, Bull. Soc. Math. France Mémoirs 48 (1976), 75–87.

A. Micali and O.E. Villamayor,
- Sur les algèbres de Clifford, Ann. Scient. Éc. Norm. Sup. 4e série 1, (1968), 271–304.
- Sur les algébres de Clifford II, J. Reine Angew. Math. 242 (1970), 61–90.
- Algebres de Clifford et groupe de Brauer, Ann. Scient. Ec. Norm. Sup. 4e serie 1 (1971), 285–310.

J. Milnor,
- Algebraic K-Theory and quadratic forms, Inventiones Math. 9 (1970), 318–344.
- Symmetric inner products in characteristic 2, in Annals of Math. Studies 70, Princeton University Press, 1971, 59–75.

J. Milnor and D. Husemoller,
- Symmetric Bilinear Forms, Springer-Verlag, Berlin, Heidelberg, New York, 1973.

G. Naber,
- The Geometry of Minkowski Spacetime, Applied Mathematical Sciences 92, Springer-Verlag, New York, Berlin, Heidelberg, 1992.

P. Nelis,
- Schur and projective Schur groups of number rings, Can. J. Math. Vol. 43(3) (1991), 540–558.
- The Schur group conjecture for the ring of integers in a number field, Proc. Amer. Math. Soc. Vol 114 (2) (1992), 307–318.
- Schur and projective Schur algebras over Dedekind domains, Ph. D. thesis, University of Antwerp, 1991.

P. Nelis and F. Van Oystaeyen,
- The projective Schur subgroup of the Brauer group and root groups of finite groups, J. Algebra 137 (2) (1991), 501–518.

J. Neukirch,
- Class Field Theory, Springer-Verlag, Berlin, Heidelberg, New York, 1985.

V. V. Nikulin,
- Integral symmetric bilinear forms and some of their geometric applications, Math. USSR Izv. 14 (1980), 103–167.
- Lectures on the Brauer group of real algebraic surfaces, University of Notre Dame, Department of Mathematics, preprint 179, 1992.

V. V. Nikulin and R. Sujatha,
- On Brauer groups of real Enriques surfaces, J. Reine Angew. Math., to appear.

M. Ojanguren,
- On Karoubi's theorem: $W(A) = W(A[t])$, Archiv Math. 43 (1984), 328–331.
- The Witt group and the problem of Lüroth, Lecture Notes, Lausanne and Pisa, 1990.

O.T. O'Meara,
- Introduction to Quadratic Forms, second edition, Springer-Verlag, Berlin, Heidelberg, New York, 1971.
- Hilbert's 11th problem: The arithmetic theory of quadratic forms, in Proceedings of Symposia in Pure Marthematics, American Math. Soc., Providence, RI, 1976, 379–400.

M. Orzech,
- On the Brauer group of algebras having a grading and an action, Can. J. Math. 28 (1976), 533–552.

- Correction to: On the Brauer group of algebras having a grading and an action, Can. J. Math. 32 (1980), 1523–1524.

M. Orzech and C. Small,
- The Brauer Group of Commutative Rings, Lectures in Pure and Applied Mathematics 11, Marcel Dekker, New York, 1975.

W. Pardon,
- A relation between Witt groups and zero-cycles in a regular ring, in Algebraic K-Theory, Number Theory, Geometry and Analysis, (A. Bak, editor) Bielefeld, 1982, Lecture Notes in Math. 1046, Springer-Verlag, New York, Heidelberg, 1984, 261–328.

R. Parimala,
- Cancellation of quadratic forms over principal ideal domains, J. Pure Appl. Algebra 24 (1982), 213–216.
- Quadratic forms over polynomial rings over global fields, J. Number Theory 17 (1983), 113–115.
- Witt groups of conics, elliptic and hyperelliptic curves, J. Number Theory 28 (1988), 69–93.

R. Parimala and V. Sridharan,
- A local global principle for quadratic forms over polynomial rings, J. Algebra 74 (1982), 264–269.
- Indecomposable rank 4 quadratic spaces of non-trivial discriminant over polynomial rings, Math. Z. 183 (1983), 281–292.
- Graded Witt rings and unramified cohomology rings of curves, Bombay 1990, preprint.

R. Parimala and R. Sujatha,
- Witt groups of hyperelliptic curves, Comm. Math. Helv. 65 (1990), 559–580.

R. Perlis, K. Szymiczek, P.E. Conner, and R. Litherland,
- Matching Witts with global fields, Contemp. Math., to appear.

M. Peters,
- Einklassige Geschlechter von Einheitsformen in totalreelen algebraischen Zahlkörpern, Math. Ann. 26 (1977), 117–120.
- Definite binary quadratic forms with class number one, Acta. Arith. 36 (1980), 271–272.

A. Pfister,
- Systeme quadratischer Formen III, J. Reine Angew. Math. 394 (1989), 208–220.
- Quadratische Formen, in Ein Jahrhundert Mathematik 1890–1990, Festschrift zum Jubiläum der Deutschen Mathematiker Vereinigung, Viehweg und Sohn, Braunschweig,Wiesbaden, 657-671.

R. Pierce,
- Associative Algebras, Graduate Texts in Mathematics, Springer-Verlag, Berlin, Heidelberg, New York, 1982.

B. Pollak,
- Orthogonal groups over global fileds of characteristic 2, J. Algebra 15 (1970), 589–595.

H. G. Quebbemann,
- Definite lattices over real algebraic function fields, Math. Ann. 272 (1985), 461–475.
- On extension of quadratic modules from affine to projective spaces, Comm. Algebra 17 (1989), 971–979.

I. Raeburn and J. L. Taylor,
- The bigger Brauer group and étale cohomogy, Pacific J. Math. 119 (1985), 445–463.

I. Reiner,
- Maximal Orders, Academic Press, New York, London, 1975.

P. Revoy,
- Sur les deux premiers invariants d'une forme quadratique, Ann. Scient. Ec. Norm. Sup. 4e serie 1 (1971), 311–319.
- Sur certaines algèbres de Clifford, Comm. Algebra 11 (1983), 1877–1891.

P. Ribenboim,
- Algebraic Numbers, Wiley-Interscience, John Wiley and Sons, New York, 1972.

C. Riehm,
- Integral representations of quadratic forms in characteristic 2, Amer. J. Math. 87 (1965), 32–64.
- The equivalence of bilinear forms, J. Algebra 31 (1974), 45–66.
- The Schur subgroup of the Brauer group of cyclotomic rings of integers, Proc. Amer. Math. Soc. 103:1 (1988), 83–87.
- The linear and quadratic Schur subgroups over the S-integers of a number field, Proc. Amer. Math. Soc. 107:1 (1989), 27–30.

K. Roggenkamp,
- A remark on separable orders, Can. Math. Bull. 12 (1969), 453–455.

P. Roquette,
- Some fundamental theorems on Abelian function fields, in Proc. Int. Cong. Math., Edinburgh 1958. Cambridge University Press, New York, 1960, 322–329.

M. Rosen,
- S-units and S-class group in algebraic function fields, J. Algebra 26 (1973), 98–108.

C.-H. Sah,
- Quadratic forms over fields of characteristic 2, Amer. J. Math. 82 (1960), 812–830.
- Symmetric bilinear forms and quadratic forms, J. Algebra 20 (1972), 144–160.

P. Samuel,
- Algebraic Theory of Numbers, Houghton Mifflin Company, Boston, 1970.

W. Scharlau,
- A historical introduction to the theory of integral quadratic forms, in Conference on Quadratic Forms – 1976, Queen's Papers on Pure and Applied Mathematics 46, 1977, 284–339.
- Quadratic and Hermitian Forms, Grundlehren der Mathematischen Wissenschaften, Vol. 270, Springer-Verlag, Berlin, Heidelberg, New York, 1985.

W. Scharlau and H. Opolka,
- From Fermat to Minkowski, Undergraduate Texts in Mathematics, Springer-Verlag, Berlin, Heidelberg, New York, 1985.

R. Schulze-Pillot,
- Remark on a note of Hsia about classification of unimodular lattices, J. Number Theory 14 (1982), 83–85.

J.-P. Serre,
- A Course in Arithmetic, Graduate Texts in Mathematics, Springer-Verlag, Berlin, Heidelberg, New York, 1973.
- Local Fields, Graduate Texts in Mathematics, Springer-Verlag, Berlin, Heidelberg, New York, 1979.

P. Shastri,
- Witt groups of algebraic integers, J. Number Theory 30 (3) (1988), 243–266.

C. Small,
- The Brauer-Wall group of a commutative ring, Trans. Amer. Math. Soc. 156 (1971), 455–491.
- The group of quadratic extensions, J. Pure Appl. Algebra 2 (1972), 83–105.

E. H. Spanier,
- Algebraic Topology, Springer-Verlag, Berlin, Heidelberg, New York, 1966.

S. Stolz,
- Simply connected manifolds of positive scalar curvature, Bull. Amer. Math. Soc. 23(2), (1990), 427–432.
- Simply connected manifolds of positive scalar curvature, Annals of Math. 136 (1992), 511-540.

R. Sujatha,
- Witt groups of real projective surfaces, Math. Ann. 28 (1990), 89–101.

A. Suslin and V. Kopeiko,
- Quadratic modules and orthogonal groups over polynomial rings, J. Soviet Math. 20 (1982), 2665–2691.

Th. Theohari-Apostolidi,
- Local crossed-product orders of finite representation type, J. Pure Appl. Alg. 41 (1986), 87–98.

J.-D. Therond,
- Le groupe des extensions quadratiques séparables libres de entiers de $Q(\sqrt{d})$, C. R. Acad. Sc. Paris, t. 281 (1975), 939–942.

- Sur deux conjectures de Small, Col. sur les Formes Quadratiques, Bull. Soc. Math. France, Mémoir 48 (1976), 117–122.

G. Thorbergsson,
- Isoparametric manifolds and their buildings. Ann. Math. 133 (1991), 429–446.
- Clifford algebras and polar planes, Duke Math. J. 67 (3) (1992), 627–632.

F. Tilborghs and F. Van Oystaeyen,
- Brauer-Wall algebras graded $\mathbb{Z}_2 \times \mathbb{Z}$, Comm. Algebra 16 (1988), 1457–1478.

J. Tits,
- Formes quadratiques, groupes orthogonaux et algèbres de Clifford, Inventiones Math. 5 (1968), 19–41.

F. Van Oystaeyen,
- On Brauer groups of arithmetically graded rings, Comm. Algebra 9 (1981), 1873–1892.

A. Wadsworth,
- Discriminants in characteristic 2, Linear Multilin. Alg. 17 (1985), 235–263.
- Merkurjev's elementary proof of Merkurjev's theorem, Contemp. Math. 55 (1986), 741–776.

C.T.C. Wall,
- Graded Brauer groups, J. Reine Angew. Math. 213, (1964), 187–199
- On the classification of hermitian forms, I. Rings of algebraic integers, Compositio Math. 22(4) (1970), 425–451.
- On the classification of hermtian forms, III. Complete semilocal rings, Inventiones Math. 19 (1973), 59–71.
- On the classification of hermtian forms, V. Global rings, Inventiones Math. 23 (1974), 261–288.

W. Waterhouse,
- Pieces of eight in the class group of quadratic fields, J. Number Theory 5 (1973), 95–97.
- Discriminants of etale algebras and related structures, J. Reine Angew. Math. 379 (1987), 209–220.

A. Weil,
- Basic Number Theory, Springer-Verlag, Berlin, Heidelberg, New York, 1967.

S. Yuzvinski,
- Composition of quadratic forms and tensor products of quaternion algebras, J. Algebra 96 (1985), 347–367.

O. Zariski and P. Samuel,
- Commutative Algebra I, II. Graduate Texts in Mathematics 28, 29, Springer-Verlag, Berlin, Heidelberg, New York, 1975, 1976.

Index

Universitext *(continued)*